AF556195

Trends in Plant Tissue Culture and Biotechnology

Trends in Plant Tissue Culture and Biotechnology

Bhupendra Kumar

RANDOM PUBLICATIONS
NEW DELHI (INDIA)

Trends in Plant Tissue Culture and Biotechnology

ISBN 978-93-5111-221-1

Published in 2014 in India by

RANDOM PUBLICATIONS

4376-A/4B, Gali Murari Lal, Ansari Road
New Delhi-110 002
Phone : +91-11-43580356, +91-11-23289044
e-mail: randomexports@gmail.com, sales@randompublications.com,
info@randompublications.com

Reprinted 2025

Type Setting by : Keystoneprintads, Delhi-110051
Digitally Printed at: Replika Press Pvt. Ltd.

Preface

Biotechnology has known since long time ago, when human beings realized that they could plant their own crops and breed their own animals, they learned to use biotechnology.

Plant tissue cultures have become important for pursuing a wide range of fundamental and applied problem in research and development. The techniques encompass a variety of procedures used for specific purposes. Biochemical and growth studies are widely employed in the growing of unorganized cells on agar or in liquid suspension. The culture of segments of stems, roots, leaves or cells provides system to study differentiation, morphogenesis and plant regeneration. Shoot apex culture methods leading to plant regeneration have been adopted for plant propagation and production of virus free stock. The culture of anthers and pollen provides new approaches to haploid plant formation. Recently the technology has been extended to include the isolation and culture of plant protoplasts which are employed in fusion and somatic cell hybridisation.

Tissue culture consists of growing plants cells as relatively on organized masses of cells on an agar medium (callus culture) or as a suspension of free cells and small cell masses in a liquid medium (suspension culture). Tissue culture is used for vegetative multiplication of many species and in some cases for recovery of virus-free plants. It has potential application in production of somatic hybrids, organelle and cytoplasm transfer, genetic transformation and germplasm storage through freeze-preservation.

Today, biotechnology is being used as a tool to give plants new traits that benefit agricultural production, the environment, and human nutrition and health. The purpose of this publication is to provide basic information about plant biotechnology and to give examples of its uses. The goal of plant breeding is to combine desirable traits from different varieties of plants to produce plants of superior quality. This approach to improving crop production has been very successful over the years. For example, it would be beneficial to cross a tomato plant that bears sweeter fruit with one that exhibits increased disease resistance. To do this, it takes many years of crossing and backcrossing generations of plants to obtain the desired trait.

It is hoped that he wide range of authoritative review and views presented by the authors in this book will provide a useful and exciting stimulus to researchers and students in the field.

I thank all members of my team who have helped in the preparation of the book. My special thanks go to "Random Publications" who have published the book.

*– **Bhupendra Kumar***

Contents

1

Plant Tissue Culture

Plants are composed of three major organ groups: roots, stems and leaves. As we know from other areas of biology, these organs are comprised of tissues working together for a common goal (function). In turn, tissues are made of a number of cells which are made of elements and atoms on the most fundamental level. In this section, we will look at the various types of plant tissue and their place and purpose within a plant. It is important to realize that there may be slight variations and modifications to the basic tissue types in special plants.

Plant tissues are characterized and classified according to their structure and function. The organs that they form will be organized into patterns within a plant which will aid in further classifying the plant. A good example of this is the three basic tissue patterns found in roots and stems which serve to delineate between woody dicot, herbaceous dicot and monocot plants.

HISTORY

Plant tissue culture is a method or technique to isolate parts of plants (protoplasm, cells, tissues, and organs) and grow them on artificial media in aseptic conditions in a controlled space so that parts of these plants can grow and develop into complete plants. Plant tissue culture techniques are essential to many types of academic inquiry, as well as to many applied aspects of plant science. In the past, plant tissue culture techniques have been used in academic investigations of totipotency and the roles of hormones in cytodifferentiation and organogenesis. Currently, tissue-cultured plants that have been genetically engineered provide insight into plant molecular biology and gene regulation. Plant tissue culture techniques are also central to innovative areas of applied plant science, including plant biotechnology and agriculture.

For example, select plants can be cloned and cultured as suspended cells from which plant products can be harvested. In addition, the management of genetically engineered cells to form transgenic whole plants requires tissue culture procedures; tissue culture methods are also required in the formation of somatic haploid embryos from which homozygous plants can be generated.

Thus, tissue culture techniques have been, and still are, prominent in academic and applied plant science.

The techniques demonstrated in these exercises range from simple ones that can easily be performed by beginning students to those done by botany or physiology students. The use of tissue culture techniques in the beginning just to prove the theory of "totipotensi" ("total genetic potential") is expressed by Schleiden and Schwann which states that the plant cell as the smallest unit can grow and thrive if kept in appropriate conditions. We have used tissue culture techniques not only as a means for studying aspects of plant physiology and biochemistry, but has developed into methods for various purposes.

ASEPTIC TECHNIQUE

The essence of aseptic technique is the exclusion of invading microorganisms during experimental procedures. If sterile tissues are available, then the exclusion of microorganisms is accomplished by using sterile instruments and culture media concurrently with standard bacteriological transfer procedures to avoid extraneous contamination.

Media and apparatus are rendered sterile by autoclaving at 15 lbs/inch 2 (121°C) for 15 minutes. The use of disposable sterile plasticware reduces the need for some autoclaving. Alternative sterilization techniques such as filter sterilization must be employed for heat-labile substances like cytokinins. Aseptic transfers can be made on the laboratory bench top by using standard bacteriological techniques (*i.e.*, flaming instruments prior to use and flaming the opening of receiving vessels prior to transfer). Aseptic transfers are more easily performed in a transfer chamber such as a laminar flow hood, which is also preferably equipped with a bunsen burner.

If experimental tissues are not aseptic, then surface sterilization procedures specific to the tissues are employed. Common sterilants are ethyl alcohol and/or chlorox with an added surfactant. Concentration of sterilants and exposure time are determined empirically. When cultured *in vitro*, all the needs of the plant cells, both chemical and physical, have to met by the culture vessel, the growth medium, and the external environment (light, temperature, etc.).

The growth medium has to supply all the essential mineral ions required for growth and development. In many cases (as the biosynthetic capability of cells cultured *in vitro* may not replicate that of the parent plant), it must also supply additional organic supplements such as amino acids and vitamins. Many plant cell cultures, as they are not photosynthetic, also require the addition of a fixed carbon source in the form of a sugar (most often sucrose). One other vital component that must also be supplied is water, the principal biological solvent. Physical factors, such as temperature, pH, the gaseous environment, light (quality and duration), and osmotic pressure, also have to be maintained within acceptable limits.

MICROBIAL

Plant cell culture is viewed as a potential means of producing useful plant products such that conventional agriculture, with all its attendant problems and variables, can be circumvented. These problems include: environmental factors (drought, floods, etc.), disease, political and labour instabilities in the producing countries (often Third World countries), uncontrollable variations in the crop quality, inability of authorities to prevent crop adulteration, losses in storage and handling.

Thus, the production of useful and valuable secondary metabolites in large bioreactors located in the consuming country is an attractive proposal. Additional advantages of such processes include: controlled production according to demand and a reduced and requirement.

However, this technology is still being developed and despite the advantages outlined above, there are a variety of problems to be overcome before it can be adopted on a wide scale for the production of useful plant secondary metabolites.

In theory, it is anticipated that such large scale suspension cultures will be suitable for industrial production of useful plant chemicals such as pharmaceuticals and food additives, in a manner similar to that of microbial fermentation.

Nevertheless, there are some significant differences between microbial and plant cell cultures that must be considered when attempting to apply plant cell cultures to the available technology. Table shows a comparison of some of the characteristics of plant and microbial cultures of relevance to fermentation.

This table serves to demonstrate some of the problems that can be encountered with plant cell cultures. The sensitivity to shear is due both to the large size of the cells and to the relatively inflexible cellulose cell wall. Thus, with normal blade impellers the cells may twist which will inhibit mitoses and, for this reason, air-lift fermentors are recommended by some researchers.

The large size of the plant cell contributes to its comparatively high doubling time (12 h - several days), which thus prolongs the time required for a successful fermentation run.

The vacuole is the major site of product accumulation, and since product secretion is uncommon, the high metabolite yields seen in microorganisms that secrete product (thereby removing product inhibition of biosynthesis) cannot be expected.

There is some ongoing research on membrane permeabilization of plant cells which may serve to relieve the constraints of product inhibition by facilitation of leakage into the extracellular medium. If this would also permit recycling of the biomass (*e.g.* via immobilization) it would help reduce production costs.

Table. Characteristic of Microbial and Plant Cell Relevant to Fermentation

Characteristics	Microorganism	Plant Cell
Size	2 u	>10 u
Shear stress	Insensitive	Sensitive
Water content	75%	>90%
Duplication time	<1 hour	days
Aeration	1-2 vvm	0.3 vvm
Fermentation time	Days	Weeks
Product accumulation	Medium	Vacuole
Production phase linked	Uncoupled	Often growth-
Mutation	Possible	Requires haploids
Medium cost ($) (MS medium)	8-9/m	65-70/m

The low aeration requirement for plant cells is an advantage over microbial cultures in general. In addition, the high cost of running a fermentation vessel over several weeks should be considered, although media costs are much less than those of animal cell cultures.

In addition to the problems outlined above, concerned with fermentation technology, there are also considerable hurdles to be overcome at the biochemical level. The two major problems concern poor expression of products and instability of cell lines.

Cultured plant cells often produce reduced quantities and different profiles of secondary metabolites when compared with the intact plant and these quantitative and qualitative features may change with time. The poor product expression is often attributed to a lack of differentiation in cultures. On the other hand, there are cases of cultures that over-produce metabolites compared with the whole plant.

There are a number of examples of cultured cells producing metabolites not observed in the plant, eg. Lithospermum erythrorhizon cultures have been observed to synthesize rosmarinic acid. It has become apparent that the choice of original plant material having high yields of the desired phytochemical may be important in establishing high-yielding cultures. Furthermore, the need to repeatedly screen for high-producing lines (due to inherent instability of cell lines) has been emphasized, although the nutritional composition of the medium is also important. Thus, a variety of approaches are being investigated by many researchers to increase productivity of useful plant metabolites in plant cell cultures as seen.

MICROPROPAGASI

Tissue culture techniques have been used to help produce crops in large scale through micropropagasi or clonal propagation of various plants. Plant tissue in very small amounts can produce hundreds or thousands of plants continuously. This technique has been used in industrial scale in various countries to commercially produce various types of plants such as ornamental plants (orchids, cut flowers, etc..), Fruit crops (like bananas), crops and forestry industries (coffee, teak, etc.). By using tissue culture methods, millions of plants with the same genetic characteristics can be obtained only from one eye buds. Therefore, this method becomes an alternative in the vegetative propagation of plants.

IMPROVED CROP

In crop improvement efforts through the glorification of the conventional methods, to obtain pure strains can take six to seven generations of self-pollination or crosses. Through tissue culture techniques, can be obtained homosigot plants in a short time by producing haploid plants through pollen culture, anther or ovaries followed by chromosome doubling. Homosigot plants can be used as plant breeding material in order to improve the nature of the plant.

PRODUCTION OF DISEASE-FREE PLANTS (VIRUS)

Tissue culture technology has contributed in a plant that is free from viruses. In plants that have been infected with the virus, the cells in the bud tip (meristem) is an area that is not infected with the virus. In this way the meristem will mengkulturkan obtained virus-free plants.

GENETIC TRANSFORMATION

Tissue culture techniques have become an important part in helping the success of plant genetic engineering (gene transfer). For example, bacterial gene transfer (such as cry genes from Bacillus thuringiensis) into the plant cells will be expressed after transgeniknya achieved plant regeneration.

The production of secondary metabolites, compounds: Plant cell culture can also be used to produce biochemical compounds (secondary metabolites) such as alkaloids, terpenoids, phenyl etc. propanoid. This technology is now available in industrial scale. For example, the commercial production of compounds "shikonin" from Lithospermum erythrorhizon cell culture. Plant cells can be grown in isolation from intact plants in tissue culture systems. The cells have the characteristics of callus cells, rather than other plant cell types. These are the cells that appear on cut surfaces when a plant is wounded and which gradually cover and seal the damaged area.

Pieces of plant tissue will slowly divide and grow into a colourless mass of cells if they are kept in special conditions.

These are:

- Initiated from the most appropriate plant tissue for the particular plant variety

- Presence of a high concentration of auxin and cytokinin growth regulators in the growth media
- A growth medium containing organic and inorganic compounds to sustain the cells
- Aseptic conditions during culture to exclude competition from microorganisms

The plant cells can grow on a solid surface as friable, pale-brown lumps (called callus), or as individual or small clusters of cells in a liquid medium called a suspension culture. These cells can be maintained indefinitely provided they are sub-cultured regularly into fresh growth medium.

Tissue culture cells generally lack the distinctive features of most plant cells. They have a small vacuole, lack chloroplasts and photosynthetic pathways and the structural or chemical features that distinguish so many cell types within the intact plant are absent. They are most similar to the undifferentiated cells found in meristematic regions which become fated to develop into each cell type as the plant grows. Tissue cultured cells can also be induced to re-differentiate into whole plants by alterations to the growth media.

Plant tissue cultures can be initiated from almost any part of a plant. The physiological state of the plant does have an influence on its response to attempts to initiate tissue culture. The parent plant must be healthy and free from obvious signs of disease or decay. The source, termed explant, may be dictated by the reason for carrying out the tissue culture. Younger tissue contains a higher proportion of actively dividing cells and is more responsive to a callus initiation programme. The plants themselves must be actively growing, and not about to enter a period of dormancy. The exact conditions required to initiate and sustain plant cells in culture, or to regenerate intact plants from cultured cells, are different for each plant species. Each variety of a species will often have a particular set of cultural requirements. Despite all the knowledge that has been obtained about plant tissue culture during the twentieth century, these conditions have to be identified for each variety through experimentation.

Table. Secondary Metabolites Produced in High Levels by Plant Cell Cultures

Compound	Plant Species	Yields (% dry wt) Culture	Yields (% dry wt) Plant	Culture Type*
Shikonin erythrorhizon	Lithospermum	20	1.5	s
Ginsenoside	Panax ginseng	27	4.5	c
Anthraquinones	Morinda citrifolia	18	0.3	s
Ajmalicine	Catharanthus roseus	1.0	0.3	s
Rosmarinic acid	Coleus blumeii	15	3	s

Ubiquinone-10	Nicotiana tabacum	0.036	0.003	s	
Diosgenin	Dioscorea deltoides	2	2	s	
Benzylisoquinoline	Coptis japonica	11	5 - 10	s	Alkaloids
Berberine	Thalictrum minor	10	0.01	s	
Berberine	Coptis japonica	10	2 - 4	s	
Anthraquinones	Galium verum	5.4	1.2	s	
Anthraquinones	Galium aparine	3.8	0.2	s	
Nicotine	Nicotiana tabacum	3.4	2.0	c	
Bisoclaurine	Stephania cepharantha	2.3	0.8	s	
Tripdiolide	Tripteryqium wilfordii	0.05	0.001	s	

* s = suspension; c = callus

As seen in many reviews, studies on the production of plant metabolites by callus and cell suspension cultures have been carried out on an increasing scale since the end of the 1950's. The large scale cultivation of tobacco and a variety of vegetable cells was examined from the late 1950's to early 1960's by Tulecke and Nickell at Pfizer Inc., Mandels *et al.* at the Natick Laboratories in the U.S. Army, Street *et al.* at the University of Leicester and Martin *et al.* at the National Research Council of Canada. Their results stimulated more recent studies on the industrial application of plant cell culture in many countries.

Since Japan has a highly developed fermentation technology, many industrial companies, in collaboration with some university groups, have tried to apply this technology for the commercial production of useful compounds. The Japan Tobacco Inc.'s interest involved around mass-production of tobacco cells as raw materials of cigarettes; the company established 20 kL fermentors which were the largest for plant cells in 1970's. Meiji Seika in Japan also elucidated the fundamentals of production of Panax ginseng cells in large volumes. Researchers reported that cultured ginseng cells stimulated physiological activities in animals in a similar fashion as elicited by native ginseng roots. The work was followed by Nitto Denko Co. which has been manufacturing cell mass of ginseng commercially. The cells are used as health foods in Japan. Researchers of Kyowa Hakko conducted extensive pharmacological screening of numerous cell cultures and found various novel products of great interest, including plasmin inhibitory proteins in Scopolia japonica cells and plant virus inhibitors in cultured cells of Phytolacca americana and other species. The virus inhibitors of P. americana are now being studied by many research groups in the world because of their activity against AIDS and other animal viruses. Other firms such as Ajinomoto and Nippon Shin-yaku also made efforts to increase the level of accumulation of alkaloids, steroids and other secondary products in cultured cells.

Groups in Germany outlined very interesting approaches to industrial application in the meeting held in 1976 at Munich, and their excellent results encouraged researchers in other countries. For example, Zenk and his colleagues presented an impressive paper in which they successfully selected high-alkaloid producing cell lines of Catharanthus roseus using a method similar to the microbial mono-colony isolation technique. A number of laboratories in industries and universities followed Zenk's approach, and in fact, some researchers could increase significantly the level of secondary products such as ubiquinone-10, biotin, and various plant pigments produced by cell cultures.

In 1982, the 5th International Congress of Plant Tissue and Cell Cultures was held in Japan and about papers presented there related to production of secondary metabolites in cultured cells and several papers seemed to be commercially promising such as production of shikonin by Fujita *et al.*) of Mitsui Petrochemical and that of several antitumor compounds by Misawa *et al.* At subsequent International Congress of Plant Tissue and Cell Cultures in Minneapolis in 1986 and that in Amsterdam in 1990 as well as other meetings such as the Meeting of Primary and Secondary Metabolism of Plant Cell Cultures held in Giessen, Germany and the 4th and International Congress on Phytotherapy held in Munich, Germany in September, 1992, many compounds were shown to be accumulated by plant cell cultures and many different strategies were presented to increase their productivity. Means for production of those compounds include not only de novo synthesis but also biotransformation processes. A biotransformation process to produce β-methyl digoxin using Digitalis lanata cells studied by Reinhard and Alfermann in Germany was evaluated by Boehringer Mannheim Co. using 4 kL bioreactors although it has not yet been commercialized.

A combination of a plant cell culture process and a simple chemical coupling reaction was invented by a Canadian company, Allelix, to manufacture vinblastine as a commercially feasible process. The technology is now being studied by Mitsui Petrochemical in Japan for commercialization. In spite of remarkable advances in plant cell culture technology, the production cost of metabolites is still high, as estimated by Zenk and by Goldstein. The former calculated that the cost would be U.S. $500 per kg of isolated product when it was accumulated in 1 g/L within a period of 15 days in 100 m batch culture. It is true that the cost has decreased since his estimation in 1974, and the producing ability of 1 g/L for 15 days is achievable in the case of some compounds such as rosmarinic acid, but it is still too expensive to produce food stuffs, food additives or pharmaceuticals which can be more easily produced by alternative ways such as chemical synthesis, extraction from plants or by microbial fermentation.

Zenk stated that industrial plant cell culture techniques would be introduced only if the plant product under consideration is produced at a price equal to or preferably lower than the field-produced product. The only factor determining the industrial realization of plant cell culture is the price by which a given product can be produced. A very detailed assessment of the costs entailed in theoretical processes has been made by Goldstein *et al.*

Four cases were considered:

A (Base state of technology)
Final biomass concentration = 200 g fresh wt./L
One day doubling time, biomass = 0.693/day
One day doubling time, product = 0.693/day
Product concentration = 0.05% fresh weight
Recycled biomass = 50%

B *As for A, but*:
One day doubling time, product = 6.93/day (10 times increase)
Recycled biomass = 90%

C *As for B, but:*
Final biomass concentration = 500 g fresh wt./L (2.5 times increase)

D *As for C, but*:
Product concentration = 0.5% fresh weight (10 times increase).

The final case (D) would translate into a production of 17.5 g product/L/day, which is well in excess of levels achieved with current technology. The rate of product formation is the major obstacle, but this could be compensated for by increased concentration of product. The calculated manufacturing costs and profit margin for each case assuming production of 10^4, 10^5 or 10^6 kg/yr are shown in Tables. Note that these figures reflect 1980 prices and will have increased since then.

Table. Manufacturing Costs for Cases A - D

Production Level (kg/yr)	10^4	10^5	10^6
	Cost ($/kg)		
Case:			
A.	551	241	106
B.	323	142	61
C.	165	72	32
D.	92	40	18

Table. Profitability of Cases A - D The Selling Price ($) and Profit (%) are Given for each of the Three Annual Production Levels.

Production Level (kg/yr)	10^4		10^5		10^6	
	$	%	$	%	$	%
Case:						
A.	1,608	5.7	683	14.7	290	13.8
B.	992	17.5	421	16.3	177	15.3
C.	528	18.7	223	17.6	94	16.6
D.	233	10.6	100	9.8	43	9.1

The selling price was calculated to take into account a 5-year payment on capital and a tax rate of 50%. The formula used was: Selling price/kg = 2 x manufacturing cost/kg + 0.4 (total invested capital/kg) = depreciation/kg.

In the case of scenario (D) with production of 10 kg/yr the selling price could be as low as $43/kg and the manufacturing cost would only be $18/kg. This assessment makes the future of commercial plant cell culture fermentation appear quite viable, but the crucial problems to be overcome are enhancing product levels, recycling the biomass (*e.g.* through immobilization) and raising the biomass concentration.

Some encouraging advances in recent years suggest that these problems may be solved eventually, *e.g.* shikonin can be produced at a level of 23% of the dry weight (roughly 2.3% or more of fresh weight) and 4 g/L; rosmarinic acid can be produced at a rate of 1 g/L/day in Coleus blumei with a final concentration of 27% of the dry weight. A biomass concentration of 40 g dry wt./L was achieved.

Batch fermentation processes are capital intensive, requiring a high initial investment and sophisticated technology. In comparison with microbial cultures, doubling times are long and product yields are low. Furthermore, since production is often not growth-linked, there are many cases where two stage production is necessary, thus increasing costs markedly. Table shows estimated costs for a hypothetical phytochemical produced by plant cell culture fermentation.

Table. Batch Fermentation Costs for a Hypothetical Plant Product as a Function of Product Concentration

Product Concentration (%)	0.1	1.0	10.0
Fixed Capital Investment ($M)	340	84	21
Total Production Cost (%/kg)	5,900	1,045	228
Assumptions:	Annual production = 20,000 kg Cell doubling time = 60 hr Batch cycle time = 15 d Cell intensity = 20 g/L		

The total production cost is the sum of direct and indirect costs, including raw material, labour, utilities, capital equipment depreciation over a 10 year period, maintenance, tax and insurance. Indirect costs were taken as 15% of capital investment

It has been calculated that if rose oil were produced in cell cultures, a fermentation process with a 20% share of the market would yield a net profit of $0.5 M/yr.

This assumes a rose oil concentration of 10% dry weight and a cell density of 20 g dry weight/L. Fowler suggests that plant products valued at $250-1,500/kg are acceptable production targets for an economically viable process. An initial market penetration of the world or U.S. market has been recommended at between 10 and 20%. This will depend on the size of the market and whether a plant cell fermentation process opens up new market.

Scragg has calculated the effect of run time on productivity. Assuming a product yield of 1% dry weight, biomass yield of 20 g dry weight/L and a bioreactor volume of 1,000 L.

A 300 day year was estimated, since time would be required for maintenance, contaminated runs, etc. Fig. shows that as productivity decreases the requisite run time increases. Similarly, as the yield of product increases the reactor volume required decreases expotentially. Under the conditions described above, a 100,000 L fermentor would be necessary to produce 300 kg/yr.

The most detailed assessment available in the literature for the production of a specific product in plant cell cultures is concerned with the production of ajmalicine from Catharanthus roseus cell cultures. In this case, a 20% market penetration was assumed, ie. 800 kg/yr and production was based on the use of a two-stage batch culture process. Stage two parameters include specific productivity of 0.26 mg/g/day (based on final dry weight), final product concentration of 0.06% dry weight and a maximum fresh weight concentration of 160 g/L.

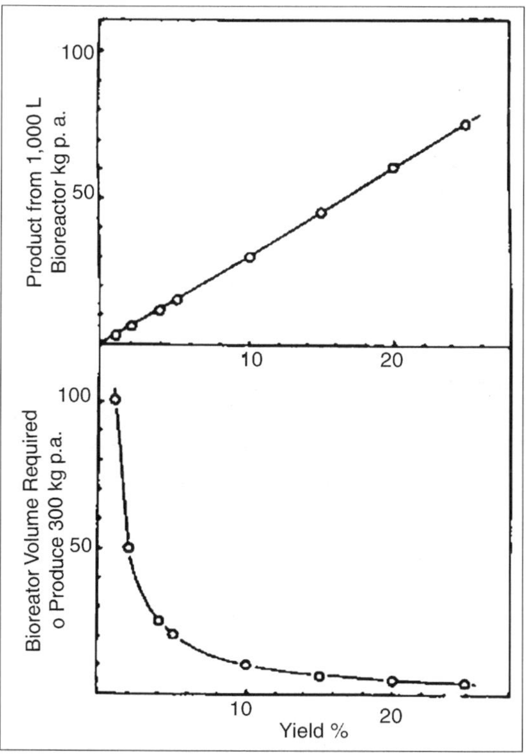

Fig. The Effect of Run Time on Bioreactor Volume and the Production from a 1,000 L Bioreactor

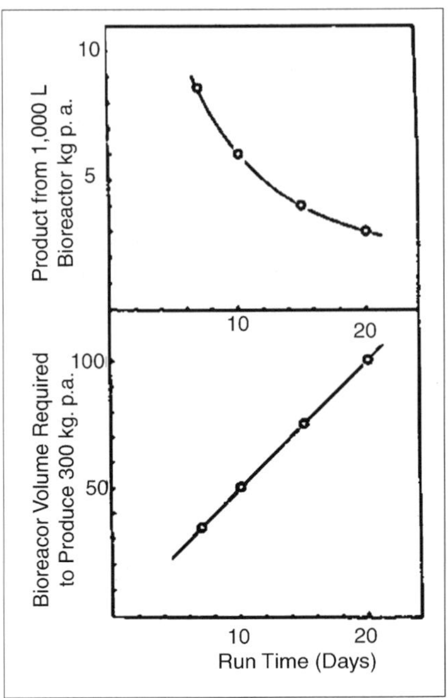

Fig. The Effect of Yield on Bioreactor Volume and the Production from a 1,000 L. Bioreactor

It was estimated that the cost of production would be $3,215/kg ajmalicine and this compares with $619/kg ($0.70/lb dry biomass) from the intact plant. Clearly, this 5-fold increase in costs is unacceptable from a commercial point of view. This example serves to illustrate the need to chose products that are particularly expensive to produce in the field. Shikonin is a good example; in that the long growing period and strict climatic requirements mean that the cost of the plant raw material is high $6.80/lb. It has been suggested that a quick way to assess the attractiveness of a cell culture method vs. conventional agriculture, is to calculate a specific biosynthesis rate "based on the final dry weight and the total time of fermentation or land occupation". With shikonin there is a 830-fold increase in plant cell culture, whilst with ajmalicine from C. roseus there was only a 24-fold improvement.

MERISTEMATIC TISSUES

Tissues where cells are constantly dividing are called meristems or meristematic tissues. These regions produce new cells. These new cells are generally small, six-sided boxlike structures with a number of tiny vacuoles and a large nucleus, by comparison.

Sometimes there are no vacuoles at all. As the cells mature the vacuoles will grow to many different shapes and sizes, depending on the needs of the cell. It is possible that the vacuole may fill 95percent or more of the cell's total volume.

There are three types of meristems:

1. Apical Meristems
2. Lateral Meristems
3. Intercalary Meristems

Apical meristems are located at or near the tips of roots and shoots. As new cells form in the meristems, the roots and shoots will increase in length. This vertical growth is also known as primary growth. A good example would be the growth of a tree in height. Each apical meristem will produce embryo leaves and buds as well as three types of primary meristems: protoderm, ground meristems, and procambium. These primary meristems will produce the cells that will form the primary tissues.

Lateral meristems account for secondary growth in plants. Secondary growth is generally horizontal growth. A good example would be the growth of a tree trunk in girth.

There are two types of lateral meristems to be aware of in the study of plants. The vascular cambium, the first type of lateral meristem, is sometimes just called the cambium.

The cambium is a thin, branching cylinder that, except for the tips where the apical meristems are located, runs the length of the roots and stems of most perennial plants and many herbaceous annuals. The cambium is responsible for the production of cells and tissues that increase the thickness, or girth, of the plant.

The cork cambium, the second type of lateral meristem, is much like the vascular cambium in that it is also a thin cylinder that runs the length of roots and stems. The difference is that it is only found in woody plants, as it will produce the outer bark.

Both the vascular cambium and the cork cambium, if present, will begin to produce cells and tissues only after the primary tissues produced by the apical meristems have begun to mature.

Intercalary meristems are found in grasses and related plants that do not have a vascular cambium or a cork cambium, as they do not increase in girth. These plants do have apical meristems and in areas of leaf attachment, called nodes, they have the third type of meristematic tissue. This meristem will also actively produce new cells and is responsibly for increases in length. The intercalary meristem is responsible for the regrowth of cut grass. There are other tissues in plants that do not actively produce new cells. These tissues are called nonmeristematic tissues.

Nonmeristematic tissues are made of cells that are produced by the meristems and are formed to various shapes and sizes depending on their intended function in the plant. Sometimes the tissues are composed of the same type of cells throughout, or sometimes they are mixed. There are simple tissues and complex tissues to consider, but we will start with the simple tissues for the sake of discussion.

SIMPLE TISSUES

There are three basic types, named for the type of cell that makes up their composition.

1. Parenchyma cells form parenchyma tissue. Parenchyma cells are the most abundant of cell types and are found in almost all major parts of higher plants. These cells are basically sphere shaped when they are first made. However, these cells have thin walls, which flatten at the points of contact when many cells are packed together. Generally, they have many sides with the majority having 14 sides. These cells have large vacuoles and may contain various secretions including starch, oils, tannins, and crystals. Some parenchyma cells have many chloroplasts and form the tissues found in leaves. This type of tissue is called chlorenchyma. The chief function of this type of tissue is photosynthesis, while parenchyma tissues without chloroplasts are generally used for food or water storage. Additionally, some groups of cells are loosely packed together with connected air spaces, such as in water lilies, this tissue is called aerenchyma tissue. These type of cells can also develop irregular extensions of the inner wall which increases overall surface area of the plasma membrane and facilitates transferring of dissolved substances between adjacent cells. Parenchyma cells can divide if they are mature, and this is vital in repairing damage to plant tissues. Parenchyma cells and tissues comprise most of the edible portions of fruit.
2. Collenchyma cells form collenchyma tissue. These cells have a living protoplasm, like parenchyma cells, and may also stay alive for a long period of time. Their main distinguishing difference from parenchyma cells is the increased thickness of their walls. In cross section, the walls looks uneven. Collenchyma cells are found just beneath the epidermis and generally they are elongated and their walls are pliable in addition to being strong. As a plant grows these cells and the tissues they form, provide flexible support for organs such as leaves and flower parts. Good examples of collenchyma plant cells are the 'strings' from celery that get stuck in our teeth.
3. Sclerenchyma cells form sclerenchyma tissue. These cells have thick, tough secondary walls that are imbedded with lignin. At maturity, most sclerenchyma cells are dead and function in structure and support. Sclerenchyma cells can occur in two forms:
   a. Sclereids are sclerenchyma cells that are randomly distributed throughout other tissues. Sometimes they are grouped within other tissues in specific zones or regions. They are generally as long as they are wide. An example, would be the gritty texture in some types of pears. The grittiness is due to groups of sclereid cells. Sclereids are sometimes called stone cells.

b. Fibers are sometimes found in association with a wide variety of tissues in roots, stems, leaves and fruits. Usually fiber cells are much longer than they are wide and have a very tiny cavity in the center of the cell. Currently, fibers from over 40 different plant families are used in the manufacture of textiles, ropes, string and canvas goods to name a few.

SECRETORY CELLS AND TISSUES

As a result of cellular processes, substances that are left to accumulate within the cell can sometimes damage the protoplasm. Thus it is essential that these materials are either isolated from the protoplasm in which they originate, or be moved outside the plant body. Although most of these substances are waste products, some substances are vital to normal plant functions. Examples: oils in citrus, pine resin, latex, opium, nectar, perfumes and plant hormones. Generally, secretory cells are derived from parenchyma cells and may function on their own or as a tissue. They sometimes have great commercial value.

COMPLEX TISSUES

Tissues composed of more than one cell type are generically referred to as complex tissues. Xylem and phloem are the two most important complex tissues in a plant, as their primary functions include the transport of water, ions and soluble food substances throughout the plant. While some complex tissues are produced by apical meristems, most in woody plants are produced by the vascular cambium and is often referenced as vascular tissue. Other complex tissues include the epidermis and the periderm. The epidermis consists primarily of parenchyma-like cells and forms a protective covering for all plant organs. The epidermis includes specialized cells that allow for the movement of water and gases in and out of the plant, secretory glands, various hairs, cells in which crystals are accumulated and isolated, and other cells that increase absorption in the roots. The periderm is mostly cork cells and therefore forms the outer bark of woody plants. It is considered to be a complex tissue because of the pockets of parenchyma cells scattered throughout.

Xylem

Xylem is an important plant tissue as it is part of the 'plumbing' of a plant. Think of bundles of pipes running along the main axis of stems and roots. It carries water and dissolved substances throughout and consists of a combination of parenchyma cells, fibers, vessels, tracheids and ray cells. Long tubes made up of individual cells are the vessels, while vessel members are open at each end. Internally, there may be bars of wall material extending across the open space. These cells are joined end to end to form long tubes.

Vessel members and tracheids are dead at maturity. Tracheids have thick secondary cell walls and are tapered at the ends. They do not have end openings such as the vessels. The tracheids ends overlap with each other, with pairs of pits present. The pit pairs allow water to pass from cell to cell. While most conduction in the xylem is up and down, there is some side-to-side or lateral conduction via rays. Rays are horizontal rows of long-living parenchyma cells that arise out of the vascular cambium. In trees, and other woody plants, ray will radiate out from the center of stems and roots and in cross-section will look like the spokes of a wheel.

Phloem

Phloem is an equally important plant tissue as it also is part of the 'plumbing' of a plant. Primarily, phloem carries dissolved food substances throughout the plant.

This conduction system is composed of sieve-tube member and companion cells, that are without secondary walls. The parent cells of the vascular cambium produce both xylem and phloem. This usually also includes fibers, parenchyma and ray cells. Sieve tubes are formed from sieve-tube members laid end to end. The end walls, unlike vessel members in xylem, do not have openings. The end walls, however, are full of small pores where cytoplasm extends from cell to cell.

These porous connections are called sieve plates. In spite of the fact that their cytoplasm is actively involved in the conduction of food materials, sieve-tube members do not have nuclei at maturity.

It is the companion cells that are nestled between sieve-tube members that function in some manner bringing about the conduction of food. Sieve-tube members that are alive contain a polymer called callose. Callose stays in solution as long at the cell contents are under pressure. As a repair mechanism, if an insect injures a cell and the pressure drops, the callose will precipitate. However, the callose and a phloem protein will be moved through the nearest sieve plate where they will for a plug. This prevents further leakage of sieve tube contents and the injury is not necessarily fatal to overall plant turgor pressure.

Epidermis

The epidermis is also a complex plant tissue, and an interesting one at that. Officially, the epidermis is the outermost layer of cells on all plant organs (roots, stems, leaves).

The epidermis is in direct contact with the environment and therefore is subject to environmental conditions and constraints. Generally, the epidermis is one cell layer thick, however there are exceptions such as tropical plants where the layer may be several cells thick and thus acts as a sponge. Cutin, a fatty substance secreted by most epidermal cells, forms a waxy protective layer called the cuticle.

The thickness of the cuticle is one of the main determiners of how much water is lost by evapouration. Additionally, at no extra charge, the cuticle provides some resistance to bacteria and other disease organisms. Some plants, such as the wax palm, produce enough cuticle to have commercial value: carnauba wax. Other wax products are used as polishes, candles and even phonographic records. Epidermal cells are important for increasing absorptive surface area in root hairs. Root hairs are essentially tubular extensions of the main root body composed entirely of epidermal cells. Leaves are not left out. They have many small pores called stomata that are surrounded by pairs of specialized epidermal cells called guard cells. Guard cells are unique epidermal cells because they are of a different shape and contain chloroplasts. There are other modified epidermal cells that may be glands or hairs that repel insects or reduce water loss.

Periderm

In woody plants, when the cork cambium begins to produce new tissues to increase the girth of the stem or root the epidermis is sloughed off and replaced by a periderm. The periderm is made of semi rectangular and boxlike cork cells. This will be the outermost layer of bark. These cells are dead at maturity. However, before the cells die, the protoplasm secretes a fatty substance called suberin into the cell walls. Suberin makes the cork cells waterproof and aids in protecting tissues beneath the bark. There are parts of the cork cambium that produce pockets of loosely packed cork cells. These cork cells do not have suberin imbedded in their cell walls. These loose areas are extended through the surface of the periderm and are called lenticels. Lenticels function in gas exchange between the air and the stem interior. At the bottom of the deep fissures in tree bark are the lenticels.

CULTURED CELLS AND TISSUES

The large scale commercial propagation of plant material based on plant tissue culture was pioneered in the USA. During the last thirty years, tissue culture-based plant propagation has emerged as one of the leading global agro-technologies. Between 1986 and 1993, the worldwide production of tissue cultured plants increased 50 per cent. In 1993, the production was 663 million plants. By 1997, production had risen to 800 million plants. During 1990–1994, the micropropagation industry declined in Europe, mainly due to production shifting to developing countries, but since then because of the demand for high quality and number, production in European countries has increased.

Since 1995, production has increased by 14 per cent in Asian countries, mainly due to the market entry of China, while the increase in South and Central America was from production in Cuba. More recently, some companies from Israel, the USA and UK have shifted their production requirements to Costa Rica and India. Tissue cultured plants have as yet to reach many growers and farmers in the developing countries. The primary advantage of micropropagation is the rapid production of high quality, disease-free and uniform planting material. The plants can be multiplied under a controlled environment, anywhere, irrespective of the season and weather, on a year-round basis. Production of high quality and healthy planting material of ornamentals, and forest and fruit trees, propagated from vegetative parts, has created new opportunities in global trading for producers, farmers, and nursery owners, and for rural employment. Micropropagation technology is more expensive than the conventional methods of plant propagation, and requires several types of skills. It is a capital-intensive industry, and in some cases the unit cost per plant becomes unaffordable. The major reasons are cost of production and know-how. During the early years of the technology, there were difficulties in selling tissue culture products because the conventional planting material was much cheaper.

Now this problem has been addressed by inventing reliable and cost effective tissue culture methods without compromising on quality. This requires a constant monitoring of the input costs of chemicals, media, energy, labour and capital. In the industrialized countries, labour is the main factor that contributes to the high cost of production of tissue-cultured plants.

To reduce such costs, some steps can be partially mechanized, *e.g.* use of a peristaltic pump for medium dispensing, and of dishwashers for cleaning containers. In the less developed countries of Africa, Asia, and Latin America, where labour is relatively cheaper, consumables such as media, culture containers, and electricity make a comparatively greater contribution to production costs. For example, the cost of medium preparation (chemicals, energy and labour) can account for 30–35 per cent of the micropropagated plant production. However, automated production processes based on pre-sterilized membrane capsules, bioreactors, mechanized explant transfer, and container sealing are not commercially viable propositions in many developing countries. Therefore, low cost alternatives are needed to reduce production cost of tissue-cultured plants.

Many of the low cost technology options described in this publication can be incorporated in various steps of plant micropropagation. The occasional tissue culturegenerated variants (somaclones) and rare spontaneous bud mutants as well as those obtained from induced mutations can also be propagated by deployment of low cost techniques described. Micropropagation, in combination with radiation-induced mutations, speeds up the recovery, multiplication and release of improved varieties in vegetatively propagated plants. Hence, low cost technology will be of great value for large scale plant multiplication of mutants of many fruits, shrubs, flowers and forest trees that are conventionally vegetatively propagated.

PLANT TISSUE CULTURE

Plant tissue culture refers to growing and multiplication of cells, tissues and organs of plants on defined solid or liquid media under aseptic and controlled environment. The commercial technology is primarily based on micropro-pagation, in which rapid proliferation is achieved from tiny stem cuttings, axillary buds, and to a limited extent from somatic embryos, cell clumps in suspension cultures and bioreactors. The cultured cells and tissue can take several pathways. The pathways that lead to the production of true-to-type plants in large numbers are the preferred ones for commercial multiplication. The process of micropropagation is usually divided into several stages *i.e.*, prepropagation, initiation of explants, subculture of explants for proliferation, shooting and rooting, and hardening. These stages are universally applicable in large-scale multiplication of plants. The delivery of hardened small micropropagated plants to growers and market also requires extra care.

Plant tissue culture refers to growing and multiplication of cells, tissues and organs on defined solid or liquid media under aseptic and controlled environment. Plant tissue culture technology is being widely used for large-scale plant multiplication. The commercial technology is primarily based on micropropagation, in which rapid proliferation is achieved from tiny stem cuttings, axillary buds, and to a limited extent from somatic embryos, cell clumps in suspension cultures and bioreactors.

EXPLANT SOURCE

Plant tissue cultures are initiated from tiny pieces, called explants, taken from any part of a plant. Practically all parts of a plant have been used successfully as a source of explants. In practice, the "explant" is removed surgically, surface sterilized and placed on a nutrient medium to initiate the mother culture, that is multiplied repeatedly by subculture. The following plant parts are extensively used in commercial micropropagation. Shoot-tip and meristem-tip culture: Shoots develop from a small group of cells known as shoot apical meristem. The apical meristem maintains itself, gives rise to new tissues and organs, and communicates signals to the rest of the plant. Shoot-tips and meristem-tips are perhaps the most popular source of explants to initiate tissue cultures. The shoot apex explant measures between 100 to 500 μ m and includes the apical meristem with 1 to 3 leaf primordia. The apical meristem of a shoot is the portion lying distal to the youngest leaf primordium, and is ca.100 μ??m in diameter and 250μ?m in length with 800-1200 cells. In practice, shoot-tip explants between 100 to 1000 μ m are cultured to free plants from viruses. Even explants larger than 1000 μ m have been frequently used. The term "meristem-tip culture" has been suggested to distinguish the large explants from those used in conventional propagation.

Nodal or axillary bud culture: This consists of a piece of stem with axillary bud culture with or without a portion of shoot. When only the axillary bud is taken, it is designated as "axillary bud" culture. Floral meristem and bud culture: Such explants are not commonly used in commercial propagation, but floral meristems and buds can generate complete plants. Other sources of explants: In some plants, leaf discs, intercalary meristems from nodes, small pieces of stems, immature zygotic embryos and nucellus have also been used as explants to initiate cultures.

Cell suspension and callus cultures: Plant parts such as leaf discs, intercalary meristems,-stem-pieces, immature embryos, anthers, pollen, microspores and ovules have been cultured to initiate callus. A callus is a mass of unorganized cells, which in many cases, upon transfer to suitable medium, is capable of giving rise to shoot-buds and somatic embryos, which then form complete plants. Such calli on culture in liquid media on shakers are used for initiating cell suspensions. Liquid suspension cultures maintained on mechanical shakers achieve fast and excellent multiplication rates. However, in commercial micropropagation, calli are cultured mostly in bottles and flasks kept on semi-solid or liquid media. To a limited extent, bioreactors have become popular for somatic embryogenic cultures. It is considered that some day robotics could be adapted to bioreactorbased micropropagation.

PATHWAYS OF CULTURED CELLS AND TISSUES

The cultured cells and tissue can take several pathways to produce a complete plant. Among these, the pathways that lead to the production of true-to-type plants in large numbers are the popular and preferred ones for commercial multiplication. The following terms have been used to describe various pathways of cells and tissue in culture.

Regeneration and Organogenesis

In this pathway, groups of cells of the apical meristem in the shoot apex, axillary buds, root tips, and floral buds are stimulated to differentiate and grow into shoots and ultimately into complete plants. In many cases, the axillary buds formed in the culture undergo repetitive proliferation, and produce large number of tiny plants.

The plants are then separated from each other and rooted either in the next stages of micropropagation or in vivo (in trays, small pots or beds in glasshouse or plastic tunnel under relatively high humidity). The explants cultured on relatively high amounts of auxin (*e.g.* (2,4-D, 2,4-dichlorophenoxyacetic acid) form an unorganized mass of cells, called callus. The callus can be further sub-cultured and multiplied.

The callus shaken in a liquid medium produces cell suspension, which can be subcultured and multiplied into more liquid cultures. The cell suspensions form cell clumps, which eventually form calli and give rise to plants through organogenesis or somatic embryogenesis. In some cases, explants *e.g.* leaf-discs and epidermal tissue can also generate plants by direct organogenesis and somatic embryogenesis without intervening callus formation, *e.g.* in orchardgrass. Dactylis glomerata L.. In organogenesis the cultured plant cells and cell clumps (callus) and mature differentiated cells (microspores, ovules) and tissues (leaf discs, inter-nodal segments) are induced to differentiate into complete plants to form shoot buds and eventually shoots, and rooted to form complete plants.

Somatic Embryogenesis

In this pathway, cells or callus cultures on solid media or in suspension cultures form embryo-like structures called somatic embryos, which on germination produce complete plants. The primary somatic embryos are also capable of producing more embryos through secondary somatic embryogenesis. Although, somatic embryogenesis has been demonstrated in a very large number of plants and trees, the use of somatic embryos in large-scale commercial production has been restricted to only a few plants, such as carrot, date palm, and a few forest trees.

Somatic embryos are produced as adventitious structures directly on explants of zygotic embryos, from callus and suspension cultures. Somatic embryos and synthetic seeds (embryos encapsulated in artificial endosperm) hold potential for large-scale clonal propagation of superior genotypes of heterogeneous plants.

They have also been used in commercial plant production and for the multiplication of parental genotypes in large-scale hybrid seed production. In many species, somatic embryos are morphologically similar to the zygotic embryos, although some biochemical, physiological and anatomical differences have been documented.

The synthetic auxin, 2,4-D is commonly used for embryo induction. In many angiosperms, *e.g.*, carrot and alfalfa, subculture of cells from 2,4-D containing medium to auxin-free medium is sufficient to induce somatic embryogenesis. The process can be enhanced with the application of osmotic stress, manipulation of medium nutrients, and reducing humidity. Selection of embryogenic cell lines has also been successfully used. For example, selection for unique morphotypes in grapevine cultures allows production of high quality embryos with predicable frequency.

A major problem in large-scale production of somatic embryos is culture synchronization. This is achieved through selecting cells or pre-embryonic cell clusters of certain size, and manipulation of light and temperature, temporary starvation or by adding cell cycle synchronizing chemicals to the medium. Cytokinins seem to play a key role in cell cycle synchronization and embryo induction, proliferation and differentiation. Abscisic acid is crucial in all the stages of somatic development, maturation and hardening.

Synthetic Seeds

The concept of production and utilization of synthetic seeds (somatic embryo as substitutes for true seeds) was first suggested by Murashige in 1977. Synthetic seeds can be produced either as coated or non-coated, desiccated somatic embryos or as embryos encapsulated in hydrated gel (usually calcium alginate).

Successful utilization of synthetic seeds as propagules of choice requires an efficient and reproducible production system and a high percentage of post-planting conversion into vigourous plants. Artificial coats and gel capsules containing nutrients, pesticides and beneficial organisms have long been thought as substitutes for seed coat and endosperm. However, this technology is still in the developmental stage, and currently cannot compete with the other methods of commercial plant propagation.

Process of Micropropagation

The process of plant micro-propagation aims to produce clones (true copies of a plant in large numbers).

The process is usually divided into the following stages:

Stage 0 – Pre-propagation step or selection and pre-treatment of suitable plants.

Stage I – Initiation of explants-surface sterilization, establishment of mother explants.

Stage II – subculture for multiplication/proliferation of explants.

Stage III – shooting and rooting of the explants.

Stage IV – Weaning/hardening.

These stages are universally applicable in large-scale multiplication of plants. The individual plant species, varieties and clones require specific modification of the growth media, weaning and hardening conditions.

A rule of the thumb is to propagate plants under conditions as natural or similar to those in which the plants will be ultimately grown ex-vitro. For example, if a chrysanthemum variety is to be grown under long day-length for flower production, it is better to multiply the material under long-day length at stages III and IV. There is a wide option to undertake production of plant material up to a limited number of stages. For example, many commercial tissue culture companies undertake production up to Stage III, and leave the remaining stages to others.

Pre-Propagation Stage

The pre-propagation stage (also called stage 0) requires proper maintenance of the mother plants in the greenhouse under disease- and insect-free conditions with minimal dust. Clean enclosed areas, glasshouses, plastic tunnels, and net-covered tunnels, provide high quality explant source plants with minimal infection. Collection of plant material for clonal propagation should be done after appropriate pretreatment of the mother plants with fungicides and pesticides to minimize contamination in the in vitro cultures.

This improves growth and multiplication rates of in vitro cultures. The control of contamination begins with the pretreatment of the donor plants. They may be prescreened for diseases, isolated and treated to reduce contamination. The explants are then brought to the production facility, surface sterilized and introduced into culture. They may at this stage be treated with antibiotics and fungicides as well as anti-microbial formulations, such as PPM. The explants are then culture indexed for contamination by standard microbiological techniques, which are occasionally supplemented with tests based on molecular biology or other techniques.

Stage I

This stage refers to the inoculation of the explants on sterile medium to initiate aseptic culture. Initiation of explants is the very first step in micropropagation. A good clean explant, once established in an aseptic condition, can be multiplied several times; hence, explant initiation in an aseptic condition should be regarded as a critical step in micropropagation. More than often, explants fail to establish and grow, not due to the lack of a suitable medium but because of contamination. The explants are transferred to in vitro environment, free from microbial contaminants. The process requires excision of tiny plant pieces and their surface sterilization with chemicals such as sodium hypochlorite, ethyl alcohol and repeated washing with sterile distilled water before and after treatment with chemicals. After a short period of culture, usually 3 to 5 days, the contaminated explants are discarded.

The surviving explants showing growth are maintained and used for further subculture. In herbaceous plants *e.g.* potato, chrysanthemum, carnation, streptocarpus, strawberry, and African violet; the explant sources are meristems, apical- and axillary buds, young seedlings, developing young leaves and petioles, and unopened floral buds. The following low cost options can be adapted to initiate explants:

Sterile Instrument Technique

This method assumes that most of the deep-seated meristems and those covered by leaves or other integuments (*e.g.* floral bracts) are sterile. In this procedure, the explant is washed with sterile water, rinsed in ethanol, and instruments are sterilized every time they touch the surface of the explant, and the explant is moved to a new location on the dissection stage.

Surface Sterilization Technique

This is by far the most commonly used method. The explants are washed in sterile water, rinsed in ethanol, and surface sterilization is achieved by using chemicals with chlorine base. Calcium or sodium hypochlorite based solutions, 1–3 per cent (v/v) are usually used for soft herbaceous materials. A cheap and ready-made sterilant is 5-7 per cent solution of 'Domestos'- a toilet disinfectant which contains 10.5 per cent v/v sodium hypochlorite, 0.3 per cent sodium carbonate, 10.0 per cent sodium chloride and 0.5 per cent (w/v) sodium hydroxide and a patented thickener). The explants are washed in sterile distilled water before and after sterilization. Other surface sterilants used include mercuric chloride (avoid its use as far as possible, since it is highly toxic), hydrogen peroxide, and potassium permanganate.

For Soft Tissues

- Wash explants from perennial plants for 1-2 hr in tap water. Eliminate this step for material from glasshouse grown plants.
- Wash in sterile distilled water three to four times for 5 to 10 minutes each.
- Dip in 95 per cent ethanol for 3 to 5 seconds.
- Wash once again with sterile distilled water for 5 minutes.
- Surface-sterilize in 5 per cent 'Domestos' (v/v) for 20-25 minutes.
- Wash with sterile distilled water three times for 10 minutes each.
- Drain water droplets by placing on pre-sterilized blotting paper.
- Transfer explants singly to the medium.

N.B.: Sterilize forceps each time to transfer explants to avoid cross-contamination.

For woody stems (e.g. roses, hardy shrubs, and trees):

- Collect stems, shoots, buds and store at 5 oC till needed.
- Rinse in ethanol for 3 to 5 seconds.
- Rinse in 1- per cent sodium hypochlorite (20 per cent bleach) for 10 minutes.
- Place lower parts of stems in flasks in 2 per cent sucrose and 200 PPM 8-hydroxyquinoline citrate at 23+2 oC. For items collected in September/October, add 50-PPM GA3.

After that l0 PPM GA will help break the dormancy:

- Re-cut the bottom of stem and replace the solution after 2 days.
- Excise the softwood from the developed shoots and use material for explants or for rooting.
- Surface-sterilize as in the above protocol.

Do not forget to sterilize forceps and scalpel every time for the transfer of explants to fresh solutions. Use sterile containers in the protocol of surface sterilization. If explants become brown or pale at the end of the protocol, reduce the strength of 'Domestos' to 2.5 per cent. Alternatively, dip explants in 10 per cent 'Domestos' for 2 minutes and then proceed to surface sterilize with 3-5 per cent 'Domestos' for 20 minutes. If basal contamination is observed after 2-3 days of culture, explants can sometimes be rescued by removing the basal end by making a single cut with a sharp scalpel and re-culturing on fresh medium.

Stage II

Stage II is the propagation phase in which the explants are cultured on the appropriate media for multiplication of shoots. The primary goal is to achieve propagation without losing the genetic stability. Repeated culture of axillary and adventitious shoots, cutting with nodes, somatic embryos and other organs from Stage I leads to multiplication of propagules in large numbers. The propagules produced at this stage can be further used for multiplication by their repeated culture. Sometimes it is necessary to subculture the in vitro derived shoots onto different media for elongation.

Stage III

The in vitro shoots obtained at Stage II are rooted to produce complete plants. If the proliferated material consists of bud-like structures (*e.g.* orchids) or clumps of shoots (banana, pineapple), they should be separated after rooting and not before. Many plants (*e.g.* banana, pineapple, roses, potato, chrysanthemum, strawberry, mint, several grasses and many more) can be rooted on half-strength-MS medium without any growth-regulators. Good sturdy well-rooted plants are essential for high survival during weaning and later transfer to soil. This stage is labour intensive and expensive. The process of in vitro rooting has been estimated to account for approximately 35–75 per cent of the total cost of production. Efforts should be made to combine rooting and acclimatization stages.

Stage IV

At this stage, the in vitro micropropagated plants are weaned and hardened. This is the final stage of the tissue culture operation after which the micropropagated plantlets are ready for transfer to the greenhouse. Steps are taken to grow individual plantlets capable of carrying out photosynthesis. The hardening of the tissue-cultured plantlets is done gradually from high to low humidity and from low light intensity to high intensity conditions. If grown on solid medium, most of the agar can be removed gently by rinsing with water.

Plants can be left in shade for 3 to 6 days where diffused natural light conditions them to the new environment. The plants are then transferred to an appropriate substrate (sand, peat, compost, etc.), and gradually hardened. Low-cost options include the use of plastic domes or tunnels, which reduces the natural light intensity and maintains high relative humidity during the hardening process. If the plants are still joined together after rooting, these should be planted as bunches in the soil and separated after 6 to 8 weeks of growth.

2

Tissue Culture Techniques

An important aspect of all biotechnology processes is the culture of either the plant cells or animal cells or microorganisms. The cells in culture can be used for recombinant DNA technology, genetic manipulations etc.

Plant cell culture is based on the unique property of the cell-totipotency. CELL-TOTIPOTENCY is the ability of the plant cell to regenerate into whole plant. This property of the plant cells has been exploited to regenerate plant cells under the laboratory conditions using artificial nutrient mediums. With the advances made in genetic engineering, it became possible to introduce foreign genes into cell and tissue culture systems. This led to the development of Genetically Modified (Gm) or Transgenic Crops which had improved traits and characteristics.

Fungi and bacteria grow on the plant surface and contaminate the culture media when they are not adequately eliminated. The process of explant introduction into in vitro conditions depends mainly on the disinfecting phase. Problems in handling produce contamination of the in vitro plant and so, a similar disinfection process to that of the in vitro introduction technique is recommended. However, it is possible to use antibiotics included in the medium, but just temporarily, while the plantlets are growing. Among the bacterial contaminants are Bacillus sp., Erwin/a sp., Pseudomonas sp., etc. In the case of systemic infections, the use of disinfectants is not effective, because the pathogens are located in the vascular system. In this case, it is recommended to use specific antibiotics and meristems or bud cuttings.

On the other hand, the process of in vitro introduction involves the use of explants in different physiological stages, as dormant buds, which require growth regulators for example, Gibberellic acid) to stimulate and accelerate bud growth. During the process of in vitro introduction there is a close connection between the greenhouse and the laboratory; that is why the necessary precautions should be taken to avoid the entrance of contaminants into the laboratory.

- In the greenhouse
  - Take cuttings from the mother plant, or buds from greenhouse tuber sprouts.

- In the laboratory
  - Immerse the cuttings for 10 minutes in a beaker conveniently labeled with the accession name, and containing a solution of 0.5 per cent acaricide with three drops/I of Tween 20.
  - Throw away the acaricide solution and rinse the nodes with tap water
  - Prepare the laminar flow chamber.
  - Immerse the cuttings in 700/o alcohol for 30 seconds and take the beakers immediately to the laminar flow chamber
- In the laminar flow chamber
  - Eliminate the alcohol and replace it with a 2.50/o solution of calcium hypochlorite. Keep the cuttings immersed during 15 minutes.
  - Throw away the hypochlorite, and rinse the cuttings three times with sterile water
  - Keep the nodes immersed in sterile water until bud extraction.
  - Dissect the buds on a sterile plate and place them in the culture medium.

In vitro plantlets, which are free of pathogens, are used as initial material for potato and sweetpotato seed programmes. The methods used in these micropropagation programmes mainly depend on their production volume and the available infrastructure.

In the case of potato micropropagation, the basic methods have been already described. They have been verified in many institutions and they are based on the rapid growth of individual node cuttings, or stems with multiple node cuttings. Afterwards, the basic micropropagation methods are described.

NODE MICROPROPAGATION

This method is based on the principle that the node of an in vitro plantlet placed in an appropriate culture medium will induce the development of the axillary bud, resulting in a new in vitro plantlet.

This type of propagation promotes the development of a pre-existent morphological structure. The nutritional and hormonal condition of the medium breaks the dormancy of the axillary bud and promotes its rapid development.

Callus formation and plant regeneration must be avoided because they tend to affect the genetic stability of the genotype. Under room-controlled conditions micropropagation is fast.

Each node planted in a propagation medium will produce a plantlet which will occupy the full length of the test tube, after approximately four weeks for potato, and six weeks for sweetpotato. The resultant in vitro plantlets may be transplanted to in vitro conditions in small pots in the greenhouse.

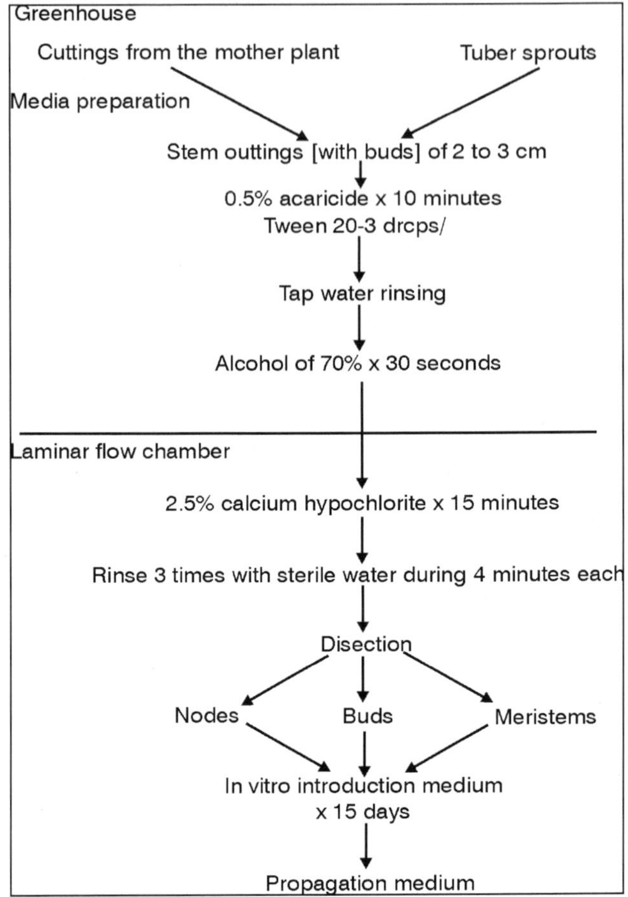

Fig. Procedure for in vitro Introducction

MICROPROPAGATION BY NODE CUTTINGS IN A LIQUID MEDIUM

This technique is applied both with potato and sweetpotato to produce a large number of nodes rapidly.

Stem cuttings with 5 to 8 nodes are prepared by removing both the apex and the root of the in vitro plant to be propagated. The stems are placed in the corresponding propagation liquid medium. It is also possible to use isolated nodes: the nodes will sprout and new plantlets will develop over a period of 3 to 4 weeks.

Micropropagation Procedure

- Sterilize petri dishes (placed in paper bags or comets) and prepare the laminar flow chamber by disinfecting the internal surfaces with alcohol.

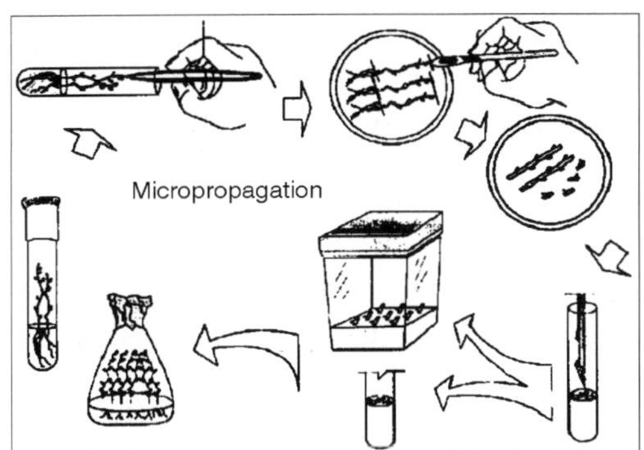

Fig. Potato Micropropagation Process Scheme

Sterilize the tools with an instrument sterilizer and place them on a sterile dish.

- Open the tube, take off the plantlet and place it on a petri dish with the help of forceps.
- Remove the leaves and cut the nodes.
- Open a tube containing fresh sterile medium and place a node inside, trying to plunge it slightly into the medium with the bud up. Close the tube.
- Seal the tube with a gas-permeable plastic tape (parafilm or saram wrap) and label it correctly. It is recommended to place two explants in 16 × 125 mm tubes, three in 18 × 150 mm tubes, five in 25 × 150 mm tubes, and 20-30 in magenta vessels.

COMMON PROBLEMS IN MICROPROPAGATION

Some problems may appear in tissue culture according to the crop or variety. To solve them, it is necessary to apply one or several preventing/ solving methods such as:

Phenolization

The explants frequently become brown or blackish shortly after isolation. When this occurs, growth is inhibited and the tissue generally dies. The young tissues are less susceptible to darkening than the more mature ones.

Prevention

The tissue darkening– mainly that of the recently isolated explants and that of the medium– may be generally prevented by:

- Removing the phenolic compounds produced by dispersion. Absorption by means of activated carbon. Absorption by polyvinilpyrolidone (PVP).
- Modifying the redox potential. Reducing agents: ascorbic acid, citric acid, L-cisteine HCL, ditriotreitol, glutation and mercaptoethanol Less availability of oxygen: stationary liquid media.

- Inactivating the phenolase enzymes. Chelating agents: NaFeEDTA, EDTA, diethyldithiocarbamate, dimethyl-dithiocarbamate
- Reducing the phenolasic activity and the availability of substrate. Low pH Darkness

Absence of Rooting

The explants can naturally form roots during propagation, without an additional rooting stage, as with the potato. However, some wild potato species may show root production deficiency. Rooting may be induced by incorporating auxins, such as IAA, NAA, and IBA, or activated carbon to the culture medium.

POTATO IN VITRO TUBERIZATION

Most of the potato microtubers are used in seed programmes in Europe, where large amounts (hundred of thousands) of pre-basic seed of a few varieties are produced. By means of this technique, microtubers are produced and stored, and it is possible to store thousands of them in a small area (in humid containers at 4°C) for long periods of time. The microtuber induction is produced through a stress effect by the CCC (chlorocholine chloride), BAP (B-benzylaminopurine) and Sucrose, which under darkness will produce from 3 to 4 microtubers per plant, according to the variety.

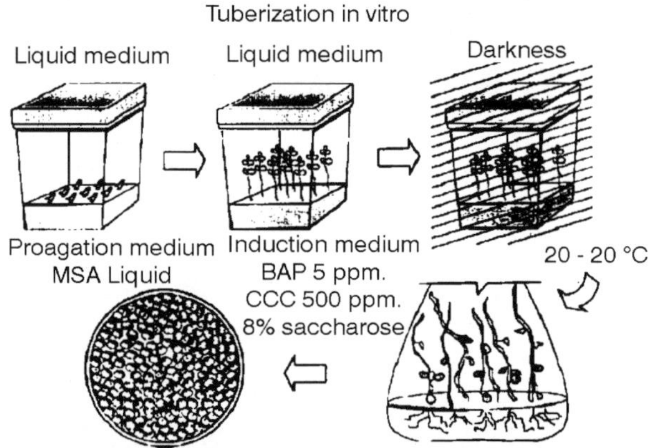

Fig. Poato in Vitro Tuberization Process Scheme

In the beginning, microtubers were used as an alternative for germplasm distribution and in vitro conservation. Curing the plantlet in vitro multiplication process, usually there is a multiplication of higher amounts to those required in the greenhouse and so, after the transference, some magentas are left over. These plantlets may be used for microtuber induction.

In accordance with the indicated procedures, the induction medium is added to the magentas and then they are placed in the dark room. After three months the microtubers will be produced. These may be harvested and transferred to sterile containers (at 4°C), where they can be maintained up to 10 months, or used in the next campaign instead of in vitro plantlets.

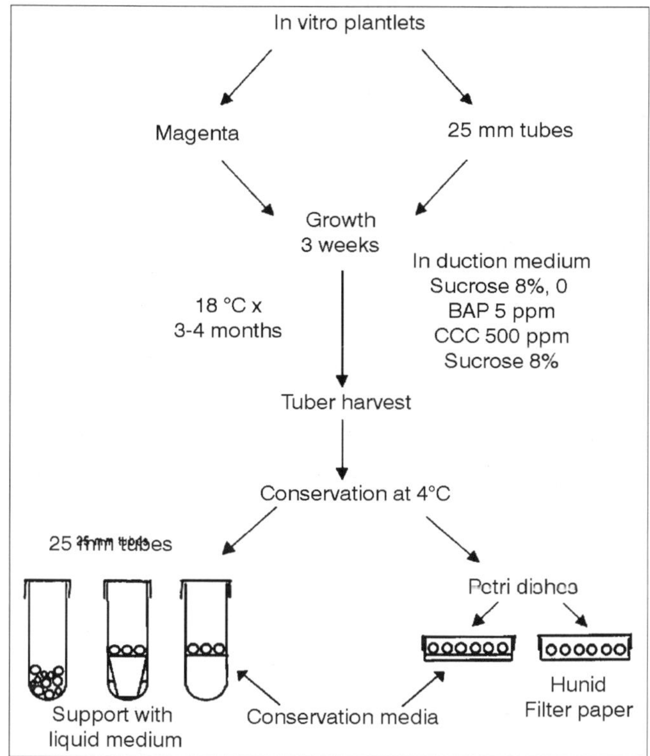

Fig. Potato in Vitro Tuberization Process Scheme

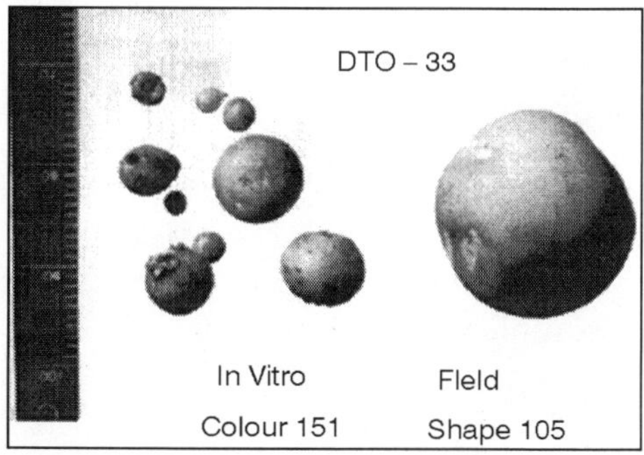

Fig. Comparison, in Size and Shape of in Vitro Microtubers with a Normal one from the Field

These microtubers can also be used as a reserve, in case in vitro material is contaminated or dies, because of temperature or handling effects. The high number and bigger size of the microtubers increases the success rates for the transfer to the greenhouse.

MEDIUM TERM GERMPLASM MAINTENANCE

The in vitro germplasm maintenance under normal growth conditions requires a series of transfers of the plantlets into a fresh medium. This leads to a consumption of time, increases the possibility of losses because of material contamination during successive sub-cultures, causes a loss of material by human error or failure of some equipment, and demands more labour A way to avoid these problems is through limitation, restriction or inhibition of growth.

This approach consists of growth speed reduction by modifying the physical or chemical conditions of the culture, and it is effective for a short or medium term period.

In Vitro Conservation Methods

These consist in maintaining the cultures (buds, plantlets derived from nodes or directly from meristems) under physical (environmental factors] or chemical (culture medium composition) stressed conditions that make it possible to extend, as much as possible, the interval of transference into the fresh media, without affecting the viability of the cultures. The methods to reduce the in vitro cultivated plant growth include the reduction of temperature and light during storage, the incorporation of growth retardants in the medium, and the induction of osmotic stress in the medium, or a combination of all these.

Temperature

Temperature reduction has been the most used way to curb culture growth. Most of the in vitro cultures are maintained at temperatures between 12 and 200C: at lower temperatures, the growth rate decreases but the reduction depends on the species.

Nutrients Concentration

The reduction of the carbohydrate concentration and the nitrogenous components of the nutritive medium may affect the growth rates. In addition, the continuous absorption of these nutrients during plantet growth will bring a nutritional defficiency, which could produce a premature death of the plants.

Use of Growth Regulators

The abscisic acid (ABA), phosphon-D, maleic hydrazide, and succinic acid are some of the most frequently used growth regulators.

Osmotic Concentration

Growth limitation caused by osmotic concentration is due to the reduction of the water and nutrients absorbed from the medium. For example, at high concentrations, sucrose acts osmotically and it is highly metabolized. No metabolized osmotics, such as Manitol and Sorbitol, are possibly more efective than sucrose in culture growth limitation.

Evaluation of in Vitro Preserved Material

To evaluate material under these conditions, some important facts about in vitro maintenance such as genetic viability and stability should be considered. The viability evaluation of the in vitro cultures must be systematic. In slow growth conditions, when the sub-culture or transfer period extends during months or years, the frequency of the culture evaluation increases. The most important characteristics to be evaluated in the low-growth cultivars storage of apical buds are: contamination, leaf senescence, the number of green sprouts, the number of viable nodes in relation to the stem length, the presence or not of roots, and callus formation.

MAINTENANCE OF ACCESSIONS IN SEED PROGRAMME

In a seed programme, a group of accessions free of virus is maintained for pre-basic seed production: however, most of them are not propagated in the greenhouse so they are maintained in vitro to be used in the future. The continuous propagation of the in vitro plantlets damages the material, mainly if we consider that environmental conditions are not adequate. The alternative is to maintain the accessions in conservation media. Each accession must be maintained in test tubes with five replications to avoid possible losses. The maintenance conditions have been tested in CIP's potato collection that consists of more than 5,000 accessions, by means of which it is possible to assure the normal recovery of the material after using stress producers.

The plantlets used in each campaign must come from the maintenance phase. Then, they will be propagated in normal media where their growth will be reestablished, and the elected procedures for plantlet multiplication in the greenhouse will continue.Each season must be initiated with this material to start off with strong plants. In addition, the plantlets taken for multiplication must be replaced, trying to maintain 5 tubes per accession in the conservation media.

The sub-cultures in the conservation medium are carried out approximately every one or two years. The conservation medium renovation must pass through a previous sub-culture in the propagation media to rejuvenate the explants.

HISTORY OF TISSUE CULTURE TECHNIQUES

The use of tissue culture techniques initiated by Gottlieb Haberlandt in 1902 in an attempt mengkulturkan hair cells of the leaf mesophyll tissue monocot plant. But the effort failed because the cells do not have cleavage, it was alleged failures because they do not use growth regulator substances needed for cell division, proliferation and induction of the embryo. In the year 1904, Hannig planting embryos isolated from several plant crucifers. In 1922, Knudson and separately each Robbin conduct investment business and culture of orchid seedlings root tip. After the 1920s, the discovery and development of tissue culture techniques continues. The following table shows the historical development of the field of plant tissue culture which was adapted from various sources. Important discoveries in the history of plant tissue culture.

Year important findings:

1902:	*Haberlandt*: The first person who tried to isolate and mengkulturkan monocot plant tissue, but failed
1922:	*Knudson*: mengecambahkan orchid seeds
	Blumenthal and Meyer 1924: The formation of callus from carrot roots eksplan
1929:	*Laibach and Hered*: Culture of embryos to overcome the incompatibility in plants Linum spp.
1934:	Culture in vitro oflGautheret woody plant tissue kambium and shrubs, but failed.
	White: Successfull culture of tomato roots in a long time
	Kogl et.al: Identification of thel first plant hormone, IAA, for the elongation of cells.
1936:	*LaRue*: Culture of embryos in some plants Gymnospermae
1939:	*Gautheret*: Successfully growing culture kambium carrot and tobacco plants
1941:	*Overbeek*: The use of coconut water for the young embryo culture menumbuhkam Datura plants
1944:	The first in vitro culture of tobacco plants to study the formation of buds adventif
	Skoog and Tsui 1948: The formation of shoots and roots of tobacco adventif
1949:	*Nitsch*: in vitro culture of fruit plants
1952:	*Morel and Martin*: meristem culture to obtain a free Dahlia plant viruses. The firstl success of micro-called grafting.
1953:	*Tulecke*: Kalus haploid pollen plants of Ginkgo biloba *Miller* 1955: The discovery of the structure and synthesis of kinetin
	Skoog and Miller 1957: Menemuan that the formation of roots and shoots in comparison tergatung culture auksin: cytokines
1958:	*Maheswaril and Rangaswamy*: Regeneration of somatic embryos Citrus nuselus ovul
	Reinert and Steward: Growth and development of carrot suspension cultures
	λ*Cocking* 1960: enzymatic degradation of cell walls to get protoplas
	Morel: λ Vegetative Propagation of orchids by meristem culture

1962:	*Murashige and Skoog*: Development of MS medium
1964:	*Guha and Maheswari*: The discovery of the first haploid plants by androgenesis Datura plant
	Erickson and Jonassen 1969: Isolation protoplas of cell suspension Hapopappus
1970:	*Power*: Figaro protoplas
1977:	*The success of T-DNA integration in plantslChilton*
	Noguchi et al: Planting of tobacco cells in Bioreaktor capacity of 20 000 L.
1978:	*Melchers et al*: Somatic hybridization between tomato and potatol plants
	Tabata et al: Production of shikonin on the industrial scale celll culture
	Zimmermann 1982: Figaro protoplas electrically (Electrofusion)
	Mitsui Petrochemicals 1983: The first secondary metabolite production in industrial scale through suspension culture at the plant Lithospermum spp.
1985-1990:	Development of gene transfer in rapidly growing crops, such as the use of Agrobacterium, particle bombardment (gene gun), electroporasi, mikroinjeksi.

Development of genetic and metabolicλ1990 - engineering challenges. growing rapidly Marketing products of geneticλ engineering

The in vitro techniques were developed initially to demonstrate the totipotency of plant cells predicted by Haberlandt in 1902. Totipotency is the ability of a plant cell to perform all the functions of development, which are characteristic of zygote, i.e., ability to develop into a complete plant. In 1902, Haberlandt reported culture of isolated single palisade cells from leaves in Knop's salt solution enriched with sucrose. The cells remained alive for up to 1 month, increased in size, accumulated starch but failed to divide. Efforts to demonstrate totipotency led to the development of techniques for cultivation of plant cells under defined conditions.

Haploid plants from pollen grains were first produced by Maheshwari and Guha in 1964 by culturing anthers of Datura. This marked the beginning of anther culture or pollen culture for the production of haploid plants. The technique was further developed by many workers, more notably by JP. Nitch, C. Nitch and coworkers. These workers showed that isolated microspores of tobacco produce complete plants. Plant protoplasts are naked cells from which cell wall has been removed. In 1960, Cocking produced large quantities of protoplasts by using cell wall degrading enzymes.

The techniques of protoplast production have now been considerably refined. It is now possible to regenerate whole plants from protoplasts and also to fuse protoplasts of different plant species. In 1972, Carlson and coworkers produced the first somatic hybrid plant by fusing the protoplasts of Nicotiana glauca and N. langsdorfii. Since then many divergent somatic hybrids have been produced.

A successful establishment of callus cultures depended on the discovery during mid-thirties of IAA (idole-3-acetic acid), the endogenous auxin, and of the role of B vitamins in plant growth and in root cultures. The first continuously growing callus cultures were established from cambium tissue in 1939 independently by Gautheret, White and Nobecourt. The subsequent discovery of kinetin by Miller and coworkers in 1955 enabled the initiation of callus cultures from differentiated tissues. Shoot bud differentiation from tobacco pith tissues cultured in vitro was reported by Skoog in 1944, and in 1957 Skoog and Miller proposed that root-shoot differentiation in this system was regulated by auxin-cytokinin ratio. The first plant from a mature plant cell was regenerated by Braun in 1959. Development of somatic embryos was first reported in 1958- 1959 from carrot tissues independently by Reinert and Steward. Thus within a brief period, the tissue culture techniques have made a great progress. From the sole objective of demonstrating the totipotency of differentiated plant -cells, the technique now finds application in both basic and applied researches in a number of-fields of enquiry.

TYPES OF TISSUE CULTURE

Tissue culture consists of growing plants cells as relatively on organized masses of cells on an agar medium (callus culture) or as a suspension of free cells and small cell masses in a liquid medium (suspension culture).

Tissue culture is used for vegetative multiplication of many species and in some cases for recovery of virus-free plants. It has potential application in production of somatic hybrids, organelle and cytoplasm transfer, genetic transformation and germplasm storage through freeze-preservation.

CLONAL PROPAGATION

Tissue culture is well suited for quick vegetative propagation of plant species. It is used for asexual propagation in many species of fruit and timber trees and also used for obtaining disease free and virus-free plants. The major difficulty in the use of this technique in clonal multiplication is the occurrence of genetic variation among the regenerated plants. This problem can be reduced to a large extent by using young tissue cultures, preferably during the first few subcultures.

MUTANT ISOLATION

Biochemical mutants are far more easily isolated from cell cultures than from whole plant populations. This is because a large number of cells, 10^6-10^9, can be easily and effectively screened for biochemical mutant cells. Biochemical mutants could be selected for disease resistance, improvement of nutritional quality, adaptation of plants to stress conditions, e.g. saline soils, and to increase the biosynthesis of plant products used for medicinal or industrial purposes.

Somaclonal Variation

Plants regenerated from tissue and cell cultures show heritable variation for both qualitative and quantitative traits; such a variation is known as somaclonal variation. Somalconal variation has been described in sugarcane, potato, tomato etc. Some variants are obtained in homozygous condition in the plants regenerated from the cells in vitro (R_0 generation), but most variants are recovered in the selfed progeny of the tissue culture-regenerated plants (R_1 generation). Somaclonal variation most likely arises as a result of chromosome structural changes, e.g., small deletions and duplications, gene mutations, plasma gene mutations, mitotic crossing over and possibly, transposons.

Somaclonal variation may be profitably utilized in crop improvement since it reduces the time required for releasing the new variety by at least two years as compared to mutation breeding and by three years in comparison to back cross method of gene transfer.

Amino Acid Analogue Resistant Mutants

Cereal grains are deficient in lysine; maize (Zea maize) is also deficient in trytophan, while wheat (T.aestivum) and rice (O.sativa) are deficient in threonine. Pulses are deficient in methionine and trytophan.

Amino acid analogue-resistant cells may be expected to show a relatively higher concentration of that particular amino acid. For e.g., carrot (D.carota) and tobacco (N.tabacum) cell lines resistant to trytophan analogue 5-methyl trytophan show a 10-27-fold increase in the level of trytophan. Similarly, rice cells resistant to lysine analogue 5-(B-aminoethyl)-cysteine, show much higher levels of lysine. This technique may prove useful in the development of crop varieties with a better-balanced amino acid content.

Disease Resistant Mutants

Many pathogenic bacteria produce toxins that ae toxic to plant cells. Plant cell cultures may be exposed to lethal concentrations of these toxins and resistant clones isolated. Plants regenerated from these resistant clones would be resistant to the disease producing pathogen. This technique should be applicable to all the pathogens, which produce the disease through the action of toxin. An e.g., an application is in the case of wildfire disease of tobacco (N.tabacum) produced by Pseudomonos tabaci.

Tobacco cells resistant to methionine sulfoximine, which is similar to the toxin produced by the pathogen, were isolated. Plants regenerated from these clones were resistant to wildfire disease, although to a somewhat lesser degree. The technique can be applied to those cases only where the disease is the result of a toxin produced by the pathogen. But many of the pathogens do not seem to produce a toxin, or the toxin does not appear to be the primary cause of the disease.

Stress Resistant And Other Mutants

Plant cells resistant to 4-5 times the normally toxic salt (NaCl) concentration have been isolated. Attempts to insolate such cells are being made. Similarly, attempts are being made to isolate clones that would produce more substances of medicinal or industrial value.

Somatic Hybridization

Protoplasts can be isolated from almost every plant species and cultured to produce callus. Protoplasts of two different species may be fused with the help of polyethylene glycol.

Genetic Transformation

There is some evidence that gene transfer may be achieved by feeding cells with DNA in case of eukaryotes, such as, Drosophila, Neurospora, cultured mammalian cells and in some plants. Genetic changes may be brought about by DNA or by radiation-killed pollen grains. This raises the possibility of genetic modification of plant cells with the help of both homologous (from the same species) and heterologous (from a different species) DNA. It is also proposed that DNA plant viruses, such as cauliflower (B.oleracea) mosic virus and potato leaf roll virus, plasmids (e.g., Ti plasmid of Agrobacterium) and transposons, may be used as the carriers of genes for genetic modification of plant cells.

Organelle Transfer

In some cases, it may be desirable to transfer only organelles or the cytoplasm into a new genetic background. This may be achieved through the use of plant protoplasts. Chloroplasts have been transferred, and other organelles including nucleus may be transferred.

Germplasm Conservation

Tissue cultures may be frozen and stored in liquid nitrogen at -196^0C for long-term storage of germplasm. This would be of great value in the conservation of germplasm of those crops which normally do not produce seeds, e.g. root and tuber crops, or where it may not be desirable to store seeds. For freeze-preservation, the cells are cooled at a slow rate and are then transferred to liquid nitrogen for storage. Thawing of the cells must be very rapid for increased survival. A cryoprotectant, such as dimethylsufoxide (DMSO), is used to protect the cells from injury due to freezing and thawing. The technique of freeze-preservation i.e., crybiology, of plant cells is still in the developing stages.

ACHIEVEMENTS AND FUTURE PROSPECTS

Tissue culture techniques are being exploited to enhance crop production and to aid crop improvement efforts. Faster clonal multiplication is being exploited on commercial scale for many horticultural species e.g. oil palm, mentha, roses, carnation etc. Tissue cultured somatic tissues are now routinely being used for conservation of those species whose seeds are recalcitrant or ones which do not produce seed at all.

Embryo culture has helped in rescuing hybrid embryos enabling the recovery of many interspecific hybrids and haploid plants. Shoot tip (meristem) culture plays a vital which is of great importance in germplasm exchange, and the development of serological techniques for the detection of viruses in plant materials is a great help to the efforts in this direction.

METHODS AND TECHNIQUES OF PLANT TISSUE CULTURE

In plant tissue culture, plants or explants such as pieces of leave, stem or root is cultured in a specific plant medium, which contains essential plant nutrients and hormones. Other plant growth factors like light and temperature are maintained and regulated by using artificial conditions. All the procedures of plant tissue culture are conducted under sterile (aseptic) conditions. The explants then develop stem, roots and leaves. The generated plantlets are hardened before planting in outdoor conditions.

To start with plant tissue culture, one should wipe his/her hands, forceps and other equipments with alcohol or other sterilizer to prevent microbial contamination. Following this, the explants are surface sterilized by using chemical solutions such as bleach or alcohol. Though mercury chloride is an effective sterilizer, it is rarely used due to its potential toxicity. After sterilization, the explants are introduced into a plant medium, which can be either solid or liquid. The cells divide and differentiate into plant parts, thus giving rise to a complete plant. Speaking about the plant medium, liquid type is prepared by mixing inorganic salts, organic nutrients (sucrose), vitamins, minerals and plant growth regulators (auxin and cytokinin). In addition to these ingredients, solid medium contains gelling agent (agar). Nutrient and plant hormones amount vary, depending on the objective of plant tissue culture. For example, in order to induce more roots, auxin amount should be high. Plant tissue culture techniques include the culture of protoplast (a cell without cell wall), meristem, node, anther, ovule, embryo and seed.

APPLICATIONS OF PLANT TISSUE CULTURE

Plant tissue culture has wide applications in agriculture. One major advantage of plant tissue culture is the production of disease and/or pest resistant varieties, thus indirectly increasing the crop yield. Commercially, it is used directly for the propagation of plants that are hard to propagate in natural conditions. Horticultural plants such as orchids, roses, banana, strawberries, potatoes, apples, etc. are successfully cultured in in-vitro conditions. Plants with valuable secondary products are grown in the controlled conditions by using plant tissue culture technique. It is also used for the propagation of medicinal herbs on a large scale. With plant tissue culture, it is possible to generate virus-free plantlet of vegetatively propagated plants.

In experimental biology such as plant breeding, cell biology, biotechnology and genetics, plant tissue culture is applied in order to solve plant related problems. It allows screening of cells for the desirable characters such as early fruit bearing, disease resistance and drought resistance. Another important application is generation of a novel hybrid by crossing two distantly related species having advantageous traits. In-vitro fertilization and/or pollination of plants is possible, irrespective of the hindrances in natural conditions. Overall, plant tissue culture is used for the conservation of germplasm.

PLANT TISSUE CULTURE MEDIA

Plant Tissue Culture refers to the technique of growing plant cells, tissues, organs, seeds or other plant parts in a sterile environment on a nutrient medium. In the plant tissue culture process, plant parts are cut into small pieces that will give rise to many more individual plants. Soil or hydroponically grown plants only need fertilizer (a source of K, N and P plus trace minerals), water, air and light to grow, because they can make their own sugar, amino acids, etc.

Plant tissue cultures need an outside source of sucrose because the tissue culture cells go into shock when the explants are made. It helps to add vitamins in addition to the fertilizer and sucrose. Pre-mixed media can be purchased in packs for preparing 1 liter. One liter usually provides enough medium to fill 60 50-ml tubes with 15 ml. Some folks like to use deep Petri plates (20 × 100mm). These need about 25 ml of medium per plate.

Is sugar in the pre-mix? Generally the pre-mixed formulations DO NOT contain sugar, nor do they contain agar. A simple way to figure out if the formulation packet contains sugar is that the total weight on a 1 liter packet would be around 30 grams. If sugar is NOT included, the total weight will be less than 5 grams. Generally, folks add 20-30 grams of sucrose to plant tissue culture medium.

What agar to use? Agar is left out because these pre-mixes are often used for suspension cultures. Note that Nutrient Agar contains lots of food for bacteria in addition to the inert agar. Be sure to use AGAR (not nutrient agar). Gel-Rite and Phytagar are more pure preparations used by professionals. Gel-Rite is used at lower concentrations (grams per liter or%) than agar. Bacteriological agar works fine for schools. Typically, agar is added at 0.7%-0.8% or 7-8 grams/liter.

PREPARATION OF PLANT TISSUE CULTURE MEDIUM

- Measure approximately 90% of the required volume of the deionized-distilled water in a flask/container of double the size of the required volume.
- Add the dehydrated medium into the water and stir to dissolve the medium completely. Gentle heating of the solution may be required to bring powder into solution.
- Add desired heat stable supplements to the medium solution.
- Add additional deionized-distilled water to the medium solution to obtain the final required volume.
- Set the desired pH with NaOH or HCl.
- Dispense the medium into culture vessels.
- Sterilize the medium by autoclaving at 15 psi (121°C) for appropriate time period. Higher temperature may result in poor cell growth.
- Add heat labile supplements after autoclaving.

CONCEPT

The 30 major soil groups of the WRB are *Acrisols, Albeluvisols, Alisols, Andosols, Anthrosols, Arenosols, Calcisols, Cambisols, Chernozems, Cryosols, Durisols, Ferralsols, Fluvisols, Gleysols, Gypsisols, Histosols, Kastanozems, Leptosols, Lixisols, Luvisols, Nitisols, Phaeozems, Planosols, Plinthosols, Podzols, Regosols, Solonchaks, Solonetz, Umbrisols,* and *Vertisols.*

Cryosols are introduced at the highest level to identify a group of soils which occur under the unique environmental conditions of alternating thawing and freezing. These soils have permafrost within 100 cm of the soil surface and are saturated with water during the period of thaw. In addition, they show evidence of cryoturbation. *Durisols* comprise the soils in semi-arid environments which have an accumulation of secondary silica, either in the form of nodules, or as a massive, indurated layer. *Umbrisols* cover the soils which have either an umbric horizon, or have a mollic horizon and a base saturation of less than 50 per cent in some parts within the upper 125 cm of the soil surface. They are a logical counterpart of the *Chernozems, Kastanozems* and *Phaeozems*. The *Plinthosols* bring together the *Plinthosols* of the Revised Legend and the soils which have a petroplinthic layer at shallow depth. In the Revised Legend the latter soils belong to the *Leptosols*. For the World Reference Base it was decided to exclude from the *Leptosols* soils with pedogenetic horizons such as indurated calcic or gypsic horizons or hardened plinthite. This necessitated the definition of a reference soil group which included these soils. Although it is realised that soils with shallow petroplinthic layers and soils having plinthite normally occupy different positions in the landscape, it was felt appropriate to group them together as they are genetically related.

Podzoluvisols are renamed *Albeluvisols*. The name *Podzoluvisols* suggests that in these soils both the processes of cheluviation (leading to *Podzols*) and subsurface accumulation of clay (resulting in *Luvisols*) take place, while in fact the dominant process consists of removal of clay and iron/manganese along preferential zones (pea faces, cracks) in the argic horizon. The name *Albeluvisols* is therefore thought to be more appropriate, expressing the presence of a bleached eluvial horizon (*"albic horizon"*), a clay-enriched horizon (*"argic horizon"*) and the occurrence of *"albeluvic tonguing"*.

WRB DIAGNOSTIC HORIZONS, PROPERTIES AND MATERIALS

Earlier it was agreed that the soil groups should be defined in terms of a specific combination of soil horizons, called *'reference horizons'* rather than *'diagnostic horizons'*. Reference horizons were intended to reflect genetic horizons which are widely recognized as occurring in soils. Unfortunately, the distinction between reference and diagnostic horizons created confusion and it was agreed to retain the FAO terminology of diagnostic horizons as well as the diagnostic properties. Additionally it appeared necessary to define diagnostic soil materials. This together resulted in a comprehensive list of WRB diagnostic horizons, properties and materials, defined in terms of morphological characteristics and/or analytical criteria. In line with the WRB objectives, attributes are described as much as possible to help field identification.

MODIFICATIONS TO DEFINITIONS OF FAO'S DIAGNOSTIC HORIZONS AND PROPERTIES

Of the 16 diagnostic horizons of the Revised Legend only *the fimic A horizon* has not been retained. It covers too wide a range of human-made surface layers and is replaced in the WRB by the hortic, plaggic and ferric horizons.

For the WRB, the definition of the *histic horizon* was broadened by reducing its minimum thickness to 10 cm and removing the maximum thickness. This is because of a second use of the definition. In the Revised Legend the histic H horizon is used to distinguish soils at second level to identify histic soil units; in the WRB it is used also at the highest level to define *Histosols*. It was agreed that *Histosols* over continuous hard rock should have a minimum thickness of 10 cm in order to avoid very thin organic layers over rock being classified as *Histosols*. The P_2O_5 content requirement for FAO's mollic and umbric A horizons has been deleted from the WRB definition of *mollic* and *umbric horizons*. This requirement cannot be considered diagnostic since thick, dark coloured, human-made horizons in, for instance, China, also have low amounts of phosphate. Other criteria have to be found to separate mollic and umbric horizons from anthropedogenic horizons.

A *chernic horizon* is defined as a special kind of mollic horizon. The present definition of the mollic horizon was felt to be too broad to reflect properly the unique characteristics of the deep, blackish, porous surface horizons which are so typical for *Chernozems*.

The definition of the *ochric horizon* is similar to the ochric A horizon. The colour requirement for the *albic horizon* have been slightly changed compared to FAO's albic E horizon, to suit albic horizons which show a considerable shift in chrome upon moistening. Such conditions are frequently found in soils of the southern hemisphere. The *argic horizon* definition differs from that of the argic B horizon of the Revised Legend in that the percentage clay skins on both horizontal and vertical ped faces and in pores has been increased from one to five per cent. This is expected to provide a better correlation with the earlier requirement of at least one per cent oriented clay in thin sections.

Guidelines to recognize a lithological discontinuity, if not clear from the field observation, were added to the description of the argic horizon. It can be identified by the percentage of coarse sand, fine sand and silt, calculated on a clay-free basis (international particle size distribution or using the additional groupings of the United States Department of Agriculture (USDA) system or other), or by changes in the content of gravel and coarser fractions. A relative change of at least 20 per cent in any of the major particle size fractions is regarded as diagnostic for a lithological discontinuity. However, it should only be taken into account if it is located in the section of the solum where the clay increase occurs and if there is evidence that the overlying layer was coarser textured.

The adjustments made in the description of the argic horizon also apply to the *natric horizon*. The definition of FAO's cambic B horizon has been slightly amended by deleting the requirement '*....and has at least eight per cent clay*'. This requirement forces some soils, which have a well-developed structural-B horizon and silt loam or silt textures with a low clay content, as found, for instance, in fluvio-glacial deposits of the nordic countries, into the *Regosols* rather than in the *Cambisols*. Because there is also no need for this requirement to separate *Cambisols* from*Arenosols* it has not been used in the definition proposed for the WRB *cambic horizon*.

Major alterations are made in the definition of the *spodic horizon*. It has been brought into line with the recent modifications in soil taxonomy (Soil Survey Staff, 1996) regarding the definition of spodic materials. Colour requirements were added, a limit of 0.5 or more in percentage oxalate extractable aluminium plus half that of iron is used, and a value for the optical density of oxalate extract (ODOE) of 0.25 or more is introduced. Moreover, the upper limit of spodic horizons has been set at 10 cm depth.

The silt-clay ratio of 0.2 or less has been deleted from the definition of the *ferralic horizon*. This criterion was felt to be too strict; the silt particle size fraction has been increased. Other values have been proposed (silt-clay ratio of 0.7 or less; fine silt-clay ratio of 0.2 or less) but, as yet, no consensus has been reached. Some alterations are made in the definitions of the *calcic* and *gypsic horizons*. For WRB purposes they are split into calcic/gypsic and*hypercalcic/hypergypsic horizons*. These latter horizons have a calcium carbonate equivalent and gypsum content of 50 and 60 per cent, respectively, but are not cemented. The definition for the *sulfuric horizon* remains the same as in the Revised Legend. In addition to these diagnostic horizons, 19 new ones are proposed. Some are adopted from FAO's diagnostic properties, others are newly formulated. Together they bring the total of diagnostic horizons recognized in the WRB to 34. The newly defined diagnostic horizons are the *andic, anthropedogenic, chernic, cryic, duric, ferric, folic, fragic, fulvic, glacic, melanic, nitic, petroduric, petroplinthic, plinthic, salic, takyric, vertic, vitric* and *yermic horizons*.

A combination of an *anthraquic horizon* at the surface with an underlying *hydragric horizon*, totalling together a thickness of at least 50 cm, defines certain*Anthrosols* which show evidence of alteration through wet-cultivation practices. It comprises a puddled layer, a plough pan and an illuvial subsurface horizon. This combination is characteristic for soils which have been used for long-term paddy rice cultivation. Newly defined diagnostic properties and materials are *albeluvic tonguing, alic* and *aridic properties*, and *anthropogeomorphic, calcaric, fluvic, gypsiric, organic, sulfidic* and *tephric soil material*.

Gleyic and *stagnic properties* have been reformulated. Slight changes are made in FAO's definitions of *abrupt textural change and geric properites*, while the definitions of *permafrost* and soft powdery lime, renamed *secondary carbonates*, have been adopted without change.

In the description of the *gleyic* and *stagnic properties* the occurrence of 'gleyic' and 'stagnic colour patterns' is introduced. These terms apply to the specific distribution pattern of Fe/Mn (hydr)oxides caused by saturation with groundwater or stagnating surface water. A gleyic colour pattern has 'oximorphic' features on the outside of structural elements, along root channels and pores, or as a gradient upwards in the soil. A stagnic colour pattern on the other hand shows these features in the centre of peas or as a gradient downwards resulting from impedance of the water flow.

The slight changes in the descriptions of *abrupt textural change* and *geric properties* refer to a different depth in which tile change in texture must occur and another way of calculating the effective cation exchange capacity (ECEC)[1], respectively. Since the inception of the Legend of the Soil Map of the World, the number of lower level units used in the Legend or soil classification has continued to grow: from 106 in 1974 to 152 in the Revised Legend of the Soil Map of the World to 209 in the first draft of the *World Reference Base for Soil Resources*. At the same time a serious effort was undertaken to expand this second level further with the introduction of third level units. Further proliferation of soil units and subunits in the World Reference Base might easily lead to a situation where it will become extremely difficult to recall and use all definitions within the main reference soil groups.

A further complication is that many soil unit names, and modifiers in the draft WRB, were inherited from the original FAO Legend and were defined depending on the grouping in which they occurred. For example, a "Dystric" soil unit may mean: "... having a base saturation of less than 75%" (in Dystric Vertisols), or "...having a base saturation of less than 50%", in different control sections (*e.g.* note the difference in control sections of Dystric Planosols and Dystric Cambisols). Another limitation inherent to the close link with the Legend of the Soil Map of the World is that, although often used as a soil classification system, the original purpose of the FAO system was to serve as a Legend for a specific map, which made certain simplifications necessary. For example, Calcic Gleysols included soil with a gypsic horizon. Similarly, Umbric Fluvisols grouped alluvial soils with an umbric horizon together with Fluvisols with a desaturated histic horizon. This resulted in loss of information due to the generalization required for the Legend.

Last but not least, it is thought that a clear-cut separation must be made between the double objectives of the World Reference Base, which on one hand should be able to serve as a soil reference system for geographers, agronomists and other users who are mainly interested in the highest level of generalization explained in non-technical terms, while on the other hand WRB must be a sophisticated tool for soil correlation able to accommodate a wide range of national soil classification systems.

CELL CULTURE

In the early 19th century, Schleiden and Schwann proposed the concept of the 'cell theory'. In 1902, Gottlieb Haberlandt, the german botanist and regarded as the father of plant tissue culture, first attempted to cultivate the mechanically isolated plant leaf cells on a simple nutrient medium. He did not succeed in achieving the growth and differentiation of the cultured cells, however, he predicted the concept of growth hormones, the use of embryo sac fluids, the cultivation of artificial embryos from somatic cells, etc.

During the period 1902 - 1930, attempts were made to culture the isolated plant organs such as roots and shoot apices (organ culture). Hanning isolated embryos of some crucifers and successfully grew on mineral salts and sugar solutions. Simon successfully regenerated a bulky callus, buds, roots from a poplar tree on the surface of medium containing IAA which proliferated cell division. Gautheret, White and Nobecourt largely contributed to the developments made in plant tissue culture. White cultured tobacco tumour tissue from the hybrid Nicotiana glauca, and N. Langsdorffii.

The period saw the development of suitable nutrient media to culture plant tissues, embryos, anthers, pollen, cells and protoplasts, and the regeneration of complete plants (in vitro morphogenesis) from cultured tissues and cells. In 1941, van Overbek and co-workers used coconut milk (embryo sac fluid) for embryo development and callus formation in Datura. Steward and Reinert first discovered somatic embryo production in vitro. Maheswari and Guha developed the anther culture for the production of haplid plants. Skoog and Miller advanced the hypothesis of organogenesis in cultured callus by varying the ratio of auxin and cytokinin in the growth medium. Muir developed a successful technique for the culture of single isolated cells wich is commonly known as paper-raft nurse technique (placing a single cell on filter paper kept on an actively growing nurse tissue). In 1952, the Pfizer Inc., New York got the US patent and started producing industrially the secondary metabolites of plants. The first commercial production of a natural product shikonin by cell suspension culture was obtained. In 1980s using Genetic engineering, for the first time, it was possible to introduce foreign genes into cell and tissue culture systems to develop plants with improved characteristics (transgenic crops) which may contribute to the path towards the second green revolution. A transgenic animal is an individual in which a gene (or genes) from another individual has been artificially inserted with the new genetic information forming a permanent part of the genome and being passed on to their offspring according to the Mendelian laws of inheritance. So far, transgenic cattle, sheep, pigs, rabbits, and chicken have been produced.

Table. List of Transgenic Animals Produced With Their Promoter, Enhancer and Structural Transgenes

Animal	Genes Tansferred (Promoter or Enhancer/Structural Transgenes)
Goat	A variant of tPA gene (LAtPA)
Sheep	mMT/hGH, mMT/TK, mMT/bGH, mMT/hGRF, oBLG/hFIX, oBLG/alpha1AT, oMT/oGH
Rabbit	mMT/hGH, hMT/hGH, rbEu/rb,c-myc
Pig	mMT/hGH, mMT/bGH, hMT/pGH, MLV/rGH, MLV/rGH, bPRL/bGH
Fish	hGH, mMT/hGh, mMT/bGal, cd-crystallin SV/hygro, AFP
Cow	BPV, lactoferrin
Chicken	ALV, REV
Mouse	mMT/rGH, mMT/bGH, mMT/oGH, mMT/hGh, mMT/hGRF, mMT/hFIX

After fertilization a mammalian embryo passes through many stages of development such as formation of pronuclei, morula formation, blastocyst formation, implantation etc. Using these early development stages of mammalian embryos, genetic engineers have used different methods to introduce foreign genes into the genome.

METHODS TO INDUCE FOREIGN GENES INTO THE GENOME

MICRO-INJECTION IN TO A PRONUCLEUS

A fertilized egg is held in a pipette by suction and several copies of the foreign genes are injected into one of the pronuclei via a micropipette. Although this has been the most widely used and most successful method, 30% of the treated embryos degenerate and die with in a few hours.

A morula is placed on a culture of fibroblasts infected with retrovirus after removing the zona pellucida layer. The virus cannot penetrate this layer therefore it is essential to get rid of it. The retrovirus, genetically engineered to carry foreign DNA, infect the cells of the morula as they are shed from the fibroblast. This technique often creates 'mosaic offspring', where some cells contain the foreign gene while others do not.

One of the limitations of this technique is that the modified virus cannot leave the transgenic species and under certain conditions it could mutate regaining its ability to cause disease in the tissues where it occurs.

In this method, stem cells, from the inner cell mass of an embryo, are infected with genetically-engineered retroviruses, then injected into the central cavity of a different blastocyst. The injected cells colonise the new embryo and participate in the formation of all the tissues, including the ovaries and testes from which cells are formed. This technique has been used to produce transgenic mice and hamsters.

The first transgenic animals produced were mice, into which a gene cloning for rat growth hormone was inserted by microinjection.

Two techniques were used to introduce the new gene into the mouse genome:

- As many as 10000 copies of the rat gene were injected into a pronucleus of a fertilized mouse egg, using a microneedle.
- After treatment with calcium chloride, to make them more permeable, the rat DNA was applied to the outside of a 2-8- celled mouse embryo in culture.

The treated eggs, or embryos were then transplanted into foster mothers. Embryos with the inserted genes were identified by DNA hybridization with small samples of DNA from white blood cells or skin cells. Transformed adult mice produced the growth hormone mainly in their livers rather than pituitary gland where it is normally synthesized.

This technique can be used successfully to transfer other genes such as those affecting the properties of hair, hides, cold tolerance, disease-resistance and milk production which might help to produce strains of farm animals with economic advantage over existing breeds and stocks. In this regard, research is going on to use techniques which can improve the quality of the farm animals. In this regard, following two techniques are of great importance.

ENVIRONMENTAL TECHNIQUES

The term "Environment" is defined as our surroundings which includes the abiotic component (the non living) and biotic component (the living) around us. The abiotic environment includes water, air and soil while the biotic environment consists of all living organisms – plants, animals and microorganisms. Environmental pollution broadly refers to the presence of undesirable substances in the environment which are harmful to man and other organisms. In the past decade or two, there has been a significant increase in the levels of environmental pollution mostly due to direct or indirect human activities. The major sources of environmental pollution are –Industries, Agricultural sources (mainly rural area), anthropogenic sources (man related activities mainly in urban areas), biogenic sources etc. The pollutants are chemical, biological and physical in nature. The Chemical pollutants include-gaseous pollutants (hazardous gases like sulfur dioxide, nitrogen oxide), toxic metals, pesticides, herbicides toxins and carcinogens

Etc. The physical pollutants are- heat, sound, radiation, and radioactive substances. The pathogenic organisms and some poisonous and dangerous biological products are the biological pollutants.

Controlling the environmental pollution and the conservation of environment and biodiversity and controlling environmental pollution are the major focus areas of all the countries around the world. In this context the importance and impact of biotechnological approaches and the implications of biotechnology has to be thoroughly evaluated. There have been serious concerns regarding the use of biotechnology products and the impact assessment of these products due to their interaction with the environmental factors.

A lobby of the environmentalists have expressed alarm on the release of genetically engineered organisms in the atmosphere and have stressed on thorough investigation and proper risk assessment of theses organisms before releasing them in to the environment. The effect of the effluents from biotechnological companies is also a cause of concern for everyone. The need of the hour is to have a proper debate on the safety of the use of the biotechnological products.

The efforts are not only on to use biotechnology to protect the environment from pollutionbut also to use it to conserve the natural resources. As we all know that microorganisms are known natural scavengers so the microbial preparations (both natural as well as genetically engineered) can be used to clean up the environmental hazards.

DEVELOPMENT OF ALTERNATE CLEANER TECHNOLOGIES USING BIOTECHNOLOGY

Biotechnology is being used to provide alternative cleaner technologies which will help to further reduce the hazardous environmental implications of the traditional technologies. *E.g.* some Fermentation technologies have some serious environmental implications. Various biotechnological processes have been devised in which all nutrients introduced for fermentation are retained in the final product, which ensures high conversion efficiency and low environmental impact.

In paper industry, the pulp bleaching technologies are being replaced by more environmentally friendly technologies involving biotechnology.

The pulp processing helps to remove the lignin without damaging valuable cellulosic fibres but the available techniques suffer from the disadvantages of high costs, high energy use and corrosion. A lignin degrading and modifying enzyme (LDM) was isolated from Phanerochaete chrysosporum and was used, which on one hand, helped to reduce the energy costs and corrosion and on the other hand increased the life of the system. This approach helped in reducing the environmental hazards associated with bleach plant effluents. In Plastic industry, the conventional technologies use oil based raw materials to extract ethylene and propylene which are converted to alkene oxides and then polymerized to form plastics such as polypropylene and polyethylene. There is always the risk of these raw materials escaping into the atmosphere thereby causing pollution. Using biotechnology, more safer raw materials like sugars (glucose) are being used which are enzymatically or through the direct use of microbes converted into alkene oxides.*e.g.* Methylococcus capsulatus has been used for converting alkene into alkene oxides.

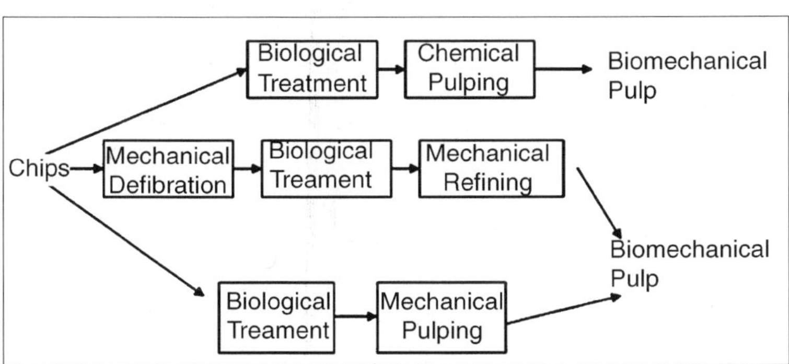

Fig. Integration of Biological Steps in Pulping Process Leading to Lignin Degradation

BIOREMEDIATION

Bioremediation is defined as 'the process of using microorganisms to remove the environmental pollutants where microbes serve as scavengers. The removal of organic wastes by microbes leads to environmental cleanup. The other names/terms used for bioremediation are biotreatment, bioreclamation, and biorestoration. The term "Xenobiotics" (xenos means foreign) refers to the unnatural, foreign and synthetic chemicals such as pesticides, herbicides, refrigerants, solvents and other organic compounds. The microbial degradation of xenobiotics also helps in reducing the environmental pollution.

Pseudomonas which is a soil microorganism effectively degrades xenobiotics. Different strains of Pseudomonas that are capable of detoxifying more than 100 organic compounds (*e.g.* phenols, biphenyls, organophosphates, naphthalene etc.) have been identified. Some other microbial strains are also known to have the capacity to degrade xenobiotics such as Mycobacterium, Alcaligenes, Norcardia etc.

FACTORS AFFECTING BIODEGRADATION

The factors that affect the biodegradation are: the chemical nature of xenobiotics, the concentration and supply of nutrients and O_2, temperature, pH, redox potential and the capability of the individual microorganism. The chemical nature of xenobiotics is very important because it was found out that the presence of halogens *e.g.* in aromatic compounds inhibits biodegradation.

The water soluble compounds are more easily degradable whereas the presence of cyclic ring structure and the length chains or branches decrease the efficiency of biodegradation. The aliphatic compounds are more easily degraded than the aromatic ones.

BIOSTIMULATION

It is a process by which the microbial activity can be enhanced by increased supply of nutrients or by addition of certain stimulating agents like electron acceptors, surfactants etc.

TISSUE CULTURE OF LILACS

Traditionally lilacs have been grown from seed, suckers or by grafting superior clones on to compatible rootstocks. Lilac seedlings of several species, privet stem cuttings as well as ash seedlings/root cuttings have all been utilized as rootstocks. In recent times there has been a move away from using rootstocks for various reasons, and nurserymen have been trying to produce cultivars on their own roots. While cuttings taken at the right time are often successful for many species, this is not always a sure recipe for thevulgaris clones. For more than two decades some of the leading nurseries in the US and Europe have been using tissue culture for propagating lilacs on their own roots. While the main reason for resorting to micropropagation is to produce large numbers of selected clones within a short period, we have discovered it to be no less effective in producing small numbers ofvulgaris cultivars, which are recalcitrant to initiating roots by conventional means.

INVITRO FLOWERING OF ROSE

The present invention is directed to compositions and methods of micropropagation of a rose plant and in vitro flowering of the rose plant. Young shoots are induced to produce buds in an enclosed vessel containing a first culturing medium comprising benzyladenine, an auxin and 2% sucrose as carbon source.

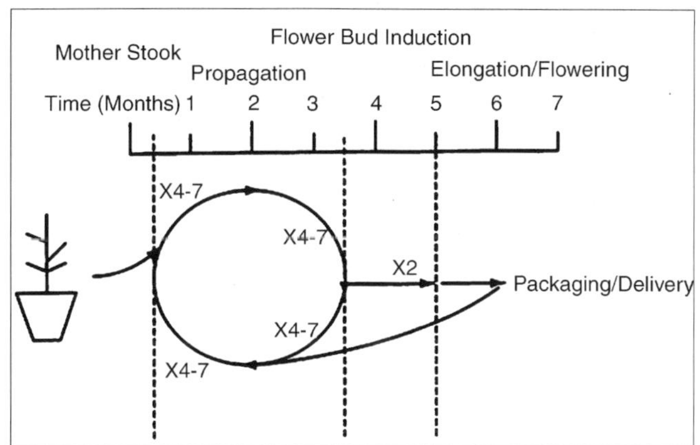

Buds are excised and cultured on a medium for propagation and multiplication of plantlets. Plantlets are then transferred to a medium comprising thidiazuron, an auxin and myo-inositol for induction of flower buds, followed by culture on a medium comprising benzyladenine and an auxin for plantlet elongation and finally culture in the fifth medium without phytohormone for flowering. Alternately, after propagation and multiplication, plantlets are transferred to a medium comprising zeatin, auxin and myo-inositol for induction of flower buds followed by culture on a medium without phytohormone for elongation and flowering.

ORGANOGENESIS

Meristematic tissues are responsible for the division of new cells... they are zones of actively dividing cells.

Before we proceed, we should first recognize that growth in plants includes two stages:

1. First the production of new cells and,
2. Secondly the expansion of these cells via uptake of water by the vacuole.

Cell division occurs solely in meristematic regions, while expansion may occur anywhere. Thus in a single plant there are zones of young dividing cells, maturing cells, and mature cells. We will go into more detail on meristematic regions when we go over the stem and root individually, but for now an introduction is sufficient

Let us recognize that there are 3 meristematic regions in the plant:

I. Apical meristems are located at the apices or tips - at root and shoot tips and are directly involved in their elongation

They create derivatives which form primary growth:

- The protoderm which forms the outer dermal layer of tissues,
- The ground meristem which forms the cortical cells and
- The procambium which forms the vascular tissue.

In shoots, plants protect their meristems with young leaves and by forming dormant reserve meristems (*i.e.*, buds), that protect their apical meristems with a root cap Ferns, Gymnosperms and Angiosperms are all terrestrial plants.

Tissues therefore arise as a result of a cell or a group of cells dividing, thus giving rise to a large number of cells. In flowering plants these cells undergo changes that change the structure and function of the cell; differentiation has take place. As a result of differentiation, a cell or group of cells can have a specialized function; *i.e.* specialization takes place. Thus, a tissue is a group of cells that have differentiated to perform a specific function.

If a group of cells are more or the less the same, then it is known as a Simple tissue *e.g.* epidermis, parenchyma, collenchyma and sclerenchyma. If, however, the tissue is composed of a number of different cells, then it is known as a Complex tissue such as xylem and phloem.

DIAGRAMMATIC REPRESENTATION OF A PLANT BODY

Ground Tissue is simple non-meristematic tissue (non-dividing tissue) made up 3 cell types:paranenchyma, collenchyma and sclerenchyma.This tissue generally forms either the pith, cortex or bulk of leaf (mesophyll).

Parenchyma Cells

- Mostabundant cells in plants;
- Spherical cells which flatten at point of contact;
- Alive at maturity;pliable, primary cell walls;
- Large vacuoles for storage of starch, fats, and tannins (denature proteins);

- Primary sites of the metabolic functions such as photosynthesis, respiration, and protein synthesis; they are "ready reserves" from which a plant makes specialized cells to meet its changing needs.

SPECIALIZED PARENCHYMA

- *Chlorenchyma*: Photosynthetic cells; have high density of chloroplasts
- *Aerenchyma*: prominent intercellular spaces that improve gas exchange capacity of the tissue; provide maximum support with a minimum metabolic requirement
- *Transfer cells*: specialized for short distance transfer of solutes between cells; have secondary cell walls; they are inner extensions of wall that increase surface area.

Transfer cells occur in areas of high solute transport, such as secretory tissues, which release substances that are produced within the protoplasm and are moved outside, *i.e.*, nectar cells, mucilage in sundews, and resins.

Collenchyma Cells- Colla = Glue, Glisten

- Living protoplasm;
- Unevenly thickened primary cell walls; elongate cells;
- Longer than wide; just beneath epidermis; function to support growing organs, grass, floral parts, and border veins; their nonlignified cell walls can stretch

Sclerenchyma Cells- (skleos = hard)

- Most are dead at maturity;
- Rigid, thick, lignified, nonstretchable secondary cell walls. there are 2 types of sclerenchyma cells:

Sclereids or stone cells we saw in lab- short; variable shape; form hard layers such as the shells of nuts and seed coats; produce the gritty texture of pears Fibres- long, slender; occur in strands or bundles; tiny cavity or lumen; the different hardnesses of fibres are used to make coarse rope, linen cloth, etc.

DIRECT SHOOT ORGANOGENESIS FROM LEAF EXPLANTS OF CHRYSANTHEMUM

Shoot regeneration from leaf and stem explants of 19 Chrysanthemum cultivars has been studied. Adventitious shoots could be regenerated from leaf and stem explants of 16 cultivars. Most cultivars showed the best shoot regeneration on Murashige and Skoog medium supplemented with 8.9 μM BAP and 2.7 μM NAA (0.5 mg/l). The number of regenerated shoots per explant was more than 10 in some of the cultivars studied. It has not been posssible to regenerate shoots from leaf explants of cultivars 170 and 260. The cultivars 140 and 8010 produced the highest number of shoots. Regenerated shoots can easily be rooted on half strength Murashige and Skoog medium supplemented with 0.98 μM IBA (0.2mg/l). After potting in soil and transfer to the glasshouse normal plants develop. Apparently no somaclonal variation has been observed among the regenerants.

Biolistic particle bombardment is used to transfer desirable genes into difficult to transform plants, but research is lacking on factors that can affect transformation success. This study was conducted to determine if manipulating culture conditions forRhododendron 'PJM Hybrid' and 'Iridon' chrysanthemum leaf explants could alter transient gene expression of β-glucuronidase (GUS) after bombardment. Both rhododendron and chrysanthemum leaf explants were bombarded with 1.0 μm gold particles coated with 2 μg of linear plasmid DNA, containing the GUS gene. Different explant treatments before bombardment were tested, including time in culture on regeneration medium, growing in the presence or absence of 0, 0.09, 0.2, 0.4, or 0.6 M sucrose, and growing in the dark for 0, 3, or 6 days. Transient GUS expression was histochemically quantified 2 days after bombardment as the percentages of GUS positive explants and the number of blue spots per GUS positive explant. Transient GUS expression was highest when 'PJM Hybrid' explants were incubated on regeneration medium 9 to 12 days before bombardment, whereas transient GUS expression by 'Iridon' explants was relatively high (>42%) for leaves incubated 3 to 18 days before bombardment.

Exposure of 'PJM Hybrid' explants to media containing different concentrations of sucrose decreased transient GUS expression by 71% compared to those grown on sucrose-free medium. Alternatively, 27% more 'Iridon' explants transiently expressed GUS compared to those on sucrose-free medium, and the mean number of spots per GUS positive explant increased from 5 to 13.

Transient GUS expression increased almost three fold when 'PJM Hybrid' explants were placed in the dark for six days compared those in the light. In contrast, the dark treatment had inconsistent effects on transient GUS expression by 'Iridon' explants. Manipulating explant culture conditions before biolistic bombardment influenced transient gene expression and illustrated that genotype-specific differences affect transformation potential.

3

In Vitro Conservation Techniques

Classical *in vitro* conservation techniques have been developed for a wide range of species, including temperate woody plants, fruit trees, horticultural species, as well as numerous tropical species.

However, a recent FAO survey indicated that only 37 600 accessions are conserved *in vitro* worldwide. Slow-growth conservation is used routinely for the conservation of only a few species, including banana, potato and cassava, in regional and international germplasm conservation centres such as INIBAP, CIP, CIAT and IITA.

Alternative medium-term conservation techniques are still at the experimental stage. Low-oxygen storage might be interesting for storing cold-sensitive tropical species since it allows growth reduction without decreasing the temperature. However, complementary experiments involving additional species and longer storage periods are still required. Medium-term storage of desiccated somatic embryos will be mainly applicable for the management of large-scale production of elite genotypes. Encapsulated apices might be employed for medium-term genetic resources conservation. Research is still needed to refine the protocols and extend the storage periods. Cryopreservation procedures have been developed for around 100 different plant species cultured in various ways including cell suspensions, calluses, apices, and zygotic and somatic embryos.

Most of this work has been performed in the framework of academic studies and has involved only one or a few genotypes. However, there are recent reports of large-scale experimentation of classical freezing techniques involving cell suspensions from several hundred genotypes. In the case of organized structures such as apices or embryos, the development of new cryopreservation procedures has also allowed experiments involving a relatively wide range of genotypes/varieties. In the case of potato, apices from more than 100 different varieties have been successfully frozen and around 25,000 apices are presently stored in liquid nitrogen.

There is an increasing number of cases where cryopreservation techniques can be considered operational. However, their routine application is mostly restricted to the conservation of cell lines in research laboratories. The only example of routine application of cryopreservation to another type of material is oil palm, where 80 clones of somatic embryos are stored in liquid nitrogen and samples thawed upon request for plant production.

GENETIC STABILITY OF IN VITRO CONSERVED MATERIAL

Genetic conservation is based on the assumption that the material is stored under conditions ensuring genetic stability. However, there are various factors linked with *in vitro* culture/conservation procedures which can be a source of variation. These factors, as well as the various approaches for assessing the genetic stability of plants recovered from *in vitro* culture, have been reviewed recently by Harding (1995). Genetic variation can pre-exist in the collected material, possibly linked to its genetic structure, such as in sugar-cane or banana where polyploids are more prone to instability than diploids. It can also arise from somaclonal variation, particularly if the propagation system includes a dedifferentiated phase or from stresses imposed by the conservation procedures. The information available on the genetic stability of plant material conserved *in vitro* is presented below.

Slow Growth

Different genotypes do not grow at the same rate when placed in culture. This induces a risk of selection when plants are placed under stress conditions, such as slow growth. It is thus important to minimize the risk of selection. One of the most efficient ways is to use, whenever possible, differentiated, organized structures such as shoot tips or embryos for which the risks of variation are minimal.

A limited number of experiments only have been performed to assess the genetic stability of plants during slow-growth storage. The most exhaustive study has been performed in the framework of a joint project between CIAT and IPGRI which aimed at monitoring the genetic stability of cassava shoot cultures stored *in vitro*. Isozyme and DNA analysis as well as examination of the morphology of plants regrown in the field did not reveal any modification after 10 years of storage under slow growth. In contrast, growth of potato shoot tips on mannitol-supplemented medium, which caused morphological changes in the material, was correlated with DNA hypermethylation, which may be an adaptive response to conditions of high osmotic stress. This clearly illustrates the importance of defining storage conditions which are likely to cause as little stress as possible to the plant material.

Cryopreservation

Cryopreservation involves a series of stresses which might lead to modifications in the recovered cultures and regenerated plants. It is therefore necessary to verify that the genetic stability of the cryopreserved material is not altered before using this technique routinely for the long-term storage of germplasm. There is now increasing evidence that, provided the cryopreservation technique applied ensures the greatest possible maintenance of the integrity of the frozen specimen, there will be no modification at the phenotypic, biochemical, chromosomal or molecular level attributable to cryopreservation; none has been reported to date.

It has been shown that cell suspensions of numerous species maintain their biosynthetic and morphogenic potential after cryopreservation. The only exception concerns lavender cell suspensions exposed to successive freeze-thaw cycles, for which the number of colonies recovered from cryopreserved cells increased with the number of freeze-thaw cycles. However, no modifications were noted in the biosynthetic capacities of cryopreserved cells, suggesting a change in population structure rather than genetic change. A similar observation was made by Bercetche *et al.* (1990), who noted that cryopreserved *Picea abies* embryogenic calluses recovered faster than nonfrozen controls; nonembryogenic tissues were killed by freezing, thus leading to the production of a more homogeneous population, with a higher embryogenic potential. This suggests that cryopreservation could be used as a tool to 'rejuvenate' cultures when their proliferation capacities are decreasing.

Plants regenerated from cryopreserved apices of strawberry and cassava were phenotypically normal. No differences were noted in the vegetative and floral development of several hundreds of oil palms regenerated from control and cryopreserved somatic embryos. In contrast, Fukai *et al.* (1994b) showed that a high percentage of plants regenerated from apices of a periclinally chimeric chrysanthemum cultivar frozen using a classical protocol had an altered flower colour. These apices had been severely harmed during freezing, leading to callusing during recovery. Regenerated plants had thus an adventitious origin which explained the disturbance in the initial chimeric structure. The same group compared the effect of freezing carnation apices using a classical protocol and the encapsulationdehydration technique. Encapsulation-dehydration ensured 100% recovery after freezing, and regrowth was rapid and direct. In contrast, after slow, controlled freezing, the recovery rate was 50% only, and callusing was observed during regrowth. The two above examples underline the importance of selecting the most appropriate freezing technique for any given material, not only as regards recovery rates but also as regards recovery pattern and genetic stability.

Electrophoretic profiles of two enzymatic systems were comparable in plants regenerated from control and cryopreserved apices of sugar-cane and sweet orange somatic embryos. The ploidy level of plants regenerated from oilseed rape somatic embryos and sensitive dihaploids of potato was not modified by cryopreservation. Restriction fragment length polymorphism (RFLP) patterns of plants regenerated from sugar-cane embryogenic cell suspensions were identical to those of unfrozen controls. Several molecular types were uncovered in plants recovered from both control and cryopreserved sugar-cane apices, thus indicating that the variation was not due to freezing but was preexisting among the *in vitro* mother plants. Finally, DNA analysis showed that a gene integrated in the genome of a navel orange cell suspension was maintained after freezing and one-year storage in liquid nitrogen.

Comparative Studies

A limited amount of work has been published on the comparative effects of slow growth and cryogenic storage on the stability of plant material. Mannonen *et al.* (1990) showed that the production of secondary metabolites by cell cultures of *Panax ginseng* and *Catharanthus roseus* decreased drastically after 6 months of culture either in standard conditions (i.e. with weekly subcultures) or in slow growth under mineral oil, whereas the productivity of cell suspensions cryopreserved and stored in liquid nitrogen during the same period was identical to that of the original cultures.

Modifications in the RFLP pattern were observed in potato plants stored for 6 months under slow growth on a medium supplemented with mannitol, whereas no such modifications were noted in plants regenerated from cryopreserved apices.

NEW CRYOPRESERVATION TECHNIQUES

New cryopreservation techniques offer practical advantages in comparison to classical ones. Vitrification-based procedures are operationally less complex than classical ones since they do not require the use of a programmable freezer. In addition, since ice formation is avoided during freezing, they are more adapted for freezing complex organs such as apices or embryos which contain a variety of cell types, each with unique requirements under conditions of freeze-induced dehydration. Finally, they have greater potential for broader applicability than classical techniques since only minor modifications are required for various cell types. Seven different vitrification-based procedures can be identified: (i) encapsulation-dehydration; (ii) a procedure actually termed vitrification; (iii) encapsulation-vitrification; (iv) desiccation; (v) pregrowth; (vi) pregrowth- desiccation; and (vii) droplet freezing. Table presents a list of species to which these new freezing techniques have been applied.

Encapsulation-dehydration

The encapsulation-dehydration technique is based on the technology developed for the production of synthetic seeds where somatic embryos are encapsulated in a bead of hydrosoluble gel. This technique has been applied mostly to apices of more than ten species of temperate and tropical origin, and to somatic embryos of several crop species as well as to a *Catharanthus* cell suspension. Before cryopreservation, specimens can be submitted to conditioning treatments which increase their survival potential. In the case of cold-tolerant species, *in vitro* mother plants or apices can be placed at low temperature for several weeks. Mulberry apices are transferred daily on media with progressively increasing sucrose concentrations to initiate dehydration.

Table. List of plant species for which new cryopreservation techniques (encapsulation-dehydration,Vitrification,encapsulation-vitrificatio,pregrowth=desiccation,pregrowth,desiocation)have been developed using different types of Specimens.

Species	Specimen	No.of Accessions	Technique
Aesculu			
Hypocastanea	Zygotic embryo	1	Desiccation
Allium Sativum	Apex	12	Vitrification
Allium wakegi	Apex	7	Vitrification
Arachis hypogaea	Zygotic embryo	6	Desiccation
Araucaria excelsa	Zygotic embryo	1	Desiccation
Armoracia rusticana	Hairy root culture	1 dehyde.	Encaps./.
		1	
Artocarpus			
Heterophyllus	Zygotic embtyo	?	Desiccation
Asparagus			
officinalis	Stem segment	1	Pregrowth/ desicc.
	somatic embryo	1	vitrification
	Cell suspension	1	vitrification
		1	
Baccaurea motleyana	Zygotic embryo	1	Desiccation
Baccaurea polyneura	Zygotic embryo	1	Desiccation
Beta Vulgaris	Apex	2	Encaps./dehydr.
Brassica campestris	Cell suspension	1	Vitrification
Brassica napus Embryo	Microspore	1 desicc.	Pregrowth/
		1	Encaps/dehyd.r

Bromus inermis	Cell suspension	1	Vitrification
Calamus manan	Zygotic embryo	1	Desiccation

Specimens are usually encapsulated in 3% calcium alginate gel. They are then submitted to the following successive steps: pregrowth, dehydration, freezing, thawing and recovery. Pregrowth is performed in liquid medium enriched with sucrose (0.3 to 1.5 M) for periods varying between 16 hours and 10 days. Replacement of sucrose with other sugars did not improve survival of grape apices. Progressive increase in sucrose concentration generally overcomes sensitivity to direct exposure to high sucrose levels which is encountered with some species such as eucalyptus, grape and coffee.

Desiccation is performed using either the air current of a laminar airflow cabinet or silica gel. The latter method is generally preferred because it ensures more precise and reproducible desiccation rates. The water content of the beads allowing optimal survival rates is around 20% (fresh weight basis). If pregrowth conditions have been well defined, only limited loss is observed after desiccation with most species. However, banana apices were found to be highly sensitive to both sucrose pregrowth treatment and to even limited desiccation, which induced a drastic survival drop.

Desiccated samples are usually frozen by direct immersion in liquid nitrogen. However, modifications in the freezing rate can have different consequences on the survival of different materials. Survival of encapsulated grape apices was higher after slow cooling down to 2100°C. In contrast, survival of sugar-cane apices was higher after rapid freezing than after slow, controlled cooling. Freezing encapsulated carnation apices at cooling rates of between 0.5 and 200°C min21 had no effect on survival.

Samples are usually stored in liquid nitrogen at 2196°C. Scottez (1993) and Tannoury (1993) showed that survival of pear and carnation apices, respectively, was not modified after 2 and 3 years of storage at 2196°C. Several authors have demonstrated that samples can be frozen and conserved in deep-freezers provided that the storage temperature is below that of ice recrystallization (250 to 270°C). Pear apices were conserved at 275°C for one year and apple, pear and mulberry apices were stored for 5 months at 2135°C.

Samples are usually placed directly under standard conditions for recovery. However, transitorily modified conditions can enhance recovery in some cases. Sugar-cane apices are placed in the dark for one week on a medium supplemented with growth hormones. Addition of the antioxidant ascorbic acid significantly improved the recovery of sugar-beet apices, which are extremely sensitive to oxidation. Extraction of apices from the beads was necessary to allow regrowth of pear and grape apices. Regrowth of material frozen using the encapsulation- dehydration technique is usually direct and rapid, without callus formation. This is due to the fact that, contrary to what is observed after classical freezing, where many cells are destroyed and callusing is frequently observed during recovery, encapsulation-dehydration preserves the structural integrity of most cells. Therefore, regrowth usually originates from the whole meristematic zone, as observed notably in the case of sugar-cane.

There are presently four crops (pear, apple, sugar-cane and potato) on which the encapsulation-dehydration technique has been successfully extended to several genotypes or varieties. In all cases, even though genotypic variations were noted, results were sufficiently high to envisage routine application of the cryopreservation protocols developed.

Vitrification

Vitrification involves treatment ('loading') of samples with cryoprotective substances, dehydration with highly concentrated vitrification solutions, rapid freezing and thawing, removal of cryoprotectants ('unloading') and recovery. Vitrification procedures have been developed for more than 20 different species using protoplasts, cell suspensions, apices and somatic embryos.

The physiological state of the explants can influence their survival potential. In some cases, the plant material is thus submitted to various treatments before the cryopreservation procedure itself. *In vitro* mother plants can be cultured at low temperature for several weeks. Explants can be placed on a medium with cryoprotectants and/or cultured at low temperature for a short period. Pepó (1994) indicated that axillary shoot tips are much more sensitive to vitrification than apical shoots. Loading consists of placing the explants in liquid medium containing cryoprotective substances (sucrose, glycerol, ethylene glycol) for a short period, varying between 5 and 90 min, depending on the material. This reduces the sensitivity of the material to the highly concentrated vitrification solutions.

Vitrification solutions are complex mixtures of cryoprotectants which have been formulated for their high ability to vitrify (i.e. form an amorphous glassy structure). Most solutions employed are derived from ones elaborated by either of two groups: that of Sakai's group, which consists of 22% glycerol, 15% ethylene glycol, 15% polypropylene glycol, 7% DMSO and 0.5 M sorbitol; and that of Steponkus' group, which comprises 40% ethylene glycol, 15% sorbitol and 6% bovine serum albumin. The duration of contact between explants and the vitrification solutions is a critical parameter, in view of their high toxicity. The dehydration period generally increases with the size of the explants used. Rye protoplasts are dehydrated for 60 s only, whereas the optimal dehydration duration of apices of pear and apple is 80 min.

Performing the dehydration step at 0°C instead of room temperature reduces the toxicity of vitrification solutions and thus broadens the window of exposure durations ensuring survival of samples. This also allows the manipulation of a large number of samples at the same time. Survival of asparagus embryogenic cell suspensions dropped rapidly after 5 min of dehydration at 25°C but high survival was obtained for dehydration periods between 5 and 60 min if it was performed at 0°C. Specimens are then cooled rapidly by direct immersion in liquid nitrogen in order to achieve vitrification of internal solutes. Reduction in the volume of cryoprotectants and the use of plastic straws (500 ml–1 ml) which have a small diameter and a large surface area of contact with the exterior allow an increase in the cooling rate, reaching 990°C min21 in the case of asparagus cell suspensions frozen in 50 ml of medium in a 500 ml straw. Apices of mint and sweet potato were frozen without cryoprotective medium, thus reaching a cooling rate of 4800°C min21.

Rewarming of samples is performed as rapidly as possible to avoid devitrification, which would lead to the formation of ice crystals detrimental to cellular integrity. Samples are immersed in a water-bath or liquid medium held at 20–40°C. Unloading aims at removing progressively the vitrification solution in order to reduce the osmotic shock. Liquid medium containing 1.2 M sucrose or sorbitol is added progressively to dilute the vitrification solution. Explants are then transferred to standard conditions. Vitrification procedures generally achieve high survival rates. Recovery is usually direct and rapid, even though a few authors have reported callusing and/or abnormal plant development. Vitrification experiments involving a large range of genotypes have been performed with several species including *Allium*, tea, *Citrus*, apple, mulberry, grape, potato and wasabi.

Encapsulation-vitrification

This technique is a combination of encapsulation-dehydration and vitrification procedures. Samples are encapsulated in alginate beads, then subjected to freezing by vitrification. It has been developed by Tannoury *et al.* (1991) with carnation apices, and applied recently by Sakai's group to apices of *Armoracia*, lily and wasabi.

In the case of carnation, encapsulated apices were pregrown for 16 h with progressively more concentrated sucrose solutions, then incubated for 6 h in a vitrification solution containing ethylene glycol and sucrose, and frozen either rapidly or slowly. Maximum survival was 100% and 92% after rapid and slow cooling, respectively. The technique employed with wasabi apices was slightly different. Apices were encapsulated in beads containing various loading solutions (glycerol or ethylene glycol and sucrose), then placed in the vitrification solution either at 25°C for 30 min or at 0°C for 70–100 min before rapid freezing. Ninety-five per cent of frozen apices recovered growth within 3 days.

Even though this technique has been used with a limited number of species only, it possesses great potential both in terms of efficiency and practicality. Matsumoto *et al.* (1995) mention that the recovery rate of apices frozen using the encapsulation-vitrification technique was 30% higher than with the encapsulation-dehydration technique. A reason for this might be that the alginate capsule reduces the toxicity of the vitrification solution. From a practical point of view, the number of manipulations is reduced as well as the total duration of the procedure.

Desiccation

Desiccation is a very simple procedure since it consists of dehydrating the plant material before rapid freezing by direct immersion in liquid nitrogen. This technique has been applied mainly to zygotic embryos or embryonic axes of various species, including numerous tropical forest trees, as well as to somatic embryos of several species and to shoot tips of mulberry.

Various parameters, such as the developmental stage of the embryos at the time of harvest, can greatly influence survival. Mature coffee embryos displayed a higher survival rate than immature ones. High variability was observed in the survival of embryos of several recalcitrant-seed-producing tree species harvested at different periods.

Desiccation is usually performed by placing the embryos in the air current of a laminar airflow cabinet. However, more precise and reproducible desiccation conditions are achieved by placing the embryos in a stream of compressed air or in an airtight container with silica gel. Optimal survival rates are obtained when the water content of the embryos is around 10–20% (fresh weight basis). Freezing is usually rapid but positive results have been obtained using slow cooling with several tropical forest tree species.

Regrowth of the frozen material usually takes place in standard conditions even though modification of the hormonal balance of the recovery medium proved beneficial with coffee zygotic embryos. Direct development of the embryos into plantlets is common but abnormal development patterns have been noted in some cases, such as the nondevelopment of the haustorium in the case of *Howea* and *Veitchia*, or callusing and/or incomplete development with *Hevea*, *Castanea* and *Quercus*, and oil palm.

Pregrowth

This technique consists of cultivating the plant material for different durations (hours to weeks, depending on the material) in the presence of cryoprotectants, then freezing rapidly by direct immersion in liquid nitrogen. It was first developed by Monnier and Leddet (1978) with zygotic embryos of *Capsella bursa-pastoris*, then applied to embryos of bean, wheat and maize, and more recently to meristematic clumps of banana.

Culture for 3 days up to 3 weeks on solid medium with high sucrose concentration (0.3 to 0.9 M) was employed with *Capsella*, maize and banana whereas a short treatment with liquid medium containing polyethylene glycol, glucose and DMSO was sufficient to ensure survival of bean and wheat zygotic embryos.

The pregrowth technique was successfully applied to five banana cultivars, resulting in survival rates ranging between 6 and 42.5%.

Modifications in the sucrose concentration and the duration of the pregrowth treatment should lead to improvements in the survival rate.

Pregrowth-desiccation

Pregrowth-desiccation is a combination of the two techniques described immediately above. In this technique, samples are treated with cryoprotectants, partially desiccated, cooled and rewarmed rapidly. This technique has been applied to a limited number of specimens only: somatic embryos of oil palm, date palm, coffee, pea and melon; stem segments of *in vitro* plantlets of asparagus; microspore embryos of rapeseed; and zygotic embryos of coconut. Culture of samples on media containing cryoprotectants usually takes place before desiccation. However, coconut zygotic embryos are dehydrated before preculture with cryoprotectants (Assy-Bah and Engelmann, 1992a). Sugars (sucrose, glucose) are generally employed for preculture. The duration of the preculture varies between 20 h for coconut and 7 days for oil palm. Various methods are employed for desiccation: coconut embryos are placed in the air current of a laminar airflow cabinet; asparagus stem segments and oil-palm somatic embryos are dehydrated using silica gel; melon somatic embryos are placed over a saturated salt solution; and date palm, coffee and pea somatic embryos are dehydrated using a stream of compressed air. The optimal water content (fresh weight basis) varies between 11.8% for melon and 25–30% for oil-palm somatic embryos.

Desiccated samples are frozen rapidly by direct immersion in liquid nitrogen. Specimens are usually stored in liquid nitrogen. However, Dumet *et al.* (1994) have demonstrated that oil-palm somatic embryos could be conserved for 6 months in a deep freezer at 280°C, i.e. below the glass transition temperature, without any modification in their recovery rate.

After rewarming, samples are usually placed directly under standard culture conditions. However, oil-palm somatic embryos are cultured on media with progressively reduced sucrose concentration, with transitory addition of 2,4-dichlorophenoxyacetic acid (2,4-D) to stimulate reproliferation.

Recovery of samples is usually rapid and direct. Alterations in the regrowth pattern of cryopreserved specimens were observed with coconut and oilseed rape microspore embryos. The haustorium of cryopreserved coconut embryos browned rapidly and did not develop further. Only half of frozen rapeseed microspore embryos developed directly into plantlets, whereas the other half produced callus and/or secondary embryos. Pregrowth-desiccation has been tested with four varieties of coconut, giving recovery rates ranging between 33 and 93%. This technique is routinely employed for the longterm storage of 80 clones of oil-palm somatic embryos.

Droplet Freezing

The droplet freezing technique has presently been applied to potato apices only. Apices are pretreated for 2–3 h in liquid medium supplemented with DMSO, placed on an aluminium foil in 5 ml droplets of cryoprotective medium and frozen rapidly by direct immersion in liquid nitrogen. This procedure is adapted from the classical procedure developed by Kartha *et al.* (1982) with cassava where apices placed in droplets of liquid medium were frozen slowly in a programmable freezer. The droplet freezing method has been successfully applied to 100 varieties of potato with recovery rates ranging between 5 and 100%. Around 25 000 apices are stored for the long term in liquid nitrogen at DSM, Braunschweig, Germany.

CRYOPRESERVATION

The development of cryopreservation for plant cells and organs has followed the advances made with mammalian species, albeit several decades later. The first report on survival of plant tissues exposed to ultra-low temperatures was made by Sakai in 1960 when he demonstrated that very hardy mulberry twigs could withstand freezing in liquid nitrogen after dehydration mediated by extra-organ freezing. More than a decade later it was shown that cultured cells of flax could resist freezing to 250°C after pretreatment with dimethylsulphoxide (DMSO).

This study was followed by a similar one in which cell cultures of carrot were shown to survive after freezing in liquid nitrogen. The methodology employed in the above experiments followed the classical procedures which had been successful with other living systems: chemical cryoprotection, slow dehydrative cooling followed by rapid immersion in liquid nitrogen, storage in liquid nitrogen, rapid thawing, washing and recovery.

The development for plant tissue cultures of so-called classical techniques which took place in the 1970s and 1980s is based on the above sequence of treatments. In recent years, several new cryopreservation techniques have been developed which allow cryopreservation to be applied to a larger range of tissues and organs, in different technical environments.

Dehydration and Freezing Injury

Most of the experimental systems employed in cryopreservation (cell suspensions, calluses, shoot tips, embryos) contain high amounts of cellular water and are thus extremely sensitive to freezing injury since most of them are not inherently freezing-tolerant. Cells have thus to be dehydrated artificially to protect them from the damage caused by the crystallization of intracellular water into ice. The techniques employed and the physical mechanisms upon which they are based are different in classical and new cryopreservation techniques. Classical techniques involve freezeinduced dehydration, whereas new techniques are based on vitrification. Vitrification can be defined as the transition of water directly from the liquid phase into an amorphous phase or glass, whilst avoiding the formation of crystalline ice. Classical cryopreservation techniques involve slow cooling down to a defined prefreezing temperature, followed by rapid immersion in liquid nitrogen. With temperature reduction during slow cooling, the cells and the external medium initially undergo supercooling; this is followed by ice formation in the medium. The cell membrane acts as a physical barrier and prevents the ice from seeding the cell interior, so the cells remain unfrozen but supercooled. As the temperature is further decreased, an increasing amount of the extracellular solution is converted into ice, thus resulting in the concentration of extracellular solutes. Since cells remain supercooled and their aqueous vapour pressure exceeds that of the frozen external compartment, cells equilibrate by loss of water to external ice. In optimal conditions, most or all intracellular freezable water is removed, thus reducing or avoiding detrimental intracellular ice formation upon subsequent immersion of the specimen in liquid nitrogen. Rewarming should be as rapid as possible to avoid the phenomenon of recrystallization, in which ice melts and re-forms at a thermodynamically favourable, larger and more damaging crystal size. New techniques are vitrification-based procedures. In such procedures, the freezable intracellular water is removed prior to freezing by exposing the samples to highly concentrated cryoprotective media and/or air desiccation. Dehydration is followed in most cases by rapid cooling. As a result, the internal solutes vitrify and deleterious intracellular ice formation is avoided. Glass transitions (i.e. changes in the structural conformation of the glass) during cooling and rewarming have been recorded with various materials using thermal analysis.

Classical Procedures

For each new material to be cryopreserved, optimal conditions have to be defined for each of the successive steps of the protocol.

The physiological state of the material can affect its survival. Cell suspensions are more likely to withstand freezing when they are employed during their exponential growth phase, i.e. when they are small and have a relatively low water content. Survival of carnation shoot tips after freezing decreased progressively with their rank on the shoot apex, starting from the terminal meristem, thus reflecting physiological and developmental differences found in explants sampled on the same plant. Harding *et al.* (1991) indicated that a long-term period in tissue culture before cryopreservation significantly reduced the ability of potato apices to survive freezing. A pregrowth period before the cryoprotective treatment on medium with osmotically active compounds such as mannitol or sorbitol can increase freeze-tolerance.

Samples are submitted to a cryoprotective treatment before freezing. It is carried out using various cryoprotective substances such as dimethylsulphoxide, sorbitol, mannitol, sucrose or polyethylene glycol.

Additional information on the role and mechanisms of action of cryoprotectants can be found in several reviews. Cryoprotectants are employed alone or in binary or ternary mixtures. Mixtures of cryoprotectants have proved especially effective with cell suspensions.

Classical cryopreservation procedures include a two-step freezing: slow, controlled cooling to a defined prefreezing temperature followed by rapid immersion of samples in liquid nitrogen. Freezing rates of between 0.5 and 2°C min21 down to prefreezing temperatures around 240°C generally give satisfactory results in most cases. However, while for some materials, such as oil palm somatic embryos, a wide range of freezing rates can be employed without modification of the recovery rate, other materials (e.g. grape cell suspensions) require very precise freezing parameters to achieve survival. Most classical freezing procedures require the use of expensive programmable freezing devices which achieve precise freezing conditions. However, sophisticated apparatus is not always necessary to obtain high survival. Withers and King (1980) have successfully cryopreserved various cell suspensions with an improvised and simple apparatus that can offer reproducible but nonlinear slow cooling. More recently, several authors have successfully employed domestic or laboratory deep-freezers to perform the slow cooling step.

Removal of cryoprotectants by washing after thawing, which was widely adopted in the past, was found to be deleterious for a number of cell cultures. A much less stressing technique for removing cryoprotectants, developed by Chen *et al.* (1984), consists of moving cultures on a filter paper through a series of Petri dishes of solid culture medium. This technique was successfully applied to various cell suspensions and calluses. In some cases, standard culture conditions have to be modified to improve recovery of cryopreserved cultures. Recovery of cell suspensions is generally improved if they are cultured for some days on solid medium before being transferred to liquid medium. Cultures are often placed in the dark or under reduced light intensity to avoid photooxidation phenomena which are harmful to the material. Finally, the culture medium can be transitorily altered by modifying its hormonal balance, mineral composition or by incorporating activated charcoal.

Classical freezing techniques have been assessed with different materials including cell suspensions, calluses, apices and embryos. They have proven to be highly successful with most cell suspensions and calluses, i.e. culture systems which consist of small units of relatively uniform morphology. However, apart from exceptions like carnation apices, these techniques are not suitable for the cryopreservation of larger units comprising a mixture of cell sizes and types, such as apices and zygotic and somatic embryos.

IN VITRO GERMPLASM EXCHANGE

In vitro culture techniques have been used extensively for the international exchange of germplasm because of their obvious advantages over *in vivo* material, notably their reduced weight and volume, and phytosanitary condition. Indeed, meristem culture, alone or employed in combination with thermotherapy, can eliminate viral pathogens. Plant material can thus be multiplied, stored and exchanged in a disease-free state.

In particular, because aseptic culture conditions are used and provided that the plant material has been cleansed of internal contaminants, problems with the international movement of germplasm are considerably reduced. Most countries will accept batches of plants *in vitro* with a phytosanitary certificate, without requiring a rigorous quarantine period.

However, an important caution in this area is that transfer to *in vitro* culture does not confer disease-free status. Indexing, using one or a combination of the various techniques available, which include symptomatology, grafting/inoculation on indicator plants, ELISA and molecular techniques such as dsRNA detection, is the only sure way of making this judgement. Routine procedures for the international exchange of *in vitro* cultures have been developed, notably for potato, cassava, yam and *Musa*. Samples are usually placed in glass, or preferably plastic, test tubes on the standard culture medium, possibly with the gelling agent modified to increase its firmness and a reduction in the carbon source to limit growth.

Heat-sealable polythene bags are sometimes employed instead of more fragile plastic or glass test tubes. With species such as potato and sweet potato, where it is possible to induce their formation *in vitro*, tubers are also used for germplasm exchange, since they are often easier to transport and to handle than plantlets. Encapsulation of plant material in alginate beads has been suggested recently as a possible way of germplasm exchange in the case of banana. Using encapsulated apices instead of *in vitro* plantlets would result in a further volume reduction.

In addition, encapsulated apices might be either grown *in vitro* after their receipt, or directly sown in the soil as seeds where they would develop into plantlets. The encapsulation technique has also been employed with nodal segments of yam. In a germplasm exchange simulation experiment, encapsulated nodal segments were conserved for two weeks in the dark in cryotubes containing culture medium, without any detrimental effect on their further growth.

SLOW-GROWTH STORAGE

Classical Techniques

Standard culture conditions can be used for medium-term storage with species which have a naturally slow-growing habit only. A wide range of *Coffea* species are thus conserved on standard medium at 27°C without subculturing for durations varying from 6 to 12 months, depending on the species.

However, for most species, growth reduction is achieved by modifying the environmental conditions and/or the culture medium. The most widely applied technique is temperature reduction, which can be combined with a decrease in light intensity or culture in the dark.

Temperatures in the range of 0–5°C are employed with cold-tolerant species. Strawberry (*Fragaria* 3 *ananassa*) plantlets have been stored at 4°C in the dark and kept viable for 6 years with the regular addition of a few drops of liquid medium. Apple (*Malus domestica*) and *Prunus* shoots survived 52 weeks at 2°C. Tropical species are often cold-sensitive and have to be stored at higher temperatures, which depend on the cold sensitivity of the species. Kiwi fruit shoots can be conserved at 8°C and taro (*Colocasia esculenta*) tolerates 3 years of storage at 9°C. *Musa in vitro* plantlets can be stored at 15°C without transfer for up to 15 months. Other tropical species such as oil palm and cassava are much more cold-sensitive: oil palm somatic embryos and plantlets do not withstand even a short exposure to temperatures lower than 18°C. Cassava shoot cultures have to be conserved at temperatures higher than 20°C.

Various modifications can be made to the culture medium in order to reduce growth. Embryogenic cultures of carrot could be conserved on a medium without sucrose for two years, and reproliferated if a sucrose solution was supplied. Kartha *et al.* (1981) could conserve coffee plantlets on a medium devoid of sugar and with only half of the mineral elements of the standard medium.

Replacement of sucrose by ribose allowed the conservation of banana plantlets for 24 months. The addition of osmotic growth inhibitors (e.g. mannitol) or hormonal growth inhibitors (e.g. abscisic acid) is also employed successfully to reduce growth.

The type of explant as well as its physiological state when entering storage can influence the duration of storage achieved. Roxas *et al.* (1995) indicate that, in the case of chrysanthemum, nodal segments showed higher survival rates than apical buds. The presence of a root system generally increases the storage capacities, as observed by Kartha *et al.* (1981) with coffee plantlets. Microtubers can be successfully employed as storage propagules, as demonstrated with potato. Preconditioning the explants by exposing them briefly to temperature and light conditions intermediate between standard and storage conditions was favourable for *Nephrolepsis* and *Cordyline* cultures. Higher survivals were obtained with shoot cultures of wild cherry, chestnut and oak if they were kept for 10 days under standard conditions after the last subculture before their transfer to the cold storage chamber.

The type of culture vessel, its volume as well as the type of closure of the culture vessel can greatly influence the survival of stored cultures. Roca *et al.* (1984) indicate that cassava shoot cultures could be stored for longer periods in a better condition by increasing the size of the storage containers. White spruce embryogenic tissues withstood a one-year storage period in hermetically sealed serumcapped flasks. Replacing cotton plugs by polypropylene caps, thus reducing the evaporation of the culture medium, increased the survival rate of *Rauvolfia serpentina* during storage. The use of heat-sealable polypropylene bags instead of glass test tubes or plastic boxes was beneficial for the storage of several strawberry varieties.

At the end of a storage period, cultures are transferred onto fresh medium and usually placed for a short period in optimal conditions to stimulate regrowth before entering the next storage cycle.

Alternative Techniques

Alternative techniques include modification of the gaseous environment of cultures, and desiccation and/or encapsulation of explants. Growth reduction can be achieved by reducing the quantity of oxygen available to the cultures.

The simplest method consists of covering the explants with paraffin, mineral oil or liquid medium. This technique was first developed by Caplin (1959), who could store carrot calluses for 5 months under a layer of paraffin oil. Augereau *et al.* (1986) and Moriguchi *et al.* (1988) applied this technique to calluses of *Catharanthus* and grape, respectively. Florin (1989) showed that 50% of a collection of 313 callus lines survived after storage under mineral oil for 12 months. However, Mannonen *et al.* (1990) reported that conservation under mineral oil did not preserve the productivity of *Catharanthus* and *Panax ginseng* callus cultures for more than 6 months.

Similar experiments performed with shoot cultures of various species led to contradictory results. After 4 months of storage under mineral oil, regrowth of surviving coffee shoot cultures was very slow and pear microcuttings did not survive. In contrast, several ginger species could be conserved for up to two years under mineral oil with high viability.

Reduction of the quantity of oxygen can also be achieved by decreasing the atmospheric pressure of the culture chamber or by using a controlled atmosphere. Tobacco and chrysanthemum plantlets could be stored under low atmospheric pressure (with 1.3% oxygen) for 6 weeks. Oil palm polyembryogenic cultures were conserved for 4 months at room temperature in a controlled atmosphere with 1% oxygen. Dorion *et al.* (1994) showed that hypoxic regimes at standard or intermediate temperature could replace low-temperature storage for shoot cultures of peach. Rice calluses could be stored for 12 weeks in a CO2 or N2 saturated atmosphere.

Desiccation of cultures as a means of achieving medium-term conservation was first reported by Nitzsche (1980), who stored dehydrated carrot callus for one year at 15°C under 25% relative humidity. Increasing interest has been paid recently to this technique, with the development of so-called synthetic seeds for various plant species. Synthetic seed is a generic term for a somatic embryo delivery system used as a means of clonal propagation. The aim is to use such somatic embryos as true seeds: embryos, encapsulated or not in alginate gel, could be stored after partial dehydration and sown directly *in vivo*.

After progressive dehydration using saturated salt solutions, naked alfalfa somatic embryos could be conserved with 10–15% moisture content at room temperature for one year and showed only a 5% decrease in their conversion rate at the end of the storage period. Carrot somatic embryos were stored for 8 months at 4°C without viability loss. Shorter storage durations were achieved with encapsulated material.

Encapsulated carrot somatic embryos survived after 3 months in liquid medium at low temperature and encapsulated shoot tips of *Valeriana wallichii* could be conserved for over 6 months at 4–6°C. According to Redenbaugh *et al*. (1991), the rapid dehydration of the encapsulating matrix limits the respiration of the encapsulated material, which is the cause for the rapid survival loss generally observed.

IN VITRO TECHNIQUES FOR COLLECTION AND EXCHANGE OF GERMPLASM

Collectors are faced with various problems when collecting germplasm of recalcitrant seed and vegetatively propagated plant species. Collecting missions often require travelling for relatively long periods in remote areas. It is thus necessary to keep the material collected in good state for some days/weeks before it can be placed in optimal growth or storage conditions. There are thus great risks that recalcitrant seeds either germinate or deteriorate before they are brought back to the genebank.

In addition, many recalcitrant seeds have considerable weight and bulk, which increases the volume of material to handle and induces additional costs, if an adequate sample of the population is to be collected. With vegetatively propagated species, the material collected will consist of stakes, pieces of budwood, tubers, corms or suckers. Not only will most of these explants not be adapted to survival once excised from the parent plant, but they will also present health risks due to their vegetative nature and contamination with soil-borne pathogens.

Difficulties can also be encountered when collecting germplasm of orthodox seed-producing species. Even with careful planning of the time of a collecting mission, there might be no or little seed available for all or part of the germplasm to be collected, or seeds might not be at the optimal developmental stage, or might be shed from the plant or eaten by grazing animals.

The problems described previously can be overcome if it is realized that the seed is not the only material which can be collected: zygotic embryos, or vegetative tissues such as pieces of budwood, shoots or apices can be sampled, transported and grown successfully if placed under adequate conditions.

Following an expert meeting organized by IBPGR in 1984 and sponsorship of various research programmes, *in vitro* collecting techniques have been developed for different materials including embryos of coconut, cacao, avocado, *Citrus*, vegetative tissues of cacao, *Musa*, coffee, *Prunus*, grape, cotton and several forage grasses.

The critical points to consider for the development of *in vitro* collecting techniques have been synthesized and analysed by Withers (1995). The techniques developed are very simple and flexible, as illustrated below with coconut and cacao.

In the case of coconut, the *in vitro* field collecting technique developed by Assy-Bah *et al.* consisted of extracting from the nut with a corkborer a plug of endosperm containing the embryo. After surface sterilization with calcium hypochlorite or commercial bleach, the embryos were dissected on the spot under the shelter of a wooden box and inoculated onto semisolid medium, or the endosperm plugs were transported to the laboratory where the embryos were dissected and inoculated onto semisolid medium in aseptic conditions.

This technique is very efficient since after approximately 6 to 9 months in culture in the laboratory under standard conditions, an average of 75% of the embryos collected developed into plantlets which could be successfully transferred to the nursery and then to the field. In addition, embryos inoculated *in vitro* in the field could be kept in the open for two months before being grown in the laboratory, without influencing their further development.

This technique has been modified by other researchers towards higher or lower sophistication. In its more sophisticated version, inoculation of the embryos onto sterile medium is performed in an inflatable glove box. The simpler procedure consists of sterilizing the endosperm plugs containing the embryos and of placing them in a cool box in sterile plastic bags for transportation. Upon arrival in the laboratory, a second disinfection is performed and the embryos are dissected and inoculated onto culture medium inside the laminar airflow cabinet.

In the case of cacao, an *in vitro* collecting method was developed for budwood. Considering that absolute sterility would be difficult to achieve in the field and would not necessarily be essential for robust, woody material, the aim was to place the samples collected in conditions which would suppress or delay deterioration. Stem nodal cuttings were disinfected with boiled water in which drinking-water sterilizing tablets had been dissolved and containing fungicides, then inoculated onto semisolid medium supplemented with fungicide and antibiotics. Explants could be maintained in a relatively clean (though not necessarily completely sterile) condition for up to 6 weeks.

4

Cloning of Plant Traits

Plant gene cloning is currently limited to simply inherited single gene traits. The transfer of genes to plants is limited to single genes or, at most, two to three genes encoding a particular, well-characterized biochemical pathway. Manipulation of multigenic traits is currently only feasible within the existing gene pool, sometimes made more efficient by markerassisted selection.

When traits are regarded in the molecular genetic sense, often the coding sequence is the major consideration. Obviously, the coding sequence is of major importance as this is the template for determining the amino acid sequence of the final protein or enzyme which determines the phenotype. However, a major factor in controlling the phenotype is the presence of regulatory sequences surrounding the coding sequence in the form of upstream (promoters and enhancers) and downstream (terminator) regulatory sequences and, in many eukaryotic genes, introns.

When traits are being manipulated via genetic engineering, the role of these regulatory sequences cannot be trivialized. Put simply, these sequences control such things as the level of gene expression, tissue/organ specificity, developmental expression and response to particular abiotic/biotic stimuli.

The regulatory sequences that have been subjected to greatest attention in plants are the upstream promoters. For proper expression of a seed storage protein, for example, the gene of interest must be under the control of a promoter which expresses in developing endosperm (monocots) or cotyledons (dicots). Expression of other genes may be required only in anthers, root hairs or photosynthetic tissues, or in response to a wound such as that caused by a chewing insect or penetration of a fungal infection structure.

In the earliest cases of transgenic plants, introduced genes were placed under the control of constitutive promoters of bacterial (octopine or nopaline synthetase) or viral (cauliflower mosaic virus (CaMV) 35S) origin. These promoters have been found to give adequate expression of selectable marker and reporter genes in dicotyledonous species.

However, expression is commonly not of a sufficient level in monocots, especially cereals, for good selection with antibiotic or herbicide resistance genes. And, as already discussed, such a constitutive pattern of gene expression is commonly not desirable for most traits. For cereals, a number of other promoters have been developed to achieve high-level constitutive expression of selectable markers or reporter genes. These include sequences cloned directly from cereal tissues such as the ubiquitin and rice actin 1 promoters and synthetic promoters involving various spliced sequences from a range of sources such as the Emu promoter.

However, to properly control gene expression such that the introduced character is most efficiently targeted without other pleiotropic effects, it is generally found that regulatory sequences of plant origin are desirable. Further to this, expression is usually best controlled when a cereal promoter is used in a cereal or a dicot promoter in a dicot.

The most widely used means of cloning such regulatory sequences is by using cDNA libraries from specific tissue types and looking for abundant signals. Naturally, cDNA libraries will not contain the regulatory sequences, hence the abundant cDNA clones must then be used to screen a genomic library so that the regulatory sequences can be identified and cloned.

METHODS FOR CLONING PLANT TRAITS

Cloning Based on Knowledge of Protein Structure/sequence

Many of the first plant traits to be cloned were those in which some degree of biochemical understanding preceded any work with DNA. The understanding may have come from plant systems or from outside the plant kingdom. Most of the cloned genes in these cases were enzymes which were part of a well-characterized biochemical pathway or had a protein structure which was otherwise well known, such as seed storage proteins.

Cloning genes based on a knowledge of the protein sequence can be achieved by working backwards from the gene product (amino acid sequence) to the coding sequence or the DNA template. Such an approach has been used for cloning many seed storage proteins, such as the 2S sulphur- rich proteins from Brazil nut.

Genes of plant origin have been used in transgenic plants to improve resistance to insect pests. One of the first to be characterized was the cowpea trypsin inhibitor, a protease inhibitor which acts as an antifeedant to a range of insect pests. The protein is naturally produced in seeds of some genotypes of cowpea (*Vigna unguiculata*), which was shown to confer lepidopteran insect resistance when expressed constitutively in tobacco. Another such report from the same group at Durham

University described the use of a snowdrop lectin gene to confer aphid resistance in transgenic tobacco, the first report of genetically engineered resistance to a sucking insect in plants. The lectin from snowdrop (*Galanthus nivalis*), known as GNA, has been demonstrated to be effective against two major rice pests, the brown planthopper and the green leafhopper, as well as the peach potato aphid. When expressed at 0.1% of total leaf protein in tobacco, GNA is antimetabolic to homopteran pests. These levels of protein were accumulated in leaf tissues when placed under the control of the CaMV 35S promoter. However, Hilder *et al.* (1995) postulate that better control would be achieved with the use of a phloem-specific promoter, such as the sucrose synthase 1 promoter from rice.

Another significant group of genes which have been cloned based on prior knowledge of the end product are the seed storage genes, including both seed storage proteins and lipids. In both cases there was considerable biochemical knowledge of the pathways involved in biosynthesis and storage. For seed storage proteins, the genes can be cloned based on the knowledge of the amino acid sequence of the mature protein and targeting signal or transit peptide.

Molecular genetic manipulation of vegetable oils is a major area of research in the developed world. While approximately 90% of vegetable oils are used for human consumption, there are emerging markets for industrial applications including lubricants, paints, cosmetics, plasticizers and soaps.

With increasing emphasis on quality and dietary considerations, there is a desire in developed countries to substantially alter the types and balances of fatty acids found in the major oil crops. Six crop species (soybean, oil palm, rapeseed/canola, sunflower, cottonseed and peanut) account for 84% of the world's vegetable oil production, with the major fatty acids produced being palmitic, stearic, linoleic, linolenic and oleic acids. However, almost 200 different fatty acids are produced by plants, mostly occurring in nondomesticated plant species. Hence, genetic manipulation of the major crop species, either via conventional breeding or genetic engineering, has the potential to modify the fatty acid types produced and stored by seeds such as soybean and canola.

The biochemistry of fatty acid biosynthesis is well established for many of the major pathways, with enzymes identified which control the chain length of fatty acids, and the degree of saturation, or number of double bonds in the fatty acid chain. The oils stored in seeds have been altered by downregulation of particular enzymes using antisense RNA techniques. The oleic acid content of rapeseed oil has been increased to 83% (from 62% in conventionally bred cultivars) by downregulating the enzyme D12-oleate desaturase using antisense RNA. A similar result has been achieved using co-suppression, which has produced rapeseed oil with 87% oleic acid content. Genetic engineering approaches have also been used to introduce totally novel fatty acids to major oilseed crops. Petroselinic acid, an isomer of oleic acid, is found in abundance in the spice plant, coriander (*Coriandrum sativum*). This oil has uses in cosmetics and pharmaceuticals, and its oxidation leads to the formation of lauric acid, used in detergents, and adipic acid, which is used in nylon production.

Hence this is a potentially valuable oil which may enable soybean and canola crops – i.e. renewable resources – to replace some of the nonrenewable carbon fuels which are currently used for many industrial polymer and lubrication applications. A cDNA clone from coriander encoding an acyl-ACP desaturase has been genetically engineered into tobacco, resulting in the accumulation of petroselinic acid. Oil crops such as rapeseed and soybean could be genetically modified to produce this fatty acid on a commercial scale. Hence, here is an example whereby a minor crop plant has become a donor of genetic diversity which will enable the commercial production of a very useful array of products for food and industrial purposes.

Cloning Genes Known Only by Phenotype

However, a more challenging task has been to move beyond proteins and pathways about which sufficient biochemistry is known into cloning genes known by phenotype only. By far the majority of genes fit into this category, which includes many of the most important plant traits such as disease resistance, adaptation to abiotic stresses, phenology and flowering behaviour and fertility.

Only in the past few years have genes for traits known only by phenotype been cloned. One of the most interesting areas of plant biology is the understanding of plant–pathogen interactions at the molecular level. The major methods used for cloning genes known only by phenotype are:

- Transposon tagging.
- Map-based or positional cloning.
- T-DNA tagging.

There have been a number of plant genes cloned using all these methods or modifications of them. There are also examples where a combination of methods has been used to expedite the rate of identification of the unknown sequence.

Transposon Tagging

The proposal that specific pieces of DNA are able to autonomously excise from a particular locus and insert, theoretically at random, into another linked or nonlinked locus was first proposed by Barbara McClintock (1949, 1950). Transposable elements are now accepted as ubiquitous, and are particularly well characterized in organisms such as *Escherichia coli*, *Drosophila* and maize.

In two plant species, maize and the snapdragon (*Antirrhinum majus*), transposable elements are sufficiently well characterized to be used for mutagenesis and gene tagging. The best understood system is the maize *Ac/Ds* (*Activator/Dissociation*) family of elements. The behaviour of the *Ac* element is such that it can excise autonomously from its locus, under the control of its own transposase, and insert into another chromosomal position. Where the *Ac* element is in or near a gene, that gene is inactivated (i.e. not expressed), and will appear phenotypically as a recessive mutation. When *Ac* transposes (excises and reinserts at another locus), the gene is commonly returned to normal function, and hence is termed a revertant.

Should this *Ac* element then insert into another gene (transcribed or regulatory sequence), then a new recessive mutation will manifest. Transposition can occur either germinally, which results in an altered phenotype for the whole plant, or somatically, which results in a mosaic mutation, and can often be detected as a sectoring of plant or seed colour. The *Ds* element is actually a group of elements derived from defective *Ac* elements, with a common feature being their lack of effective *Ac* transposase.

Table. Plant Transposable Elements which have been used for Transposon-Taggine

Element	*Species*	*Type of Element*	*Size*
Ac	Maize	Autonomous	4.6
Ds	Maize	Nonautonomous	0.5–30]
Tam1	Snapdragon	Autonomous	15
Tam2	Snapdragon	Nonautonomous	5
En/Spm	Maize	Autonomous	8.3

Hence these elements are nonautonomous, and will only transpose in the presence of an *Ac* element capable of producing transposase. Equipped with sequence and map information of an *Ac* element in maize, geneticists have been able to clone important maize genes using the transposon-tagging approach.

When a gene is inactivated by an *Ac* element, proof of which can be demonstrated by a low frequency of trait reversion, the *Ac* element can be used as a probe for 'fishing-out' the tagged gene.

A tagged gene can be cloned by performing Southern blot analysis with the *Ac* element used as probe, and any polymorphism represents an excision/ insertion event. The technique is described more fully in the review by Walbot (1992).

The *Ac* system has proven extremely useful for cloning maize genes for traits such as plant pigment genes (*bronze*; the anthocyanin pathway), starch biosynthesis (*opaque2*), and abscisic acid (AbA) insensitivity (McCarty *et al.*, 1989). Other maize transposable elements such as *Mutator* (*Mu*) and the *En/ Spm* system have been used to clone maize genes, as has the *Tam* system in snapdragon. As a result, the search began for transposable element systems in other species so that a similar approach could be taken.

However, it was soon demonstrated that the well-characterized *Ac/Ds* system could be genetically engineered into other plant species including tobacco, *Arabidopsis*, tomato and petunia. It was subsequently shown that the *Ac* element acted in a similar manner in these heterologous species, undergoing germinal and somatic excision and insertion, and maintaining the propensity to reinsert at a linked site (usually within 4 cM). As a result, the frequency of mutation in the surrounding area approaches 1024. It was not until 1993 that a plant gene was cloned using the heterologous system, that of the *Ph6* gene in petunia, which is involved in floral pigmentation.

Using the transposon-tagging approach, a number of plant disease resistance genes (R genes) have now been cloned. The first of these was the tomato *Cf-9* gene, which confers resistance to race 9 of the fungal pathogen *Cladosporium fulvum,* causal agent of leaf mould. The gene was tagged using a modified *Ac/Ds* system.

The system relied on the use of a stabilized *Ac* element (s*Ac*), which was effectively a disabled *Ac* element that nevertheless produced *Ac* transposase and hence enables *Ds* transposition. The two elements are maintained in separate lines, and hence are both stable until controlled crossing is performed. This gives a significant advantage in that *Ds* insertion mutants can be maintained indefinitely in the homozygous or hemizygous form, and revertants can be isolated by crossing lines with s*Ac* tomatoes.

The transposon-tagging approach used to isolate the *Cf-9* gene, which was known by phenotype only, did have considerable advantages in that much was known of the pathogen. The hypersensitive response elicited by the virulent strain fitted into the gene-for-gene hypersensitive response proposed by Flor (1971), with a dominant avirulence gene (*Avr9*) interacting with the *Cf-9* gene to form an incompatible or hypersensitive response. *Avr*9 had previously been cloned and fully characterized, and was known to specify a 28 amino acid peptide which alone could elicit a necrotic response in resistant tomatoes.

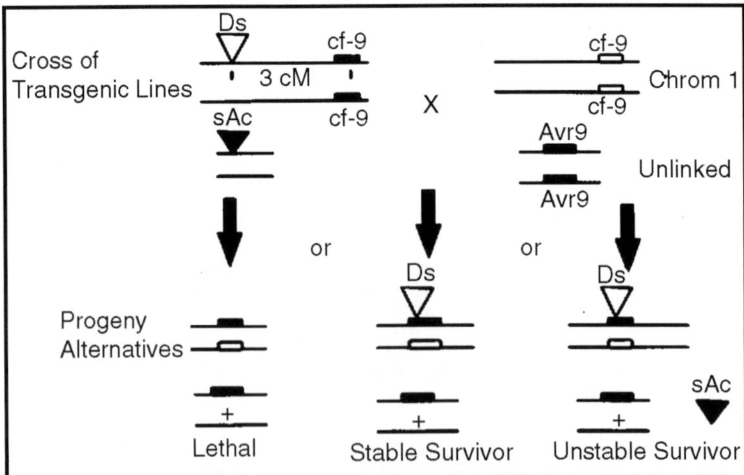

Fig. Graphical Representation of the Crosing Strategy of Transgenic Ac/Ds Tomato Lines.

Crosses were made to develop a transgenic tomato line homozygous for *Cf-9*, but heterozygous for *Ds*. Crossing was performed as outlined in Fig.. As can be seen, most seedlings died from systemic hypersensitive response, with *Avr9* and *Cf-9* present in all cells. The only survivors would be individuals where *Ds* had inserted into the *Cf-9* gene. Where this happened in plants carrying s*Ac*, the seedlings were variegated for the hypersensitive response (somatic excision), and in the absence of s*Ac*, the survivors were stable (germinal excision).

This work was a major undertaking, involving the germination of approximately 160 000 progeny, yielding 118 survivors, representing 63 independent mutations. After considerable effort to characterize these, the *Cf-9* gene was cloned and characterized, with an open reading frame encoding an 863 amino acid protein.

Map-based Cloning

The previously described transposon-tagging scheme utilized to clone *Cf- 9* had an element of the map-based approach by utilizing the phenomenon of *Ac/Ds* transposition to linked sites, with a *Ds* element within 3 cM of the locus of interest, which greatly enhanced the likelihood of insertional mutagenesis. However, map-based cloning requires a much tighter genetic linkage (in centimorgans or recombination fraction), which in turn needs to be converted into a physical distance (kbp of DNA sequence).

The map-based, or positional cloning concept relies on building a well-saturated genetic map of DNA sequences of the organism in question. The genetic map is built up in a polymorphic mapping population, commonly an F2, backcross (BC) or recombinant inbred, with markers such as RFLPs (restriction fragment length polymorphisms). The first RFLP map to be made was of human. Based on such maps, important human genes responsible for disorders such as cystic fibrosis and muscular dystrophy have been cloned in this manner.

Genetic maps of plant species have been made with DNA markers in virtually all important plant species, particularly the major cereals, grain legumes, oilseeds and important vegetable crops including tomato and potato. The RFLP marker system is based on Southern blot analysis, while many of the newer technologies use the polymerase chain reaction (PCR).

These include random amplified polymorphic DNA (RAPDs), amplified fragment length polymorphisms (AFLPs), and a number of different marker systems based on simple sequence repeats (SSRs), which may use PCR, such as inter simple sequence repeat (ISSR)-PCR, or Southern analysis. The applications of DNA markers to plant genetic analysis and breeding are many, and the subject has been reviewed by various authors.

The population for saturation mapping should be segregating in a Mendelian manner for the trait of interest. This requires a lot of tedious but demanding work, and it may be that particular populations such as near isogenic lines, or DNA pooling strategies, such as bulked segregant analysis, will allow the mapping effort to be more targeted to the region of interest, hence improving the rate of map saturation.

It is desirable to obtain extremely close genetic linkage to the locus of interest (<1 cM), with flanking markers if possible. It can be of considerable advantage to identify a marker which co-segregates (say within 0.1 cM) with the trait. The next step is to convert the genetic map into a physical map of the chromosome region, translating distances from centimorgans to nucleotides.

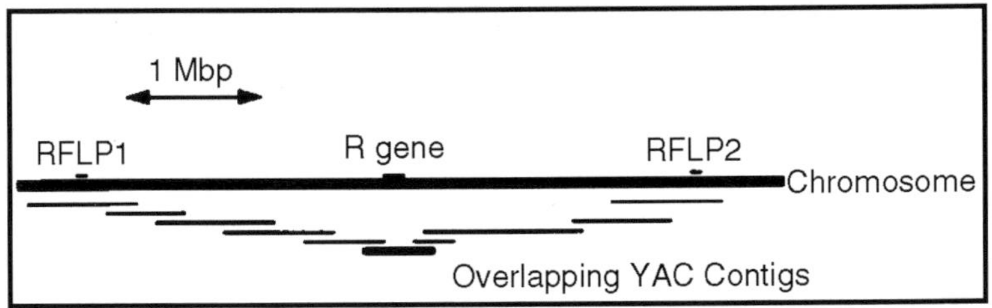

Fig. Generalized Strategy of Chromosome Walking Used in Map-based gene Cloning.

The Gene of Interest (in this illustration,R) is Flanked by Markers such as RFLPs or SSRs.A Series of Overlapping YAC Contigs is used to cover th Chromosomal Region Containing the Gene.The YAC Which Contains the Complete Gene is Marked in Bold.This is then Subcloned or Matched to a cDNA Library to Idenfify the Gene(s) Contained the Region, and Ultimately Identify the R Gene.

Physical mapping then involves the use of YAC or BAC clones which span the region of interest. These YAC/BAC contigs will span the region between two flanking DNA marker sequences as shown in Fig., a process known as chromosome walking. In this example, two YAC/BAC clones contain the gene of interest. Methodology is being developed to enable BAC clones to be directly transformed into plants, to find which one contains the gene of interest (in the example, the *R* gene). Alternatively, the YAC/BAC must be subcloned (for example as a cDNA library), and these subclones can be put into the susceptible plant and tested for the R phenotype.

The first R gene to be cloned in plants via map-based cloning was the *Pto* gene, which confers resistance to *Pseudomonas syringae* pv. *lycopersici*, the causal agent of bacterial speck in tomato. The starting points for the work were:

- A high-density RFLP map of tomato.
- A tomato YAC library.

The *Pto* locus had been previously mapped to tomato chromosome 5 and one RFLP marker co-segregated with *Pto*. Hence this was a useful starting point to screen the YAC library. A 400 kb YAC was identified, and via inverse

PCR, some end-specific probes were generated. The left arm of the YAC was 1.8 cM from *Pto,* with the right arm perfectly co-segregating. This YAC was then used to screen a 920 000 plaque leaf cDNA library, from which 30 were selected. One of these was shown to co-segregate with *Pto*. This cDNA was then transformed into a susceptible tomato cultivar under the control of the CaMV 35S promoter, where it conferred resistance. Hence the gene was identified, and could then be sequenced and characterized.

The success of this approach was partly due to the small genome size of tomato (950 Mbp), and a relatively low level of repeat sequence. Such an approach is much more difficult in species with larger genomes such as maize (2300–2700 Mbp) and wheat (16 000 Mbp). One such example of this problem is the 79 000 clone maize YAC library produced by Zeneca Seeds. As reported by K. Edwards, only 15% of the YACs have unique ends, with the remainder terminating in repeat sequences, which makes chromosome walking very difficult. The synteny of genomes is a major advantage in this instance. It is now well established that there is a high degree of synteny of genome organization, as seen in the co-linearity of maps of the most important grass genomes. This has resulted in the great interest in using this co-linearity to facilitate cloning of genes in large genomes such as maize and wheat, by starting with the smaller genomes, particularly rice, and to a lesser extent, sorghum. One particular study of maize YACs demonstrated that a single YAC with the *Adh1* sequence contained 36 different repetitive sequences, whereas these repetitive sequences were largely absent in corresponding areas of the rice and sorghum genomes, a pattern seen also for the *a1–sh2* region in all three genomes. In a similar manner, many genes of interest for *Brassica* improvement are being identified in the smaller *Arabidopsis thaliana* genome.

Paterson *et al.* (1995) have recently demonstrated the potential of this approach, by identifying quantitative trait loci (QTLs) for seed mass and phenology which were co-linear on molecular marker maps of maize, sorghum and rice. They proposed that, as the maize genome is four and six times larger than the sorghum and rice genomes, respectively, the genes of interest in maize may be cloned in sorghum/rice more expediently. It must be remembered that QTLs are subject to genotype 3 environment interaction, which will be a major complicating factor in attempts to clone and characterize a QTL.

T-DNA Tagging

When T-DNA is inserted into the plant genome, it does so via illegitimate recombination. In the same way as a transposon, the insertion of T-DNA into a gene will inactivate it, which phenotypically will appear as a mutation. The comparative advantage of such a method is that stable mutations will be conferred. This can be a disadvantage, however, as reversions, such as those made possible by the *Ac/Ds* system are not possible. It is also the case that each independent transformation event will result in just one mutation event (or more if multiple independent insertions are made), whereas the heterologous transposon approach allows multiple mutation events to take place from a single initial transformant. Hence, mutated genes with T-DNA insertions can be cloned, in much the same manner as those with transposon insertions, by using part of the T-DNA as a probe in Southern blots.

The procedure for selecting out the tagged gene is quite straightforward, particularly when the plant contains a single T-DNA copy. However, this means that a large number of independent transformants must be generated. This can be achieved with species such as *Arabidopsis*, where Feldmann (1991) demonstrated that a seed-based transformation system using *Agrobacterium* can be quite efficient. Feldmann (1991) demonstrated that by inoculating approximately 1000 seeds, approximately 300 000 selfed T2 seed can be collected. They estimated that approximately 8000 transformants could be recovered from an experiment of this scale and, using this approach, cloned a gene involved in trichome development and a dwarf gene from *Arabidopsis*.

A regulatory gene in the ethylene biosynthesis pathway of *Arabidopsis* has also been cloned using the T-DNA tagging strategy. With modification, T-DNA tagging methodology can be used to clone plant promoter sequences. Koncz *et al*. (1989) designed a vector which contained a promoterless antibiotic selection gene (aminoglycoside phosphotransferase II) and transformed this into *Nicotiana* and *Arabidopsis*. In this way, promoters could be 'trapped' as plants that were regenerated on selective antibiotic were antibiotic resistant because the T-DNA was inserted downstream of a promoter such that the gene was expressed. The most serious limitation to the T-DNA tagging approach is that large numbers of independent transformation events must be generated, which is not currently possible with most species. This is really only feasible with a system such as the seed transformation technique developed for *Arabidopsis*, or possibly the 'whole-leaf' transformation system developed for tobacco.

CLONING

Genetically identical plants derived from an individual are called clones. Processes that produce clones can be put under the term 'cloning'. This includes all the methods of vegetative propagation such as cutting, layering, and grafting. Propagation by tissue culture also helps in producing clones. Using the shoot tip, it is possible to obtain a large number of plantlets. This technique is used extensively in the commercial field for micropropagation of ornamental plants like chrysan-themum, gladiolus, etc. and also crops such as sugar cane, tapioca, and potato. Thus an unlimited number of plants that are genetically similar or are clones can be produced in a short span of time by tissue culture.

CLONAL PROPAGATION OR MICROPROPAGATION

It has been demonstrated that a variety of plant species can be conveniently propagated through the techniques of cell, tissue or organ culture. This is popularly described as clonal propagation ormicropropagation.

The major benefits of this method include the following;

- Rapid multiplication of superior clones and maintenance of uniformity;
- Multiplication of disease free plants and
- Multiplication of sexually derived sterile hybrids.

In most cases, clonal propagation is achieved by placing sterilized shoot tips or auxillary buds onto a culture medium that is sufficient to induce formation of multiple buds. This method has already been used to propagate a large number of marketable ornamentals.

Following stages are involved in the method of clonal propagation:

- Stage I involves establishment of tissue in vitro.
- Stage II involves multiplication of shoots (often media are not changed between stage I and stage II). (iii)
- Stage III concerns root formation and conditioning of propagules prior to transfer to the green house; this stage requires alteration of media for promotion of root formation.
- Stage IV involves growth in pots followed by field trials. The number of steps in micropropagation may sometimes be reduced to three or two as shown in diagram.

Three Different Methods of Micropropagation of Plant Material

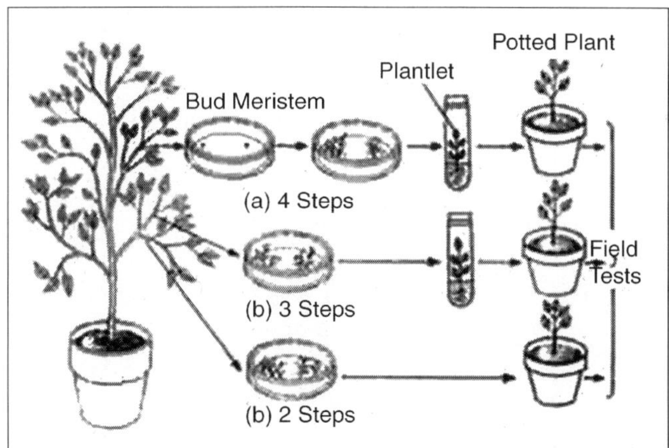

A wide range of plants have now been regenerated through technique of tissue culture. This has been found particularly useful for propagation of tree species, so that a large number of plant species have been successfully grown by tissue culture.

LARGE-SCALE PROPAGATION

To bridge the gap between research and application, the Department of Biotechnology, Government of India sponsored the setting-up of two pilot-scale facilities for large-scale propagation of elite planting material of forest trees through tissue culture. One of these facilities has been established at TERI's 36-hectare-campus in Gual Pahari, Haryana with an annual capacity of a million plantlets. Research at these facilities focuses exclusively on developing new protocols for mass cloning of elite planting material, mainly of trees.

Till date, over 4 million plants have been dispatched for field plantation from these facilities. The tissue culture raised plants are presently being evaluated under field conditions. This is being done in tandem with the forest departments of Haryana, Uttar Pradesh, Madhya Pradesh, Bihar, Jammu and Kashmir, and Orissa. For initial screening for phenotypically superior trees only a few hundred plantlets of the same are raised and tested under various agroclimatic zones. The best clones are then mass multiplied and monitored regularly for their performance. Field data suggest a survival percentage of more than 90% even in the harsh conditions of Aravalis without the life-saving irrigation. At half the rotation age some of the selected clones of Eucalyptus are showing a significant increase in productivity as compared to the conventional seed raised progenies.

ADVANTAGES OF CLONAL PROPAGATION

Clonal forestry is often seen as modern biotechnology. It is true that some techniques used for creating clones are new. The technical capacity to produce clones is improving. The current paper aims to review the possible advantages of clones in forestry and forest tree breeding and to give information about the occurrence of clones in nature and forestry. Some specific attention is given to focus on matters, which are relevant for cloning as a tool to get quality birch.

No effort is done to discuss possible disadvantages and problems in full, but this has been done many times before. Neither are the possibilities of other "competing" alternatives to get some of the advantages (like using controlled crosses) fully explored.

Clones in Nature and Forestry

Clones are common in Nature, and they have been used as a tool of domestication for very long time and are used in large-scale forest operations. It seems to be important to remember that to put clonal forestry into perspective,

The earliest land plants lacked vascular tissue and had inefficient systems of sexual reproduction and dispersal and clonal reproduction was the typical way of propagation in nature. Clonal growth was advantageous enabling physical dominance of large areas through horizontal growth.

Clonal propagation is still important in nature for many species. Many of the commonest clones are herbaceous but the habit is also found in woody plants throughout the world. In temperate broad-leaved woodlands, root suckering is a common form of clonal growth and there are good examples in the genera Populus and Prunus. Coppice is a natural growth form that has been exploited for centuries in European forests. Betula, Carpinus, Corylus, Quercus, Salix and Tilia are all genera with the ability to self-coppice.

The use of vegetative propagation of trees as a tool in their domestication has a long history. An early use was to propagate good clones of fruit trees and this has been done for thousands of years, thus humanity has long experiences of cloning trees.

For specific cases clonal forestry is often regarded as a "new technique", and "biotech" leads associations to the public to something more radical than most cases of clonal forestry. To put things into perspective it has to be remembered that many of the techniques used in modern "seedling forestry" are also "new", and the problems and uncertainties with that is often larger than the uncertainties connected to clones.

Sugi

The first known use of vegetative propagation for forestry purposes was in Japan with Sugi in the 15th century. Since then reforestation with monoclones or clonal mixtures has been widely used in Japan. Cryptomeria japonica is the most widely cultivated tree species in Japan comprising nearly half of the ten million ha of Japanese plantations. Almost half of the plants used are clones.

The extreme uniformity of the plantations has probably contributed to greater than average damage from typhoons and heavy snow. Attacks from Sugi bark borer and bark midge have increased in severity as the area of Sugi plantation has grown, but there is no evidence of altered disease or insect virulence related to clonal forestry, despite careful monitoring.

Clonal forestry is most common on the south-western island of Kyushu where about 100 cultivars are in use. Many foresters use only 1-3 cultivars of local origin in their plantations so that a single clone may cover as much as 2 hectares.

Allergy to Sugi pollen is a health problem in Japan as a whole, but is only a minor problem on Kyushu with its large use of clones. Cloning by cuttings for several generations has reduced the incidence of male flowers, so these plantations produce little pollen. Analogously one may speculate in that a reversion of North European birch forestry to clonal forestry in a very remote future may reduce allergic reactions with birch pollen.

Eucalypts

Clonal forestry is more common with broadleaf species than with conifers. Probably the most successful example is the eucalypt plantations in countries in tropical and subtropical climates like South America, South Africa and

Portugal. The total area of eucalypt plantations today is about 30 million hectares and about half of this area is planted with cloned material.

The eucalypt plantations in Brazil cover an area of approximately 3 million hectares. A succession of eucalypt species has been grown in Brazil due primarily to disease problems. Clonal forestry started in the late 1960:s and is today the dominant form of eucalypt forestry in the country. Progeny from an Eucalyptus grandis female (male sterile) growing in an arboretum were found to be natural hybrids which were resistant to pests and diseases. The offspring of this tree are believed to be Eucalyptus grandis ? Europhylla hybrids.

This knowledge was rapidly adopted by the forest industry companies and they selected clones within the open-pollinated offspring from this single tree, which means that a small number of clones originating from the same family are widely grown over large areas in Brazil. This has and is still changing with most companies having good breeding programmes backing their clonal programmes. The clones are deployed in huge monoclonal blocks. This situation seems favourable for a pathogen, but nothing has happened yet - despite the possible risks of a disease in a monoculture. Today the companies have become more risk-conscious and are developing breeding and clonal testing programmes that include strategies for clonal diversity in space and time. A typical programme is now deploying about 20 clones each year and these clones will be replaced with time as new clones are selected. The production landscape forms a genetic mosaic. New genetic material has been imported to widen the genetic base for breeding.

Poplars and Willows in Europe

Vegetative propagation of poplars (Populus sp.) has a long history in Europe. Organised clonal forestry started in the beginning of the 20th century. Monoclonal plantations of poplars have become a common land use especially on river plains in southern European countries like Italy, Spain and France. Some of the best poplar clones in use were selected as early as during the first half of the 20th century. Clonal forestry with poplars is common in countries with subtropical and temperate climates like Italy, Spain, France, Belgium, USA and Canada. Single clones of poplars have been propagated extensively. There have been disease problems and the overuse of single clones and use of large monoclonal plantations has been questioned. Poplar has covered over 100000 ha in one country and a single clone may comprise one third of this area. The focus has been to develop disease resistance within individual clones, but these clones have proved to be highly susceptible to new varieties of disease.

The disease problem seems to be growing and has caused Germany to more or less abandon poplars and other countries may follow. Nevertheless these incidents have not provided poplar growers with sufficient incitement to focus on more diverse alternatives. It is unlikely that monoclonal cultures and the wide use of some clones are the only explanation to increased disease, but they act as contributing factors. Legal and commercial reasons often favour the use of single clones. Even willows and short rotation forestry in Sweden are threatened by disease. Resistance is broken down within a single rotation (it is decades in between replanting even when harvesting is done at intervals of a few years). Two evaluations of the Swedish willow breeding programme have rather strongly recommended an increased number of selections to be made (more than a single selection released per year). Nevertheless, the growers′ organisations still do not encourage clone mixtures.

Hybrid aspen in Finland

A Finnish forest industry company has recently started a clonal forestry programme with hybrid aspen. This raises the question if something similar could be done with birch. It may be noted, however, that suggested that micro propagation costs are high and that the technique needs development.

Overoptimistic Statistics

It is a clear tendency that statistics reported about the use of vegetative propagation, at least there it is regarded as modern hitech biotech, gives the impression of more rapid progress than actually occurs. It is reported, based on a survey, that four Swedish operators produces 7.3 millions rooted Norway spruce cuttings annually. It is estimated the annual production to 5 millions and believed in a raise to 10 millions. However, looking backwards, less than 20 million were planted in total till 2000 and now only some hundred thousands are planted annually.

Crop Uniformity

As genetic diversity and possible uniformity is an evident and possible important aspect of clonal forestry. For evaluation of possible ecological impact diversity seems central. I will thus start to discuss disadvantages with a low genetic diversity in a stand. In other respects this paper focuses on the advantages of clonal forestry.

Disadvantages of a Uniform Crop

There are reasons to believe that the biological production can be higher in a genetically diverse crop than in a uniform:

- A single genotype demands the same things at the same time, and thus utilises the site worse than a mixture of genotypes.
- In a mix, another genotype may take over the ecological space left by a failed genotype.
- A disease spreads faster in a uniform crop.

These statements are supported by the average of a very large number of agricultural mixing experiments and a few experiments and experiences with forestry. On average, mixes performed some per cent better and were less susceptible and more stable than the average of its components, but usually the production of the mix was not better than its best component grown pure.

Even if mixes seem more productive on an average, there are many examples and experiments, when no advantage was found and some experiments where diversity actually seems negative for production. I conclude that from a production point of view an advantage with some degree of diversity is expected, but that does not imply that production is higher the higher the diversity is. But the superiority in biological production is not expected to be large and it is not certain for individual cases, but just a rough generalisation. This general statement ought to apply to birch also, thus biological production is assumed to be lower and more uncertain in a uniform crop, but the effect is probably small and practically unimportant (a few per cent of biological production).

Advantages of a Uniform Crop

There must be considerable advantages in uniform monoclonal blocks as they are used frequently. Although the advantages with diversity are repeated and reiterated and very politically correct, it is seldom applied in intensive agriculture or husbandry. The main reason is that maximizing biological production and maximizing economical production is not the same thing. It makes it easier to manage the crop if it behaves uniformly. For a forester it may not matter very much if individual birches drop their leaves on different days, but a farmer wants the crop to become ripe the same day, as the whole crop is harvested the same day. Diversity is a usually a problem for the customer, who wants a uniform raw material for further processing. For administrative processes, like describing the crop, it is an advantage if it is uniform and does not change from specification over time and space. These problems are evident for many types of agricultural crops, and therefore these crops are usually uniform even when biological production would be higher with a diverse crop.

The dream of many foresters is to find the best genotype and use only that. This has been successfully done for other domesticated species. The uniformity itself in a clonal plantation has a considerable advantage from a management point of view. It is easier in the nursery to raise plants of a single genotype. It is easier to manage and harvest a uniform forest. It has advantages to handle a single type of material. A customer or processor of a product knows more about a lot if it is a specified clone than if just the species is known. A batch of uniform logs is worth more and easier to handle than the same amount of very variable logs.

Even if interest is only in quantitative production and that is lower in a uniform crop, still the higher genetic performance of the best clones compared to the best seedlings available usually much more than compensates for the production loss by a uniform crop. Thus uniform clonal crops produce more than diverse seedling crops.

My opinion is that genetic diversity often is overemphasised in comparison with the higher genetic gain achievable at the cost of a reduction in genetic diversity for future forest production plantations in many countries (like Sweden) both for propagation with seedlings and cuttings. To get genetic diversity many clones are used in seed orchards and clonal mixtures and there constraints of relatedness of clones may be seen as desirable. This was not a disadvantage at an early stage of forest tree improvement, as it does not influence gain much if many phenotypically selected plus trees are placed in seed orchards, and in the early stages of cutting propagation there were physical constraints on the number of ramets which can be produced by each clone, so it is necessary to have many clones. But now this constraint often means selection is not intensive and the best genotypes are not exploited to the extent, which would be possible.

However, clonal forestry is unlikely to deal with a single clone only. That does not seem a likely scenario for birch. Many clones will be tested and test results as well as experiences will accumulate of the best clones, so the opinion of what is the best clone will change over time. It is also dependent on the locality, the desires of the user and the availability. Even if uniform plantations are planted, there will usually be a mosaic of clones on the landscape level. From the risk management point of view a clonal deployment philosophy offers advantages.

If a clone is sensitive to a pest or pathogen the problem can be identified and managed. If silvicultural means to deal with a problem are insufficient, the stand can be salvage harvested and replanted. The clone which turns out inferior can quickly be taken out of production. It is an advantage for an operator which uses monoclone blocks to use several clones in the same time that it will become more evident what the strong and weak point of the clones are. A disadvantage with mosaics is that the uniform lots of wood from a stand are often too small to be practical as a specific variety at the end user, and actually the end user may get trouble with larger between lot differences if clonal forestry is used. Alternatively the clones can be grown in intimate mixtures, in that they most of the possible advantages of the diversity of seedlings may be gained by a mix of rather few clones.

When we are talking quality birch - and not quantity birch - the reasons for monoclonal plantings become stronger. When high quality of the product is concerned, it often becomes more important to have uniform quality also, while in a low quality product - as firewood - the quantity is usually relatively more important and a uniform and predictable product is less important. Attention on quality instead of quantity also generally coincides with a higher level of intensity, and at a higher level of intensity, clonal forestry often becomes more interesting. Thus considering the aim the project arranging this symposium has in mind; uniform clonal plantings becomes relatively more important than for forestry in general.

Improved Clones can be Preferable to Seedlings

There are a number of arguments for forestry with clones. It takes less time to multiply a genotype as a clone than to build a seed production unit with that clone as one of the parents and wait till that works. The time lag is actually much longer than till the first seed production, because a seed production unit is usually designed to last for many years. The risk for getting the seed storage empty means that forestry often want to have five to ten years foreseen seed need in storage. That is expensive, many seeds may never be used and the seeds actually used are less improved than possible.

A seed production unit is designed to meet a certain need of plants with certain characteristics many years ahead. If too many seeds are produced it means that the investment was unnecessary large. If too few seeds are produced it may mean that foresters must use genetically inferior unimproved seeds. If the propagation is done fast by vegetative propagation by *e.g.* micro propagated clones or somatic embryogenesis, it takes only a year to multiply a desirable genotype in the desired number of copies. Thus it is fairly easy to adapt to the latest news from breeders and customers.

There are large annual variations in seed production both in the forest and in seed orchards. In a seed orchard the reproductive output is variable among genotypes and the reproductive phenology may make clones incompatible, so the actual genetic set up of the progeny is a bit uncertain. These uncertainties can be avoided by clonal forestry. Many of our forest trees - like birch - can both spread pollen and produce seeds. That means that selfing can occur and the occurrence of selfing in a plant material will decrease the forest production. With the use of clonal forestry selfing can be completely eliminated.

In a wind pollinated seed orchard wild uncontrolled and unimproved pollen may contribute to pollination, clonal forestry is a way to eliminate this problem. However, for birch, seed orchards are small enough to be managed in plastic green houses, where the inflow of unimproved pollen is low, so for birch it is not fair to see this pollen contamination problem as an argument against open pollinated seed orchards.

In a seed orchard it is important to avoid related clones, as relatives get inbred progeny when they mate, which means that forest production drops because of inbreeding depression. Mating between relatives may be a worse problem than selfing in future seed orchards, as self-zygotes usually die; self-seedlings are often culled before planting; or early out-competed in forest, while milder inbreeding is more directly transferred to production loss. If clones are related in a clonal forest, that does not cause inbreeding depression, and thus there is little harm if related clones occur in a stand (actually this is very natural, trees in a natural stands are often related). The clones in the breeding population will be increasingly more related as breeding goes on, it will be easier to handle the consequences of this for the production forest with clonal forestry.

A seed orchard is a big investment intended to meet the seed need for a long period for a specific species used in a specified area, thus a considerable local market is needed to justify the investment. Conifer seed orchards are usually designed for a need of many millions plants per year over some decades. Multiplication of a clone can be seen as a smaller investment during a shorter period, thus the local market is allowed to be much smaller with clonal forestry, *e.g.* micro propagation for curly birch in Finland serves a market of 200 000 seedlings a year. A seed orchard has a stiff structure, its composition can be only slightly modified once it is established and it is not easy to justify several orchards heading for different characteristics for the same area. Clonal propagation is more flexible; an individual customer can choose the clones, which fits the particular needs for the customer and occasion. Sometimes a combination of rather rare characters is desired, a clone can be found which combines several desirable characters in a much more efficient way than dealing with seed orchards.

It is often stated that clonal forestry reduces genetic diversity. This is a generalisation, which need not always be true. Actually, clonal forestry provides a tool to choose genetic diversity at will, while when a seedling material is used, the genetic diversity is essentially left to chance. Probably those using intensively managed birch plantation for producing high quality timber with special characteristics would choose the option to reduce or eliminate genetic diversity on the stand level, because they found it economically sound to do that. It seems likely this is how the tool will be used. But clones in a clonal mix can be chosen so that the genetic diversity is larger than in a seedling material.

Clones can be mixed with seedlings, that may be a good strategy not only for diversity, but economically in cases where cloned plants are much better, but also much more expensive. If the clones succeed well, their share at the final harvest will be larger than at establishment. Over time and space different clones will be used even in a system with monoclone cultures (like Eucalypts in Brazil). In such a system the genetic diversity among stands is likely to be much than in nature. Sometimes clones have been used for natural conservation in situation when good seeds could not be obtained. Clonal forestry may strengthen the trend to produce the needed wood at a limited acreage of intensively managed production forests, allowing larger areas to be set aside for other purposes. Thus the question how clones may affect diversity has many aspects and it is an oversimplification to state that it must be severely reduced.

A common situation is that some good seeds exists or can be obtained, but quantitatively insufficient for the plant need. If such a situation occurs unexpected and unforeseen, vegetative propagation may be the solution. A more common situation is that a few superior seeds can be produced by controlled crosses, but too few to meet the plant demand. Clonal propagation is often a way to make forestry with controlled crosses practical feasible. Clonal propagation may be seen as a way to amplify the impact of a few good parents or a tested cross by mating the parents and when multiply the seeds.

The genetic value of a genotype depends on the breeding value of its parents, but also how well the genes of the parents fit together. This may be called dominance. The later reason for genetic differences cannot be exploited in an ordinary seed orchard, but it can be exploited with clonal forestry.

Hybrids sometimes result in good seeds. It may be difficult to make many of those seeds by controlled crosses, but hybrid seeds (or trees) can be multiplied by vegetative propagation. This is used in the Eucalypts and Aspen programmes mentioned above. Personally I am not that convinced that hybrid superiority is so common as it is often stated. When hybrids are better than the parents it may be because the parents were not well adapted for the conditions where they were tested or that the hybrid combines characters of the parents for the environment and purpose of the forester in a better way than tested alternatives. Hybrid maize is a best seller, but the reason is more that farmers are unable to produce hybrid seeds them self, but has to buy them, and when it is good economy of a seed company to invest in hybrids, so they can market the seeds. To get a genetically superior forest, it is desirable to use the genes of the very best clones. In a seed orchard it has to be rather many parents to avoid selfing and problems with lack of overlap in flowering time. This means a drop in breeding value. For clonal forestry the very best clone can be used. This is true even if in the complete absence of dominance. As dominance is seldom the major cause of genetic variation for the most important characters in forest tree breeding, this more effective exploitation of breeding value is usually a more important reason for clonal forestry than the exploitation of dominance variation, at least as long as we do not talk about hybrids of different species or races. Quantitatively the low importance for the superiority of cloning of a limited amount of dominance variation is demonstrated.

We have also special features for clonal selection. We can intentionally select clones, which do not waist resources on sexual activities but spend resources on stem wood production. Such clones are desirable for forestry. In a seed orchard it is not desirable with clones, which do not produce seed and pollen, so such clones would not be chosen for a seed orchard, and if they were, they would be unable to transmit their genes to the seed crop.

Cloning methods have different characteristics. They are often very depending on the state of the material which is cloned, cuttings usually do not work well on physiologically old material, for somatic embryogenesis other tissue than immature seeds are often a difficult starting material. With micro propagation it is usually possible to multiply old trees or clones, and this is often an advantage. Cloning and the cloning method can have effects on the physiological status of the tree. *E.g.* conifer grafts have not thick bark at the bottom of the trunk. The protocol for making clones can have effects on the plants. These effects sometimes look desirable and can be exploited. For example Norway spruce cuttings seems to suffer a lower mortality from pine weevil damage than seedlings. Clones usually origin from older and more mature tissues than seedlings and this affects the characters of the plants and may have effects also on the mature trees. I guess these differences to seedlings are not dangerous and nursery raised seedlings can also be claimed to be unnatural, and they my be different because of different growing regimes.

DNA CLONING

DNA cloning a method to manipulate DNA in vitro and in vivo or is a logical extension of the ability to manipulate DNA molecules with restriction endonucleases and ligases. Imagine that an plant gene has been obtained as a single restriction fragment after digestion of a larger molecule with the restriction enzyme *Bam*HI, which leaves 5-GATC-3 sticky ends. Imagine also that a small *E. coli* plasmid has been purified and treated with *Bam*HI, which cuts the plasmid in a single position.

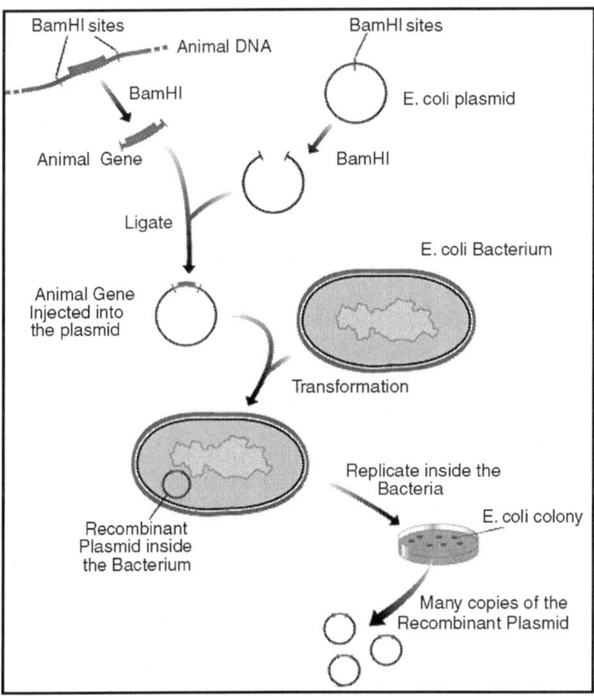

Fig. An Outline of Gene Cloning.

The circular plasmid has therefore been converted into a linear molecule, again with 5-GATC-3 sticky ends. Mix the two DNA molecules together and add DNA ligase. Various recombinant ligation products will be obtained, one of which comprises the circularized plasmid with the animal gene inserted into the position originally taken by the *Bam*HI restriction site. If the recombinant plasmid is now re-introduced into *E. coli,* and the inserted gene has not disrupted its replicative ability, then the plasmid plus inserted gene will be replicated and copies passed to the daughter bacteria after cell division. More rounds of plasmid replication and cell division will result in a colony of recombinant *E. coli* bacteria, each bacterium containing multiple copies of the animal gene. This series of events, constitutes the process called DNA or gene cloning.

Vector that Works with Plant Cells

Not all aspects of the genetic manipulation of plant cells are readily accomplished. Not only do plants usually have a great deal of chromosomal material and grow relatively slowly as compared with single cells grown in the laboratory, but few cloning vectors can successfully function in plant cells. While researchers working with animal cells can choose among a wide variety of cloning vectors to find just the right one, plant cell researchers are currently limited to just a few basic types of vectors.

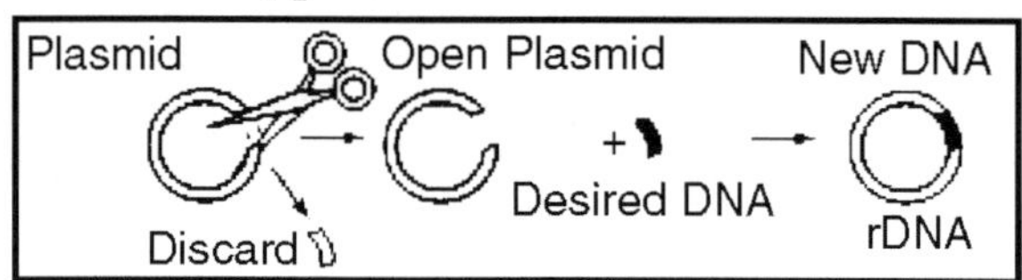

Fig. Insertion of Foreign DNA in Plasmid

Perhaps the most commonly used plant-cloning vector is the "Ti" plasmid, or tumor-inducing plasmid. This plasmid is found in cells of the bacterium known as Agrobacterium tumefaciens, which normally lives in soil. The bacterium has the ability to infect plants and cause a crown gall, or tumorous lump, to form at the site of infection. The tumor-inducing capacity of this bacterium results from the presence of the Ti plasmid.

The Ti plasmid itself, a large, circular, double-stranded DNA molecule, can replicate independently of the A. tumefaciens genome. When these bacteria infect a plant cell, a 30,000 base-pair segment of the Ti plasmid - called T DNA - separates from the plasmid and incorporates into the host cell genome. This aspect of Ti plasmid function has made it useful as a plant cloning vector.

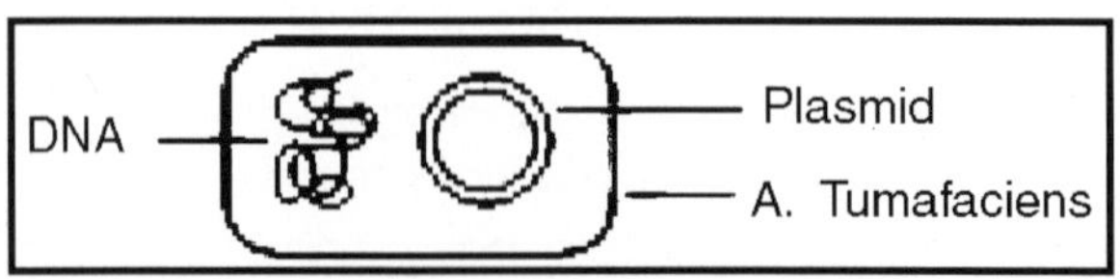

Fig. A. Tumefiecian

The Ti plasmid can be used to shuttle exogenous genes into host plant cells. This type of gene transfer requires two steps: 1) the endogenous, tumor-causing genes of the T DNA must be inactivated and, 2) foreign genes must be inserted into the same region of the Ti plasmid. The resulting recombinant plasmid, carrying up to approximately 40,000 base pairs of inserted DNA and including the appropriate plant regulatory sequences, can then be placed back into the A. tumefaciens cell. That cell can be introduced into plant cell protoplasts either by the process of infection or by direct insertion. Once in the protoplast, the foreign DNA, consisting of both T DNA and the inserted gene, incorporates into the host plant genome. The engineered protoplast - containing the recombinant T DNA - regenerates into a whole plant, each cell of which contains the inserted gene. Once a plant incorporates the T DNA with its inserted gene, it passes it on to future generations of the plant with a normal pattern of Mendelian inheritance.

One of the earliest experiments that involved the transport of a foreign gene by the Ti plasmid involved the insertion of a gene isolated from a bean plant into a host tobacco plant. Although this experiment served no commercially useful purpose, it successfully established the ability of the Ti plasmid to carry genes into plant host cells, where they could be incorporated and expressed.

A. Limition of A. Tumifecian

The fact that only certain types of plants were naturally susceptible to infection with the host bacterial organism initially limited the usefulness of the Ti plasmid as a cloning vector. In nature, A. tumefaciens infects only dicotyledons or "dicots" - plants with two embryonic leaves.

Dicotyledenous plants, divided into approximately 170,000 different species, include such plants as roses, apples, soybeans, potatoes, pears, and tobacco. Unfortunately, many important crop plants, including corn, rice, and wheat, are monocotyledons - plants with only one embryonic leaf - and thus could not be easily transfected using this bacterium.

Overcoming the Limition

Research efforts in the past few years have reduced the limitations of A. tumefaciens. Scientists discovered that by using the processes of microinjection, electroporation, and particle bombardment, naked DNA molecules can be introduced into plant cell types that are not susceptible to A. tumefaciens transfection. Microinjection involves the direct injection of material into a host cell using a finely drawn micropipette needle.

Electroporation uses brief pulses of high voltage electricity to induce the formation of transient pores in the membrane of the host cell. Such pores appear to act as passageways through which the naked DNA can enter the host cell. Particle bombardment actually shoots DNA-coated microscopic pellets through a plant cell wall. These developments, important in the commercial application of plant genetic engineering, render the valuable food crops of corn, rice, and wheat susceptible to a variety of manipulations by the techniques of recombinant DNA and biotechnology.

PLANT TRANSFORMATION

Genetic engineering of plants is much easier than that of animals. There are several reasons for this:

- There is a natural transformation system for plants (the bacterium *Agrobacterium tumefaciens*),
- Plant tissue can redifferentiate (a transformed piece of leaf may be regenerated to a whole plant), and
- Plant transformation and regeneration are relatively easy for a variety of plants.

The soil bacterium *Agrobacterium tumefaciens* ("tumefaciens" meaning tumor-making) can infect wounded plant tissue, transferring a large plasmid, the Ti plasmid, to the plant cell. Part of the Ti (tumor-inducing) plasmid apparently randomly integrates into the chromosome of the plant.

The integrated part of the plasmid contains genes for the synthesis of (1) food for the bacterium, and (2) plant hormones. Genes from the Ti plasmid that are integrated in the plant chromosome are expressed at high levels in the plant. Overproduction of the plant hormones leads to continuous growth of the transformed cells, causing plant tumors. Rapid, cancerous growth of the transformed plant tissue obviously is advantageous to the bacterium: more food gets produced.

The Ti plasmid has been genetically modified ("disarmed") by deleting the genes involved in the production of bacterial food and of plant hormones, and inserting a gene that can be used as a selectable marker. Selectable marker genes generally are coding for proteins involved in breakdown of antibiotics, such as kanamycin. Any gene of interest can be inserted into the Ti plasmid as well.

In principle, one can thus transform any plant tissue, and select transformants by screening for antibiotic resistance. However, unfortunately, there are some complications: (1) it has proven difficult to transform some monocots (grasses, etc.) by *Agrobacterium*, and (2) regeneration of plants from tissue culture or leaf discs is not always possible.

A number of genetically engineered plant varieties have been developed. Traits that have been introduced by transformation include herbicide resistance, increased virus tolerance, or decreased sensitivity to insect or pathogen attack. Traditionally, most of such genetically engineered plants were tobacco, petunia, or similar species with a relatively limited agricultural application. However, during the past decade it now has become possible to transform major staples such as corn and rice and to regenerate them to a fertile plant. Increasingly, the transformation procedures used do not depend on *Agrobacterium tumefaciens*.

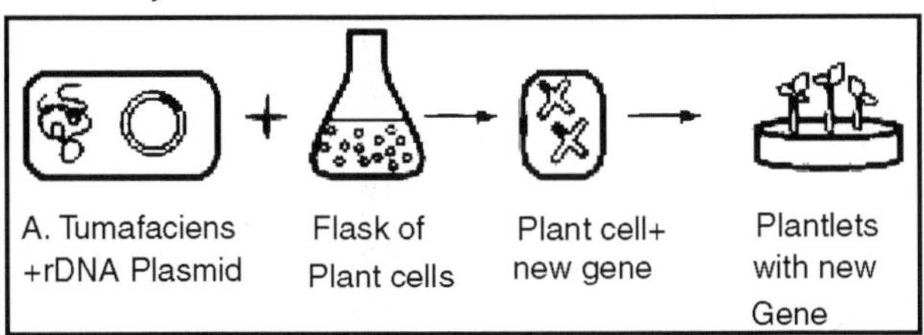

Fig. Transformation Using Ti Plasmid

Instead, DNA can be delivered into the cells by small, μm-sized tungsten or gold bullets coated with the DNA. The bullets are fired from a device that works similar to a shotgun. The modernized device uses a sudden change in pressure of He gas to propel the particles, but the principle of "shooting" the DNA into the cell remains the same. This DNA-delivery device is nicknamed "gene gun", and has been shown to work for DNA delivery into chloroplasts as well.

Over the last several years, use of the "gene gun" has become a very common method to transform plants, and has been shown to be applicable to virtually all species investigated. For example, transformation of rice by this method is now routine. This is a very important development as rice is the most important crop in the world in terms of the number of people critically dependent on it for a major part of their diet.

Another method to get foreign genes into cereals is by electroporation: a jolt of electricity is used to puncture self-repairing holes in protoplasts (*i.e.,* the cell without the cell wall), and DNA can get in through these holes. However, it is often very difficult to regenerate fertile plants from protoplasts of cereals. Nonetheless, significant advances in overcoming these practical difficulties have been made over the years.

Now even transgenic trees have been created: for example, the gene for a coat protein of the plum pox virus has been introduced into apricot. The plum pox virus leads to the feared Sharka disease, for which there is no cure. The resulting transgenic tree shows a markedly decreased sensitivity to this virus. The reason why continuous exposure of the tree to the viral coat protein leads to tolerance against viral infection is not yet understood, however. Thus, now there are a number of different techniques to introduce foreign genes into plants. Essentially all major crop plants can be (and have been or are being) genetically engineered, the procedures are now routine and the frequency of success is very high. Even though genetically engineered crops are more costly than the usual ones, US farmers have rather readily accepted them provided that tangible benefits can be demonstrated.

However, it is questionable whether the farmer in poorer countries can come up with the funds to "try out" and use the new crops. Another issue in this respect is how genetically engineered crops are perceived by the consumer. Even though in the US there is little resistance to such crops as long as the products can be shown to be safe and advantageous, in other countries genetically modified foods are received poorly by the consumer.

It is unlikely that there is a rationally sound basis for this rather hostile reaction of the consumer, as most of the crops are the results of human manipulation (such as centuries of breeding) and may have been treated with harmful herbicides and pesticides. Time and education will need to be invested to provide consumers and consumer advocates with a balanced opinion on the acceptability of the origin of their foods.

On the other hand, as so many plants (soybean, corn, etc.) are genetically modified and the nature of the genetic modification is not necessarily easy to explain, it may be simpler to label those foods that are guaranteed free of "genetically modified organisms" or their products. However, keep in mind that traditional breeding has genetically modified essentially all agricultural products, so it may be difficult to define what is actually free of genetically modified organisms.

Transgenic Crops

Transgenic crops: The varieties created through the use of a biotechnological approach commonly known as recombinant DNA, genetic engineering (GE), transgenic modification, or genetic modification (GM). The products of genetic engineering are often called genetically modified organisms, or GMOs. All these terms refer to methods of recombinant DNA technology by which biologists splice genes from one or more species into the DNA of crop plants to transfer chosen genetic traits.

Genes are segments of DNA that controlled the information that determines the structure and function of a living organism. Genetic engineers are molecular biologists who manipulate this information, typically by taking genes from one species—an animal, plant, bacterium, or virus—and inserting them into another species, such as an agricultural crop. With the advent of genetic engineering of plants around 1983, it appeared that this new biotechnology would benefit and even revolutionize agriculture.

The transfer of desirable genetic traits across species barriers has shown promise for solving problems in the management of agricultural crops. Potential benefits include reduced toxic pesticide use, improved weed control resulting in less tillage and soil erosion, and water conservation.

Gene Flow to Neighbouring Crops and to Related wild Species

Ecological scientists have little doubt that gene flow from transgenic fields into conventional crops and related wild plants will occur. Gene flow from transgenic to conventional crops is of concern to farmers because of its potential to cause herbicide resistance in related conventional crops. For example, in western Canada, three different herbicide-resistant canola varieties have cross-pollinated to create canola plants that are resistant to all three types of herbicide. Through what is called gene-stacking, this new triple resistance has turned volunteer canola into a significant weed problem.

Gene flow from transgenic crops to wild relatives creates a potential for wild plants or weeds to acquire traits that improve their fitness, turning them into "super weeds." For example, if jointed goatgrass—a weedy relative of wheat—acquires the herbicide-tolerant trait of Roundup Ready wheat, it will thrive in crop fields unless applications of other herbicides are made. There is already evidence of such outcrossing from herbicide-resistant wheat to jointed goatgrass. Frank Young and his colleagues at Washington State University found that imidazolone-resistant wheat (not a transgenic variety) outcrossed to goatgrass in one season. Other traits that wild plants could acquire from transgenic plants that would increase their weediness are insect and virus resistance.

Because of their experience with classically bred plants, few scientists doubt that genes will move from crops into the wild: seven of the world's thirteen most important crop weeds have been made weedier by genes acquired from classically bred crops. Because gene flow has the potential to affect farmers' crop and pest management, crop marketability, and liability, more research needs to be done to determine the conditions under which gene flow from transgenic plants is likely to be significant.

Pesticide Resistance in Insect Pests

Bt has been widely used as a microbial spray because it is toxic only to caterpillars. In fact, it is a pest management tool that organic farmers depend on—one of the few insecticides acceptable under organic rules. Unlike the commercial insecticide spray, the Bt engineered into crop plants is reproduced in all, or nearly all, the cells of every plant, not just applied on the plant surface for a temporary toxic effect. As a result, the possibility that transgenic Bt crops will accelerate insect pests' development of resistance to Bt is a serious concern. Pest resistance to Bt would remove this valuable and environmentally benign tool from farmers' and forest managers' pest control toolbox. Antibiotic resistance The use of antibiotic-resistant marker genes for the delivery of a gene package into a recipient plant carries the danger of spreading antibiotic-resistant bacteria. The likely result will be human and animal health diseases resistant to treatment with available antibiotics. Research is needed on antibiotic resistance management in transgenic crops. Already the European Commission's new rules governing transgenic crops stipulate phasing out antibiotic-resistant marker genes by the end of 2004.

Effects on Beneficial Organisms

Evidence is increasing that transgenic crops—either directly or through practices linked to their production—are detrimental to beneficial organisms. New studies are finding that Bt crops exude Bt in concentrations high enough to be toxic to some beneficial soil organisms. In addition, a study by University of Arkansas scientists has shown that "root development, nodulation and nitrogen fixation are impaired" in some varieties of Roundup Ready soybeans. The reason is that the beneficial rhizobium responsible for nitrogen fixation in soybeans is sensitive to Roundup. It also appears that disruption of beneficial soil organisms can interfere with plant uptake of phosphorus, an essential plant nutrient.

Insecticidal crops in two ways can affect beneficial insects that prey on insect pests. First, the Bt in transgenic insecticidal crops has been shown in some laboratory studies to be toxic to ladybird beetles, lacewings, and monarch butterflies. The extent to which these beneficials are affected in the field is a matter of further study. Second, because the insecticidal properties of Bt crops function even in the absence of an economic threshold of pests, Bt crops potentially can reduce pest populations to the point that predator species are negatively affected.

Reduced Crop Genetic Diversity

As fewer and larger firms dominate the rapidly merging seed and biotechnology market, transgenic crops may continue the trend towards simplification of cropping systems by reducing the number and type of crops planted. In addition, seed saving, which promotes genetic diversity, is restricted for transgenic crops.

Food Safety

Because food safety is predominantly a concern of consumers, However, because food safety issues do impact marketing and international trade, here is a short list of consumer safety concerns about transgenic food:

- Possibility of toxins in food
- Possibility of new pathogens
- Reduced nutritional value
- Introduction of human allergens
- Transfer of antibiotic resistance to humans
- Unexpected immune-system and genetic effects from the introduction of novel compounds

Farm Management Issues

The most widely planted transgenic crops on the market today can simplify short-term pest management for farmers and ranchers. In the case of herbicide-tolerant crops, ideally farmers can use a single broad-spectrum herbicide for all their crop weeds, though they may need more than one application in a season. By planting insecticidal crops, farmers can eliminate the need to apply pesticides for caterpillar pests like the European corn borer or the cotton bollworm, though they still have to contend with other crop pests. While these crops offer handy pest control features, they may complicate other areas of farm management. Farmers who are growing both transgenic and conventional varieties of the same crop will need to segregate the two during all production, harvesting, storage, and transportation phases if they are selling into differentiated markets or plan to save their own seed from their conventional crops.

To minimize the risk of gene flow from transgenic to adjacent conventional crop fields, federal regulations require buffer strips of conventional varieties around transgenic fields. Different transgenic crops require different buffer widths. Because the buffer strips must be managed conventionally, producers have to be willing to maintain two different farming systems on their transgenic fields. Crops harvested from the buffer strips must be handled and marketed as though they are transgenic. Planted refuges—where pest species can live outside fields of insecticidal and herbicide-tolerant transgenic crops—are also required to slow the development of weed and insect pest resistance to Bt and broad-spectrum herbicides. These refuges allow some individuals in the pest population to survive and carry on the traits of pesticide susceptibility. Requirements governing the size of refuges differ according to the type of transgenic crop grown.

Farmers growing herbicide-tolerant crops need to be aware that volunteer crop plants the following year will be herbicide-resistant. This herbicide resistance makes no-till or direct-seed systems difficult because volunteers can't be controlled with the same herbicide used on the rest of the crop. In a no-till system that relies on the same broad-spectrum herbicide that the volunteer plants are resistant to, these plants will contaminate the harvest of a following conventional variety of the same crop, a situation farmers need to avoid for two reasons. First, the contamination means a following conventional crop will have to be sold on the transgenic market. This leads to the second reason. If farmers grow and market a transgenic crop for which they do not have a technology agreement and did not pay royalty fees, they face possible prosecution by the company that owns the transgenic variety. Many farmers have been charged with "theft" of a company's patented seed as a result of contamination in the field. Farmers growing insecticidal crops need to recognize that insect pressure is difficult to predict and may not warrant the planting of an insecticidal variety every year. In a year when pest pressure is low, the transgenic seed becomes expensive insurance against the threat of insect damage.

Farmers growing transgenic crops need to communicate with their neighbours to avoid contaminating their fields and to ensure that buffers are adequate. In Maine, farmers growing transgenic crops are now required by law to be listed with the state agriculture department, to help identify possible sources of cross-contamination when it occurs. The new law, L.D. 1266, also "requires manufacturers or seed dealers of genetically engineered plants, plant parts or seeds to provide written instructions to all growers on how to plant, grow, and harvest the crops to minimize potential cross-contamination of non-genetically engineered crops or wild plant populations". Farm management issues common to all transgenic crops include yield, cost, price, profitability, management flexibility, sustainability, market acceptance, and liability.

Crop Yield, Costs, and Profitability

Some farmers will get higher yields with a particular transgenic crop variety than with their conventional varieties, and some will get lower yields. The same is true for costs. Some producers will see their overall profitability rise, and some will see it drop. If the results of research studies so far on yield, chemical use, costs, crop prices, and profitability of these crops seem contradictory, it is because farmers and farms are not all alike. Neither are the conventional varieties used for comparison.

Some yield, cost, and profitability trends do appear to be emerging from the growing body of research data for transgenic crops, however. As noted in the Wallace Centre report, Roundup Ready soybeans were designed simply to resist a particular chemical herbicide, not to increase yields. In contrast, Bt corn and cotton, by resisting insect pests, may result in higher yields from reduced pest pressure.

Yield: Herbicide-tolerant Crops
—soybeans, Cotton, Canola

Herbicide-tolerant soybeans appear to suffer what's referred to as a "yield drag." Again, in some areas and on some farms this tendency of Roundup Ready soybean varieties to yield less than their comparable, conventional counterparts varies, but overall, they appear to average yields that are five to ten per cent lower per acre.. Drought conditions worsen the effects. The bacterium that facilitates nodulation and nitrogen fixation in the root zone apparently is sensitive to both Roundup and drought.

Yield: Insecticidal Crops—corn, Cotton

Bt cotton, especially, outpaces yields of conventional cotton by as much as 9 to 26 per cent in some cases, though not at all in others. Yield increases for Bt corn have not been as dramatic. Time will tell whether farmers can expect yield increases or decreases in the long run with these and other transgenic crop varieties.

Changes in Chemical Pesticide use

One of the promises of biotechnology is that it will reduce pesticide use and thereby provide environmental benefits and reduce farmers' costs. The herbicide-tolerant and insecticidal varieties are designed specifically to meet these goals. Increased future pesticide use resulting from the buildup of resistance to heavily used herbicides is a long-term concern. Pesticide use depends on the crop and its specific traits; weather; severity of pest infestations; farm management; geographic location of the farm; and other variables.

As a result, conclusions drawn by various studies analysing pesticide use on transgenic crops remain controversial. According to the Wallace Centre report, in a review of the data available up through 2000, crops engineered to contain Bt appear to have decreased the overall use of insecticides slightly, while the use of herbicide-resistant crops has resulted in variable changes in overall herbicide use, with increases in use of some herbicides in some places and decreases in others

Profitability

Farmers need to consider all the factors that determine profitability. No single factor can tell the whole story. Transgenic crop seeds tend to be more costly, and farmers have the added expense of a substantial per-acre fee charged by the owners of transgenic varieties. These costs have to be considered along with input cost changes—whether herbicide or insecticide use and costs go down, go up, or stay the same. Market price is another factor: Prices for some transgenic crops in some markets are lower than prices for comparable conventional crops, though rarely are they higher.

Farmers need to watch the markets. Some buyers will pay a premium for a non-transgenic product, though as transgenic seeds find their way into conventional transportation, storage, and processing streams, these premiums may disappear along with confidence that "GMO-free" products are in fact truly free of engineered genes.

Marketing and Trade

Buyer acceptance is a significant marketing issue for farmers raising transgenic crops. Farmers need to know before they plant what their particular markets will or won't accept.

Since most grain handlers cannot effectively segregate transgenic from non-transgenic crops in the same facility, many companies are channeling transgenic crops into particular warehouses. Farmers need to know which ones and how far away those are.

While the European Union officially lifted its two-year moratorium on the introduction of new transgenic crops in February 2001, during the debate over labeling and traceability regulations the moratorium remains in effect. Under the proposed new EU requirements, "all foods and animal feed derived from GMOs have to be labeled and, in the case of processed goods, records have to be kept throughout the production chain allowing the GMO to be traced back to the farm".

Liability

The need to separate transgenic crops from conventional and organic ones opens farmers to liability for their product at every step from seed to table. Effective systems for segregation do not exist at present, and will be costly to develop and put into place. Farmers may well end up bearing the added costs of crop segregation, traceability, and labeling. In the meantime, farmers who grow transgenic varieties—and, ironically, those who do not—are liable for transgenic seeds ending up where they aren't wanted: in their own non-transgenic crop fields, in neighbours' fields, in truckloads of grain arriving at the elevator, in processed food products on retail shelves, and in ships headed overseas.

Farmers who choose not to grow transgenic varieties risk finding transgenic plants in their fields anyway, as a result of cross-pollination via wind, insects, and birds, which may bring pollen from transgenic crops planted miles away. Besides pollen, sources of contamination include contaminated seed and seed brought in by passing trucks or wildlife. Those farmers whose conventional or organic crops are contaminated, regardless of the route, risk lawsuits filed against them by the companies that own the proprietary rights to seed the farmer didn't buy. Likewise, farmers who grow transgenic crops risk being sued by neighbours and buyers whose non-transgenic crops become contaminated.

Because contamination by transgenic material has become so prevalent in such a short time, all farmers in areas of transgenic crop production are at risk. Insurance, the most common recourse for minimizing potential losses because of liability, is not available to the nation's farmers for this risk because insurance companies do not have enough information to gauge the potential losses. A full risk assessment and legal clarification of the distribution of liability among farmers, seed companies, grain handlers, processors, and retailers is needed before farmers can rest assured that transgenic crops won't result in lawsuits against them.

For farmers who have adopted transgenic technology—and for those who have not—in the words of Brian Leahy, executive director of California Certified Organic Farmers, "This technology does not respect property rights". In essence, farmers who grow transgenic crops on some of their fields, and farmers who grow none, risk bearing tremendous liability. This situation won't change until farmers either gain legal protection or until they all grow transgenic crops on all their fields.

Influences on Public Research

While transgenic crop varieties are generally the property of private corporations, those corporations often contract with public-sector agricultural research institutions for some of their development work. In fact, private investment in agricultural research, including germplasm development, has surpassed public investment in recent years. With this shift in funding sources, the following questions become important: Is the private sector unduly influencing the public research agenda? Are corporations directing public research in less socially valuable directions while research on, for instance, sustainable agriculture goes wanting? Are the outcomes of corporate-funded transgenic research and development by our public institutions equitable across the food and agricultural sectors? Is equity even a consideration of our public institutions when they accept this work?

Because intellectual property rights (patents) apply to living organisms, making them private property, the free flow of scientific information that has historically characterized public agricultural research is being inhibited. What are the implications for the future of agriculture and society of the secrecy that now surrounds so much of what was formerly shared public knowledge? For a brief history of intellectual property rights as they apply to living organisms. These and other questions need to be addressed by citizens and their public institutions. These issues are of particular concern to farmers and consumers who would benefit from research into alternative technologies that are less costly (in every way), less risky, and more equitable.

Equity requires that the economic benefits and risks of transgenic technology be fairly distributed among technology providers, farmers, merchants, and consumers. For long-term sustainability, farmers need research that focuses on farms as systems, with internal elements whose relationships can be adjusted to achieve farm management goals. In contrast, transgenic crop research so far has focused on products that complement toxic chemical approaches to control of individual pest species. These are products that can be commercialized by large agribusiness or agri-chemical interests and that farmers must purchase every year. This research orientation only perpetuates the cost-price squeeze that continues to drive so many out of farming.

Industry Concentration and Farmers' Right to Save Seed

The broadening of intellectual property rights in 1980 to cover living organisms, including genes, has resulted in a flurry of mergers and acquisitions in the seed and biotech industries. According to the Wallace Centre report, "Relatively few firms control the vast majority of commercial transgenic crop technologies. These firms have strategically developed linkages among the biotechnology, seed, and agri-chemical sectors to capture as much market value as possible. However, these tightly controlled linkages of product sectors raise serious issues of market access, product innovation, and the flow of public benefits from transgenic crops. Unlike Plant Variety Protection—which does not allow for the patenting of individual genes, but only of crop varieties—Intellectual Property Rights prohibit farmers from saving their seed and undertaking their own breeding programme, and prohibit plant breeders from using the material to create a new generation of varieties adapted to specific regions or growing conditions. Farmers can't save proprietary seed for planting and so must purchase new seed every year. In addition, farmers choosing transgenic varieties must sign a contract with the owner of the variety and pay a substantial per-acre technology fee, or royalty. The anticipated commercial introduction of transgenic wheat represents a dramatic shift in an industry in which farmers still widely save their own seed. As non-transgenic varieties become contaminated with transgenic ones, even those farmers who choose to stick with conventional varieties will lose the right to replant their own seed. This loss has already occurred in Canada's canola industry, with Monsanto winning its court case against farmer Percy Schmeiser for replanting his own canola variety that had become contaminated with Monsanto's Roundup Ready canola.

The adoption of transgenic crop varieties has brought with it an increasing prevalence of contract production. While contract production can lead to increased value and reduced risk for growers, farmers are justified in their concern about their loss of control when they sign a contract with a private company. Issues associated with contract production of transgenic crops must be considered within the broader context of a sustainable agriculture to include ownership, control, and social equity.

TRANSGENIC PLANTS WITH BENEFICIAL TRAITS

During the last decades, a tremendous progress has been made in the development of transgenic plants using the various techniques of genetic engineering. The plants, in which a functional foreign gene has been incorporated by any biotechnological methods that generally are not present in the plant, are called transgenic plants.

As per estimates recorded in 2002, transgenic crops are cultivated world-wide on about 148 million acres (587 million hectares) land by about 5.5 million farmers. Transgenic plants have many beneficial traits like insect resistance, herbicide tolerance, delayed fruit ripening, improved oil quality, weed control etc.

STRESS TOLERANCE

Biotechnology strategies are being developed to overcome problems caused due to biotic stresses (viral, bacterial infections, pests and weeds) and abiotic stresses (physical actors such as temperature, humidity, salinity etc).

Abiotic Stress Tolerance

The plants show their abiotic stress response reactions by the production of stress related osmolytes like sugars (*e.g.* trehalose and fructans), sugar alcohols (*e.g.* mannitol), amino acids (*e.g.* proline, glycine, betaine) and certain proteins (*e.g.* antifreeze proteins). Transgenic plants have been produced which over express the genes for one or more of the above mentioned compounds.

Such plants show increased tolerance to environmental stresses. Resistance to abiotic stresses includes stress induced by herbicides, temperature (heat, chilling, freezing), drought, salinity, ozone and intense light. These environmental stresses result in the destruction, deterioration of crop plants which leads to low crop productivity. Several strategies have been used and developed to build ressitance in the plants against these stresses.

HERBICIDE TOLERANCE

Weeds are unwanted plants which decrease the crop yields and by competing with crop plants for light, water and nutrients. Several biotechnological strategies for weed control are being used *e.g.* the over-production of herbicide target enzyme (usually in the chloroplast) in the plant which makes the plant insensitive to the herbicide. This is done by the introduction of a modified gene that encodes for a resistant form of the enzyme targeted by the herbicide in weeds and crop plants. Roundup Ready crop plants tolerant to herbicide-Roundup, is already being used commercially.

The biological manipulations using genetic engineering to develop herbicide resistant plants are:

- Over-expression of the target protein by integrating multiple copies of the gene or by using a strong promoter.,
- Enhancing the plant detoxification system which helps in reducing the effect of herbicide.,
- Detoxifying the herbicide by using a foreign gene., and
- Modification of the target protein by mutation.

Some of the examples are:

Glyphosate resistance - Glyphosate is a glycine derivative and is a herbicide which is found to be effective against the 76 of the world's worst 78 weeds. It kills the plant by being the competitive inhibitor of the enzyme 5-enoyl-pyruvylshikimate 3- phosphate synthase (EPSPS) in the shikimic acid pathway. Due to it's structural similarity with the substrate phosphoenol pyruvate, glyphosate binds more tightly with EPSPS and thus blocks the shikimic acid pathway.

Certain strategies were used to provide glyphosate resistance to plants:

- It was found that EPSPS gene was overexpressed in Petunia due to gene amplification. EPSPS gene was isolated from Petunia and introduced in to the other plants. These plants could tolerate glyphosate at a dose of 2- 4 times higher than that required to kill wild type plants.
  - *By using mutant EPSPS genes*: A single base substitution from C to T resulted in the change of an amino acid from proline to serine in EPSPS. The modified enzyme cannot bind to glyphosate and thus provides resistance.
  - The detoxification of glyphosate by introducing the gene

(isolated from soil organism- Ochrobactrum anthropi) encoding for glyphosate oxidase into crop plants. The enzyme glyphosate oxidase converts glyphosate to glyoxylate and aminomethylphosponic acid. The transgenic plants exhibited very good glyphosate ressitance in the field.

Another Example is of Phosphinothricin Resistance

Phosphinothricin is a broad spectrum herbicide and is effective against broad-leafed weeds. It acts as a competitive inhibitor of the enzyme glutamine synthase which results in the inhibition of the enzyme glutamine synthase and accumulation of ammonia and finally the death of the plant. The disturbace in the glutamine synthesis also inhibits the photosynthetic activity. The enzyme phosphinothricin acetyl transferase (which was first observed in Streptomyces sp in natural detoxifying mechanism against phosphinothricin) acetylates phosphinothricin, and thus inactivates the herbicide. The gene encoding for phosphinothricin acetyl transferase (bar gene) was introduced in transgenic maize and oil seed rape to provide resistance against phosphinothricin.

OTHER ABIOTIC STRESSES

The abiotic stresses due to temperature, drought, and salinity are collectively also known as water deficit stresses. The plants produce osmolytes or osmoprotectants to overcome the osmotic stress. The attempts are on to use genetic engineering strategies to increase the production of osmoprotectants in the plants. The biosynthetic pathways for the production of many osmoprotectants have been established and genes coding the key enzymes have been isolated. *E.g*. Glycine betaine is a cellular osmolyte which is produced by the participation of a number of key enzymes like choline dehydrogenase, choline monooxygenase etc. The choline oxidase gene from Arthrobacter sp. was used to produce transgenic rice with high levels of glycine betaine giving tolerance against water deficit stress.

Scientists also developed cold-tolerant genes (around 20) in Arabidopsis when this plant was gradually exposed to slowly declining temperature. By introducing the coordinating gene (it encodes a protein which acts as transcription factor for regulating the expression of cold tolerant genes), expression of cold tolerant genes was triggered giving protection to the plants against the cold temperatures.

INSECT RESISTANCE

A variety of insects, mites and nematodes significantly reduce the yield and quality of the crop plants. The conventional method is to use synthetic pesticides, which also have severe effects on human health and environment. The transgenic technology uses an innovative and eco-friendly method to improve pest control management.About 40 genes obtained from microorganisms of higher plants and animals have been used to provide insect resistance in crop plants

The first genes available for genetic engineering of crop plants for pest resistance were Cry genes (popularly known as Bt genes) from a bacterium Bacillus thuringiensis. These are specific to particular group of insect pests, and are not harmful to other useful insects like butter flies and silk worms. Transgenic crops with Bt genes (*e.g.* cotton, rice, maize, potato, tomato, brinjal, cauliflower, cabbage, etc.) have been developed. This has proved to be an effective way of controlling the insect pests and has reduced the pesticide use. The most notable example is Bt cotton (which contains CrylAc gene) that is resistant to a notorious insect pest Bollworm (Helicoperpa armigera).. There are certain other insect resistant genes from other microorganisms which have been used for this purpose. Isopentenyl transferase gene from Agrobacterium tumefaciens has been introduced into tobacco and tomato. The transenic plants with this transgene were found to reduce the leaf consumption by tobacco hornworm and decrease the survival of peach potato aphid.

Certain genes from higher plants were also found to result in the synthesis of products possessing insecticidal activity. One of the examples is the Cowpea trypsin inhibitor gene (CpTi) which was introduced into tobacco, potato, and oilseed rape for develping transgenic plants. Earlier it was observed that the wild species of cowpea plants growing in Africa were resistant to attack by a wide range of insects. It was observed that the insecticidal protein was a trypsin inhibitor that was capable of destroying insects belonging to the orders Lepidoptera, Orthaptera etc. Cowpea trypsin inhibitor (CpTi) has no effect on mammalian trypsin, hence it is non-toxic to mammals.

VIRUS RESISTANCE

There are several strategies for engineering plants for viral resistance, and these utilizes the genes from virus itself (*e.g.* the viral coat protein gene). The virus-derived resistance has given promising results in a number of crop plants such as tobacco, tomato, potato, alfalfa, and papaya. The induction of virus resistance is done by employing virus-encoded genes-virus coat proteins, movement proteins, transmission proteins, satellite RNa, antisense RNAs, and ribozymes.

The virus coat protein-mediated approach is the most successful one to provide virus resistance to plants. It was in 1986, transgenic tobacco plants expressing tobacco mosaic virus (TMV) coat protein gene were first developed. These plants exhibited high levels of resistance to TMV.

The transgenic plant providing coat protein-mediated resistance to virus are rice, potato, peanut, sugar beet, alfalfa etc. The viruses that have been used include alfalfa mosaic virus (AIMV), cucumber mosaic virus (CMV), potato virus X (PVX), potato virus Y (PVY) etc.

RESISTANCE AGAINST FUNGAL AND BACTERIAL INFECTIONS

As a defence strategy against the invading pathogens (fungi and bacteria) the plants accumulate low molecular weight proteins which are collectively known as pathogenesis-related (PR) proteins.

Several transgenic crop plants with increased resistance to fungal pathogens are being raised with genes coding for the different compounds. One of the examples is the Glucanase enzyme that degrades the cell wall of many fungi. The most widely used glucanase is beta-1,4-glucanase. The gene encoding for beta-1,4 glucanase has been isolated from barley, introduced, and expressed in transgenic tobacco plants. This gene provided good protection against soil-borne fungal pathogen Rhizoctonia solani.

Lysozyme degrades chitin and peptidoglycan of cell wall, and in this way fungal infection can be reduced. Transgenic potato plants with lysozyme gene providing resistance to Eswinia carotovora have been developed.

DELAYED FRUIT RIPENING

The gas hormone, ethylene regulates the ripening of fruits, therefore, ripening can be slowed down by blocking or reducing ethylene production. This can be achieved by introducing ethylene forming gene(s) in a way that will suppress its own expression in the crop plant. Such fruits ripen very slowly (however, they can be ripen by ethylene application) and this helps in exporting the fruits to longer distances without spoilage due to longer-shelf life.

The most common example is the 'Flavr Savr' transgenic tomatoes, which were commercialized in U.S.A in 1994. The main strategy used was the antisense RNA approach. In the normal tomato plant, the PG gene (for the enzyme polygalacturonase) encodes a normal mRNA that produces the enzyme polygalacturonase which is involved in the fruit ripening.

The complimentary DNA of PG encodes for antisense mRNA, which is complimentary to normal (sense) mRNA. The hybridization between the sense and antisnse mRNAs renders the sense mRNA ineffective. Consequently, polygalacturonase is not produced causing delay in the fruit ripening. Similarly strategies have been developed to block the ethylene biosynthesis thereby reducing the fruit ripening. *E.g.* transgenic plants with antisense gene of ACC oxidase (an enzyme involved in the biosynthetic process of ethylene) have been developed. In these plants, production of ethylene was reduced by about 97% with a significant delay in the fruit ripening.

The bacterial gene encoding ACC deaminase (an enzyme that acts on ACC and removes amino group) has been transferred and expressed in tomato plants which showed 90% inhibition in the ethylene biosynthesis.

MALE STERILITY

The plants may inherit male sterility either from the nucleus or cytoplasm.

It is possible to introduce male sterility through genetic manipulations while the female plants maintain fertility.

In tobacco plants, these are created by introducing a gene coding for an enzyme (barnase, which is a RNA hydrolyzing enzyme) that inhibits pollen formation. This gene is expressed specifically in the tapetal cells of anther using tapetal specific promoter TA29 to restrict its activity only to the cells involved in pollen production. The restoration of male fertility is done by introducing another gene barstar that suppresses the activity of barnase at the onset of the breeding season. By using this approach, transgenic plants of tobacco, cauliflower, cotton, tomato, corn, lettuce etc. with male sterility have been developed.

GENETIC TRANSFORMATION TECHNIQUES

A number of genetic transformation techniques have been developed since the first report of a transgenic plant. With few exceptions, these methods rely on the ability to transfer DNA to a single cell or small group of cells, such as nondifferentiated leaf tissue or meristems. Then some form of tissue culture is required to regenerate whole, fertile transgenic plants for the process to be completed.

Therefore, the first requirement for plant transformation is a reliable plant regeneration system using explants/tissues which are amenable to the DNA delivery system.

This requirement was, more than any other factor, the greatest impediment to genetic engineering of members of the cereal and grain legume families – the two most important crop groups. Regeneration in most of these species is difficult, particularly for species like soybean and chickpea.

At best, regeneration is extremely genotype-dependent for species such as rice, where the *japonica* rices are easier to manipulate *in vitro* than the *indica* rices, which are more important on a worldwide basis. However, international effort in most of the important cereal species has largely overcome the problem, although there are still genotypes of a number of species which remain recalcitrant to regeneration.

The cereals and pulses (grain legumes) proved somewhat recalcitrant to genetic transformation until the late 1980s, when rice became the first cereal to become reliably transformed at high frequency, with soybean first successfully transformed via *Agrobacterium* in 1988, although repeatable transformation was only achieved later using a microprojectile approach.

There are three basic transformation systems being routinely used in most genetic engineering laboratories internationally.

These are:

- *Agrobacterium*-mediated transformation.
- DNA uptake via protoplasts.
- Microprojectile bombardment.

Agrobacterium-mediated Transformation

Agrobacterium tumefaciens, the causal agent of crown gall disease, transfers a portion of its own DNA (T-DNA) to the plant, where it is stably incorporated into the plant genome. The T-DNA is delimited by specific 25 bp repeat sequences, known as the border sequences. None of the T-DNA is actually required for transfer, and whatever DNA is bound by the borders will be transferred, hence the pathogenic DNA can be effectively replaced with genes of interest, allowing delivery of these desirable genes to cells from which whole plants can be regenerated.

The ability of *Agrobacterium* to transfer genes to plants is known as virulence, and is largely controlled by the *vir* genes on the Ti plasmid, and to a lesser extent, the chromosomal virulence (*chv*) genes. The *vir* genes are induced by the plant's wound response, in particular the production of phenolic compounds such as acetosyringone.

The Vir proteins then facilitate the identification, copying, transfer and targeting to the host nucleus, and incorporation on a plant chromosome of the T-DNA. Generally the simplest manner by which to replace the T-DNA is to remove it, leaving a 'disarmed' Ti plasmid with the *vir* genes, then introducing a binary plasmid which contains the gene(s) of interest within the border sequences. Hence, with the induction of virulence, the Vir proteins will effect the transfer of the engineered T-DNA.

Most dicotyledonous species are hosts to *Agrobacterium* and as such can be genetically engineered in this manner. *Agrobacterium*-mediated transformation remains the method of choice for most species of the *Solanaceae* and *Brassicae*, and other top 20 most important plants such as grape, orange, cotton, sugarbeet and apple. While there are a number of reports of successful transformation of important grain legumes including soybean, chickpea and *Phaseolus* bean, there is generally great genotypedependence of such methods. It has been shown repeatedly that plant genotype 3 *Agrobacterium* strain interactions can vary from complete absence of transformation to 100% transformation within many species, including soybean, pea and sugarbeet.

Most important monocotyledonous species were once regarded as being outside the host range of *Agrobacterium*. In the late 1980s it became evident that some monocot species, including asparagus, yam and onion, were transformable using *Agrobacterium*.

Early reports of *Agrobacterium*-mediated transformation of rice, maize and sorghum remained largely unrepeatable in these and other labs, until the first report of a high-frequency, repeatable, and painstakingly substantiated system was published for three *japonica* rice cultivars Chan *et al*., (1993) first reported stable transformation of rice using *A. tumefaciens*.

Progeny analysis and Southern hybridization demonstrated that it was possible to generate transgenic rice via *Agrobacterium*, albeit at low frequency. Hiei *et al*. (1994) improved the efficiency greatly by co-cultivating callus originating from scutella of immature embryos with a super binary vector then selecting on hygromycin, and produced over 400 primary transgenics of three cultivars.

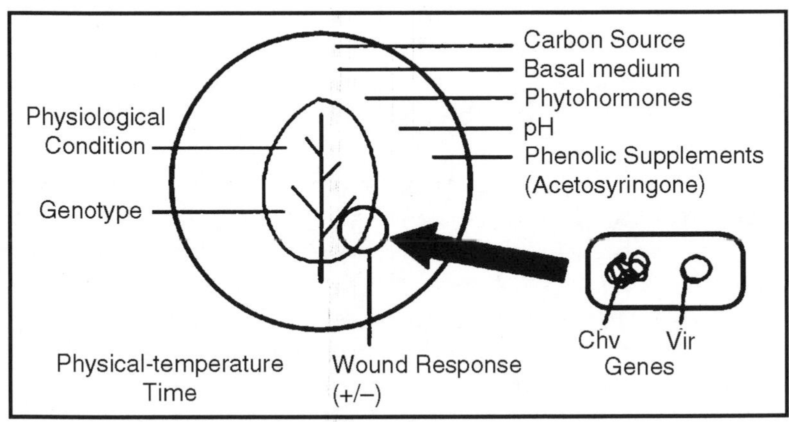

Fig. A Summary of the Plant,Co-cultivation Medium,Bacterial and Physical Factors which Affect the Efficiency of Virulence Induction and Arobacterium-mediated Gene Transfer.

Detailed analyses of 29 plants and their progeny revealed that stable transformation was possible, and in most cases marker genes segregated in the expected Mendelian fashion.

In addition, these authors sequenced T-DNA/plant DNA junctions and found that incorporation occurred at similar sites to those found in transgenic tobacco plants. The 'super binary' plasmid contained *virB* and *virG* genes from the hypervirulent pTiBo542 plasmid, and, although it enhanced transformation frequencies, was not an absolute requirement for transformation. Basic requirements for *Agrobacterium*-mediated transformation are host recognition, virulence gene induction, the ability to select and proliferate transgenic cells, and the ability to regenerate whole, fertile plants from these cells. Host recognition and virulence gene induction require the correct carbon source and pH, and the presence of virulence-inducing molecules. Many wounded plant tissues produce signal molecules such as acetosyringone, which set in train the events of T-DNA transfer and integration into the host nuclear genome.

Protoplasts

Direct gene transfer has been the method of choice for monocotyledons, predominantly because most of these species have proved recalcitrant to *Agrobacterium*-mediated transformation. Protoplasts are cells with the cell wall enzymatically removed. In the absence of a cell wall, macromolecules such as proteins and nucleic acids may be taken up by such cells. The efficiency of uptake of DNA is greatly enhanced by the application of physiological shock, such as an ionic, osmotic or electrical current, which causes reversible permeabilization of the cell membrane. DNA has been shown to transfer to cells of most major cereals using the electric shock, or electroporation, methodology. If the DNA enters the nucleus and is stably incorporated into the plant genome, the cell is transgenic. This has been achieved with all major cereals (e.g. sorghum). One of the advantages of this methodology is that a large number of theoretically competent cells can be exposed to DNA at one time.

Regeneration of whole, fertile plants from cereal protoplasts has remained the greatest hurdle to electroporation-mediated transformation. Even when protoplasts are regenerable and DNA transfer is effected, it is difficult to combine the two. The physiological shock necessary for DNA uptake may often be enough to prevent subsequent cell wall synthesis or substantially reduce regenerability. As protoplast cultures usually require greater time *in vitro*, there may also be an increased frequency of somaclonal variation. Success has been achieved with regeneration of rice, wheat and maize but protoplasts remain a widely used method for rice only, with microprojectile-mediated transformation overtaking it in efficiency.

Microprojectile Bombardment

The potential for genetic transformation with microprojectiles was first reported by Klein *et al.* (1987), who achieved strong transient expression of a reporter gene in intact onion tissues. The first report of fertile transgenic plants and their progeny described stably transformed tobacco plants with Mendelian inheritance among selfed progenies. However, much of the focus with microprojectile bombardment has concentrated on cereals, due to the apparent inability of *Agrobacterium* to transfer genes to most cereal species, and the recalcitrance and genotype-specificity of cereals (with the exception of rice) to regeneration of fertile plants from transformed protoplasts.

The microprojectile bombardment, or 'Biolistic' (commercial description used by Bio Rad Labs), approach is a simple concept, based on the introduction of DNA-coated small inert projectiles into plant cells, not unlike the action of a microscale shotgun. The system is a modification of one used by virologists, whereby high-velocity microprojectiles are used to facilitate viral infection. To adapt this for transformation, the major alterations required were to reduce microprojectile size to reduce the extent of cell damage, to be able to perform the procedure under sterile conditions to exclude contamination by bacteria or fungi, and most importantly, to ensure that the target tissue be regenerable to produce whole, fertile plants.

Here lies one of the greatest advantages of the microprojectile approach over protoplast and *Agrobacterium*-mediated techniques. Effectively, any target tissue may be used for bombardment, so the most regenerable tissues can be used.

In addition, there does not appear to be any genotype specificity in the ability to deliver DNA, which is demonstrably the case with *Agrobacterium*mediated transformation (e.g. Owens and Smigocki, 1988). The genotype dependence of plant regeneration remains the greatest hurdle in most species, although this is by no means the barrier that protoplast regeneration presents.

The microprojectiles, which are commonly an inert metal such as tungsten or gold, range in diameter from 1 to 4 mm; a velocity of approximately 250 m s-1 is required to penetrate cell walls and membranes. For successful stable transformation, the DNA must be delivered intact to the nucleus where the DNA will then be incorporated onto a chromosome. This requires that the velocity of microprojectiles be sufficient to penetrate cells, but not cause excessive damage. For most of the cereals, the target tissue of choice is the immature zygotic embryo, particularly the scutellum, or embryogenic callus cultures, usually derived from zygotic embryos. Other targets such as embryogenic suspension cultures have also been used. As can be seen, the most important cereal species have been recently transformed using this approach. In addition, transgenic oats and rye have been produced with microprojectiles. Early microprojectile apparatus was based on a gunpowder charge for particle acceleration. Improvements in safety and reliability were made by using electrical discharge or helium gas to provide acceleration, and most laboratories currently use helium-based guns, such as the commercially available Biolistic gun.

Tissue sterility is maintained by performing all operations in a laminar flow cabinet with bombardment actually taking place in a sealed chamber under a slight vacuum. One of the drawbacks of the microprojectile approach is the cost of the Biolistic gun, hence other laboratories have developed simpler apparatus, such as the particle inflow gun (PIG), which relies on a solenoid to regulate helium gas inflow. The great advantage of the PIG is that it can be built for around US$1000. There are potential questions about the use of such apparatus and whether there is any contravention of international patents.

Alternative Transformation Methods

While the great majority of transgenic plants worldwide are being produced via *Agrobacterium,* microprojectile bombardment and, to a lesser extent, protoplasts, a number of alternative transformation methods have been trialled. Although not intended to be an inclusive list, these include:

- Virus-mediated transformation.
- Pollen pathway or intact inflorescence methods.
- Microinjection.
- Silicon carbide fibres.
- Electroporation of intact tissues.
- Electrophoresis of intact tissues.

While it is not possible to discuss the success, advantages and disadvantages of these methods, the subject has been recently reviewed by Songstad *et al.* (1995).

The pollen pathway has long been recognized as a potentially useful means of introducing genes to the germline, with its greatest advantage being the avoidance of tissue culture and the need to regenerate undifferentiated tissues such as callus or protoplasts. The absence of any time in plant tissue culture should also totally avoid the problem of somaclonal variation. There have been many difficulties with this approach, however, and there are no substantiated instances of genetic transformation via the pollen pathway.

More promising methods are the use of silicon carbide fibres and electroporation/ electrophoresis of intact tissues. Stable tissue transformation of tobacco and corn (maize) suspension cultures has been achieved following agitation with silicon carbide fibres. Stable cereal transformants have been produced by electroporating germinating rice seeds and immature embryos of corn, and by electrophoresis of germinating barley seeds. However, until some of these alternative systems can be made more reliable and efficient, it would appear that the two most widely used transformation systems will remain *Agrobacterium* and, in the case of cereals, microprojectile bombardment.

GENETIC ENGINEERING TECHNOLOGY

Genetic manipulation for biotechnological plant improvement relies on the ability to modify genes as DNA sequences. This requires utilization of standard, well-documented techniques. Enzymes are the basic tool kit of the genetic engineer, and can be used for basic functions of manipulating nucleic acids in such a way as to copy (DNA polymerase), transcribe (RNA polymerase), cut at specific sites (restriction endonucleases), cut and degrade in a nonspecific nature (nucleases) and join pieces together (DNA ligase).

DNA libraries are another important resource for genetic manipulation as they represent a means of storing and screening sequences from an entire organism. The two basic types of library are the cDNA library and the genomic library. A cDNA (complementary DNA) library is based on expressed sequences. Such a library is made by extracting mRNA from a particular plant tissue of interest, then making first single-stranded DNA (via reverse transcriptase) then double-stranded DNA (via DNA polymerase). A cDNA library will then represent only those genes which are expressed in the tissue of interest (or in response to a stress of some sort), and will be devoid of introns and regulatory sequences.

A genomic library represents the entire genome (often including the organellar genomes), and will include all sequences whether they are expressed in the donor tissue or not, and even sequences which are not expressed at all. Genomic libraries will therefore include highly repeated sequences such as rRNA genes and nonexpressed simple sequence repeats

(SSRs), as well as genes and their regulatory sequences and introns. Genomic libraries may be cloned and stored in lambda bacteriophage, cosmids or larger fragments which will often include multiple genes in bacterial artificial chromosomes (BACs) and yeast artificial chromosomes (YACs). BAC library inserts may average 150 kb, whereas YAC inserts are larger on average, at up to 1 Mb of DNA.

Overall, the ability to genetically engineer a plant relies on:

- A genetic transformation system for transferring DNA into plant cells.
- A plant regeneration system to produce plants from the transgenic cells.
- A gene construct which can modify the plant phenotype in a desirable and controlled manner.

THE MAINTENANCE OF GENETIC VARIATION

In applied plant breeding there is continuous interest in the introduction of new genetic variation. Sources for extending the genetic variation with regard to some crop species are natural populations of the same species or of related species. (Genetic transformation is a rather recently developed way for extending the genetic variation to be exploited for crop improvement.)

Often such natural populations appear to accommodate a wealth of genetic diversity. Genetic variation may also be maintained in breeding populations of cultivated crops. This is remarkable, because natural (and/or artificial) selection occurs generation after generation and one might speculate that this implies a continuous reduction of genetic variation. In the absence of human intervention genetic variation is/was, however, often maintained, notwithstanding the continuous selection. With regard to cultivated crops one might even state that plant breeding has stimulated the development and maintenance of a wide genetic diversity. It seems that human interference promotes an increase of the genetic diversity in the involved crop. (In contrast to this, wild plant and animal species suffer from genetic erosion because of annihilation of ecological niches due to human activities. In recent times many species have become completely extinct.)

Ecological population genetics studies the mechanisms responsible for the maintenance genetic diversity. In this section four mechanisms (tentatively) explaining this seemingly paradoxical situation are elaborated, namely

- Overdominance,
- Frequency-dependent fitness,
- Recurrent mutations and
- Immigration of pollen or plants.

Overdominance

Crumpacker (1967) and Allard, Jain and Workman (1968) have presented, for cross-fertilizing and self-fertilizing crops respectively, examples of overdominance with regard to traits controlled by a single locus. Reduced probability of recombination alongside a certain chromosome segment gives rise to a gene cluster. If the loci belonging to the cluster control the same trait, an oligogenic basis for overdominance is present. (In humans such a gene cluster has been shown to control the immune system). These few examples do not represent the common situation.

A more realistic concept is pseudo-overdominance, due to alleles linked in repulsion phase. An example is a chromosome segment behaving as a single allele (because recombination within the segment hardly ever occurs). Crossing of two homozygous genotypes, differing for such segment, yields an offspring heterozygous for this segment which, consequently, may exceed both homozygous parents.

In 1917 Jones had already stated that hybrid vigour could be due to the assembling of favourable alleles from each of both parents in one genotype. Linkage of such favourable alleles to unfavourable alleles hampers fixation of the superior heterozygous F1-genotype into an equivalent homozygous genotype. However, it does not exclude such fixation. Results of experiments using electrophoresis substantiate the concept of pseudo-overdominance. Notwithstanding the previous remarks, many population genetical models, aimed at explaining genetic polymorphisms, have been developed on the basis of a single locus. Population genetic theory shows that for loci with overdominance, *i.e.* $s0 > 0$ and $s2 > 0$, a stable equilibrium of the genotypic composition may occur, notwithstanding the selection. Thus a genetic polymorphism is maintained, and – in contrast to what was said at the beginning of this chapter – the genotypic composition may be stable, notwithstanding selection. The equilibrium allele frequencies can be derived to be,

$$q_e = \frac{S_2}{S_0 + S_2} \; and \; P_e = \frac{S_0}{S_0 + S_2}$$

thus $0 < pe < 1$

Note. One may criticize the derivation undertlying Equation on two grounds:

- It is based on the assumption that the preceding generation had the Hardy–Weinberg genotypic composition. This composition applies in the case of mass selection occurring before pollen distribution. Selection with regard to vitality is thus, implicitly, assumed not to occur.
- Overdominance with regard to a single locus is a rare event.

Frequency-Dependent Fitness

The concept of frequency-dependent fitness is based on the fascinating observation that it is, under constant ecological conditions, both rare for plants (or animals) with a certain genotype to be completely extinct as well as rare that the frequency of plants with the considered genotype grows unrestricted. Apparently, there are mechanisms regulating the number of individuals with a certain genotype in such a way, that the number increases if it is low and that it decreases if it is high.

Example.Two examples of frequency-dependent fitness are mentioned here:

- The seed-set of male sterile barley plants (with genotype *mm*) may depend on the frequency of such plants. Refers to the relation $w0 = 0.6 - f0$.
- In the case of self-incompatibility, a low frequency of a genotype for the incompatibility locus/loci tends to be associated with a higher fitness of the genotype than the fitness of a genotype with a higher frequency.

A tentative explanation for genotypes to have a frequency-dependent fitness is as follows. Plants with the same genotype tend to have similar demands, at the same time. These demands are specific for the genotype. Among the plants with a certain genotype, more plants will survive the 'struggle' for the same, restrictedly available resources, as the genotype's frequency is lower. Plants with a genotype with a relatively low frequency may thus tend to have a relatively high fitness. This phenomenon might apply to genotypes adapted to rare environmental conditions. Such genotypes are favoured by selection. Mather (1973) called such selection disruptive selection. It may lead to distinct types or it may be balanced by stabilizing selection, for example by the genotype adapted to rare environmental conditions becoming increasingly common.

Recurrent Mutations

Mutations are, in fact, the ultimate source of all genetic diversity. However, their frequencies are generally very low. Thus in the equilibrium between the production of a new allele and its elimination, if it does not give rise to a better adapted phenotype, the new allele will have a (very) low frequency. It is concluded that recurrent mutations should not be considered as a quantitatively important factor for maintenance of genetic diversity.

Note:The frequency of the occurrence of a mutation is very low. Furthermore, one should realize that a mutant allele is not transmitted to the next generation when the mutation occurs outside the chain of cells connecting two generations, the so-called germ-line. Such mutations have no population genetical implications. This concerns mutations in cells of roots, stems, leaves, style, stigma, seed coat, connectivum, *etc.*

Immigration of Pollen or Plants

The effect of immigration of pollen or plants on the genotypic composition of the considered population depends on

- The difference in the allele frequencies of 'donor' and 'recipient' and
- The extent of the immigration

Both factors may play a role in legislation concerning mutual isolation distances required at the multiplication of seed of varieties of cross-fertilizing crops. It is emphasized here that introgression means the incorporation by crossing and repeated backcrossing of alleles originating from a *different* species. This may occur spontaneously or as a breeding activity.

Alleles may immigrate into a population in different ways:

- Flow of pollen, transported by wind or by insects
- Mixing, intended or not, of seed lots representing different varieties

Flow of Pollen

We define q as the frequency of allele a in the recipient, qm as the frequency of a among the immigrating pollen, and m as the proportion of immigrating pollen among the effective male gametes. The frequency, $q\_$, of the effective pollen grains with haplotype a is,

$$q' = (1 - m)q + mq_m$$

The case of immigrating pollen situation can be considered as a form of bulk crossing. According to Equation the frequency of a in the 'hybrid' population will be,

$$q_1 = \frac{1}{2}(q + q') = \frac{1}{2}[q + (1 - m)q + mq_m] = q + \frac{1}{2}m(q_m - q)$$

Thus

$$\Delta q = q_1 - q = \frac{1}{2}m(q_m - q)$$

This expression contains both factors mentioned before. For $qm = q$ or for $m = 0$ the allele frequency will not change. For $m > 0$ the expression yields of course $\Delta q > 0$ if $qm > q$ and $\Delta q < 0$ if $qm < q$.

If immigration occurs generation after generation, selection aiming at the elimination of allele a will never succeed. Then, notwithstanding selection, a genetic polymorphism is maintained.

Mixing of Seed

This case is considered as immigration of sporophytes. For a diploid crop one can then derive:

$\Delta q = m(qm - q)$

In certain situations immigration of sporophytes is applied intentionally, *e.g.* as a remedy against genetic erosion in populations of a small size.

5

Plant Cell Structure and Parts

Plant cells like animal cells are eukaryotic cells i.e. they contain membrane bound nuclei and cell organelles. Plant cell differs from the animal cell in having three distinctive structures – cell wall, vacuoles and plastids. Plant cells lack centrioles and intermediate filaments, which are present in animal cells.

PARTS OF A PLANT CELL STRUCTURE

Plant cells are classified into three viz. parenchyma cells, collenchyma cells and sclerenchyma cells based on the structure and function. Now let us see the different parts of a plant cell.

Cell Wall

Cell wall is the outermost rigid layer composed of cellulose, hemicellulose, pectin and sometimes lignin. The function of cell wall is protection, structural support and also it helps in filtering mechanism.

Cell Membrane

Cell membrane also called as plasma membrane is present inside the cell wall and surrounds the cytoplasm. It connects the intracellular components with the extracellular environment and helps in protection and transportation.

Plasmodesmata

Plasmodesmatas are small openings, which connect plant cells with each other enabling transport and communication between them.

Vacuole

Vacuoles are large membrane bound compartments, which stores compounds and provides storage, excretory and secretory functions. The membrane surrounding vacuole is called tonoplast.

Cytoplasm

Cytoplasm is filled up by cytosol, which is a gelatinous, semitransparent fluid.

Nucleus

Nucleus is a specialized organelle, which contains the plant's hereditary material i.e. DNA (Deoxyribonucleic Acid). It also contains structures, which regulates the cell cycle, growth, protein synthesis and reproductive function.

Plastid

Plastids are organelles responsible for the photosynthetic activity and for the manufacture and storage of chemical compounds in plants. Chloroplast is an important form of plastid containing chlorophyll pigment, which helps in harvesting light energy and converting it to chemical energy.

Mitochondria

Mitochondria are oblong shaped organelles that are also known as "the powerhouse of the cell". They are responsible for breaking down the complex carbohydrate and sugar molecules to simpler forms that the plants can use.

Endoplasmic Reticulum

Endoplasmic reticulum is an organelle responsible for the manufacturing and storage of chemical compounds like glycogen and steroids, translation and transportation of protein. It is also connected to the nuclear membrane so as to make a channel between the cytoplasm and the nucleus.

Golgi Apparatus

Golgi apparatus also known as golgi complex is an organelle responsible for the processing and packaging of macromolecules such as proteins and fats, which are synthesized by the cell and prepares them for transportation.

Ribosome

Ribosomes are organelles, which are made up of 60% RNA (Ribonucleic Acid) and 40% protein and play an important role in protein translation.

Microbodies

Microbodies are single membrane bound organelles, globular in shape and contains degradative enzymes. Most common microbodies are peroxisomes.

Microtubules

Icrotubules are straight, hollow, tubular cylinders, which make up the cytoskeleton. They are responsible for structural support and transport of the cell.

Microfilaments

Microfilaments are thin filaments of the cytoskeleton and are responsible for structural support of the cell.

PLANT CELLS AND THE PLANT KINGDOM

Since the time of Aristotle, biologists have sought to classify organisms. At first the purpose was ease of identification ("artificial" classification schemes). Carolus Linnaeus (1707–1778), arguably the greatest of the pre-modern Naturalists, sought to classify plants and other organisms according to affinity groups that reflected the mind of the Creator. Later, after Darwin, the goal of classification was to show evolutionary relationships ("natural" classification schemes). For the past 150 years, biologists have emphasized natural systems of classification and have attempted to define morphological criteria that reveal evolutionary relationships.

We now know that morphology, the form and structure of organisms, is the end product of the actions of genes. Virtually all of the information needed to form a complete organism is encoded in its DNA sequences, both nuclear and cytoplasmic (mitochondria and chloroplasts). DNA sequence analysis has thus provided evolutionary biologists with a powerful new tool for arriving at a truly natural classification system

On the basis of phylogenetic analyses of highly conserved DNA sequences, living organisms have been divided into three major domains: Bacteria, Archaea, and Eucarya.

- The common ancestor of all the organisms first gave rise to the Bacteria and the common ancestor of the Archaea and the Eucarya.
- The Archaea branch off from the Eucarya lineage.
- The Eucarya common ancestor acquires mitochondrial endosymbiont (an alpha-proteobacterium-like cell).
- A heterogeneous group of eukaryotes called protists branch off the lineage leading to plants, fungi, and animals.
- The common ancestor of fungi and animals form a branch, followed by a divergence into the fungal and animal lineages.
- The common ancestor of plants, green algae, red algae, and glaucophytes acquires chloroplast endosymbiont (a cyanobacterium).
- The three lineages of glaucophytes, red algae, and green algae diverge.
- Various lineages of protists acquire chloroplasts via green or red algal endosymbionts.
- The earliest branch of green algae (Green Algae I) diverge.
- The later branch of green algae (Green Algae II, Characeae, Coleochaetales) diverge.
- The remaining lineage leads to plants.

The Eucarya include the eukaryotes, organisms whose cells contain a true nucleus. The Bacteria, or eubacteria, which include the cyanobacteria, lack a true nucleus and are therefore prokaryotic. The Archaea, or archaebacteria, are also prokaryotic, but they differ from the Bacteria: Besides their morphological and biochemical differences, they are often adapted to extreme environments, such as sulfur hot springs or saline ponds. Phylogenetic studies have indicated that the Archaea and Eucarya split after the Bacteria separated from the common ancestor.

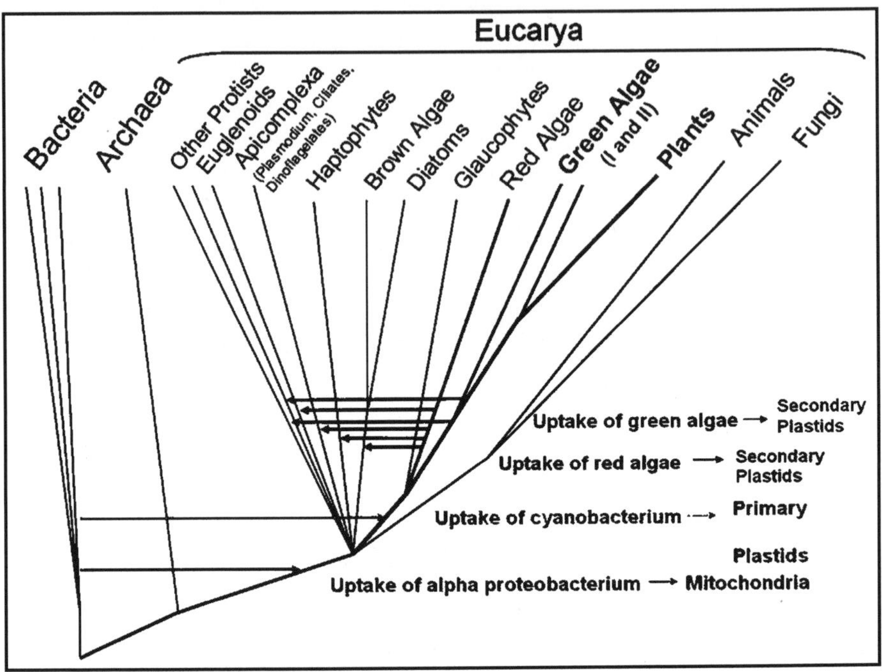

Fig. Natural Classification Scheme and Phylogeny of living Organisms, Including Endosymbiotic Events.

Thus Archaea and Eucarya represent sister groups. This closer relation between Archaea and Eucarya is reflected in their similar promoter structures and RNA polymerases, the presence of histones, and many other characteristics. Fungi were formerly classified as algal-like plants that had lost their chloroplasts. However, as the phylogenetic tree in Figure.A shows, fungi and animals branched off from the Eucarya lineage *before* the appearance of choloroplasts. They are thus more closely related to animals than to plants. Fungi are heterotrophic; that is, they depend on other organisms for their food, and they satisfy their nutritional needs by absorbing inorganic ions and organic molecules from the external environment. Most fungal species are filamentous and possess cell walls made of chitin, the same substance that is found in insect exoskeletons.

A recent phylogenetic tree of the plant kingdom is shown in Figure.

Bryophytes are small (rarely more than 4 cm in height), very simple land plants, and the least abundant in terms of number of species and overall population. Bryophytes do not appear to be in the direct line of evolution leading to the vascular plants; rather, they seem to constitute a separate minor branch. Bryophytes include mosses, liverworts, and hornworts. These small plants have life cycles that depend on water during the sexual phase.

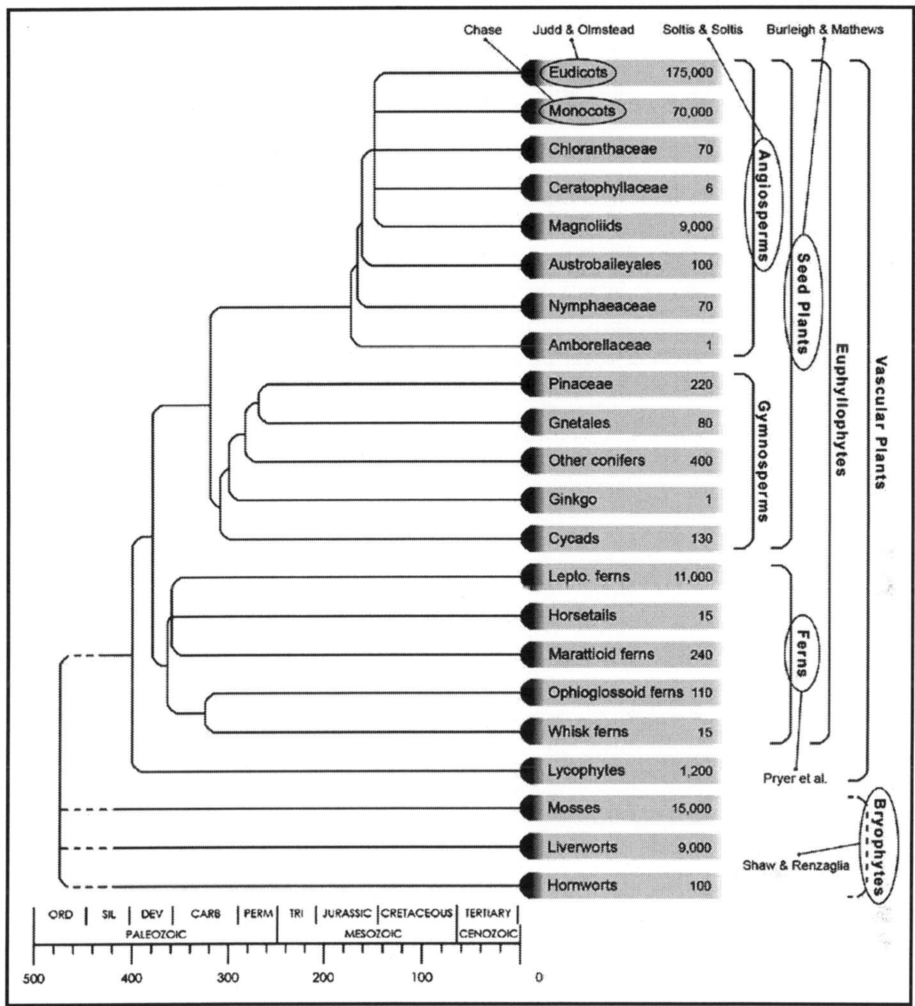

Fig. Phylogenetic Tree of the Plant kingdom, with Approximate time scale on the Horizontal axis.

Water facilitates fertilization, the fusion of gametes to produce a diploid zygote, a feature also seen in the algal precursors of these plants. Bryophytes are like algae in other respects as well: They have neither true roots nor true leaves, they lack a vascular system, and they produce no hard tissues for structural support. The absence of these structures that are important for growth on land greatly restricts the potential size of bryophytes, which, unlike algae, are terrestrial rather than aquatic.

The ferns represent the largest group of spore-bearing vascular plants. In contrast to the bryophytes, ferns have true roots, leaves, and vascular tissues, and they produce hard tissues for support. These architectural features enable ferns to grow to the size of small trees. Although ferns are better adapted to the drying conditions of terrestrial life than bryophytes are, they still depend on water as a medium for the movement of sperm to the egg. This dependence on water during a critical stage of their life cycle restricts the ecological range of ferns to relatively moist habitats.

The most successful terrestrial plants are the seed plants. Seed plants have been able to adapt to an extraordinary range of habitats. The embryo, protected and nourished inside the seed, is able to survive in a dormant state during unfavorable growing conditions such as drought. Seed dispersal also facilitates the dissemination of the embryos away from the parent plant. Another important feature of seed plants is their mode of fertilization. Fertilization in seed plants is brought about by wind- or insect-mediated transfer of pollen, the gamete-producing structure of the male, to the sexual structure of the female, the pistil. Pollination is independent of external water, a distinct advantage in terrestrial environments. Many seed plants produce massive amounts of woody tissues, which enable them to grow to extraordinary heights. These features of seed plants have contributed to their success and account for their wide range.

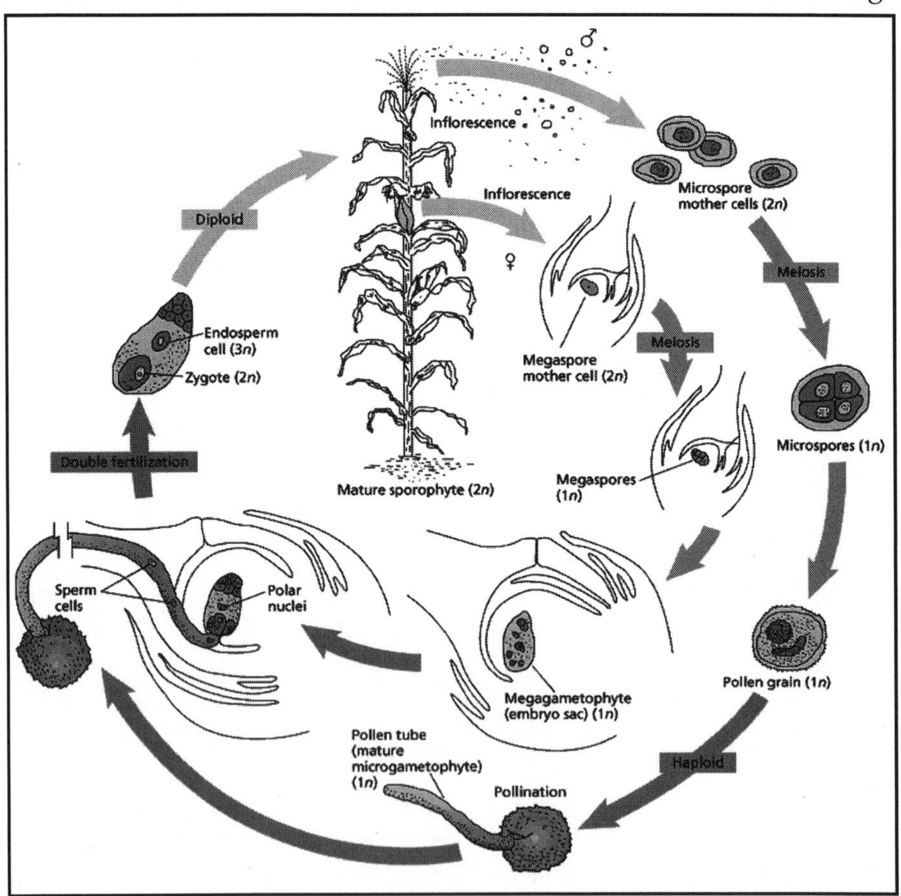

Fig. Life Cycle of corn (*Zea mays*), a monocot.

There are two categories of seed plants: gymnosperms (from the Greek for "naked seed") and angiosperms (based on the Greek for "vessel seed," or seeds contained in a vessel). Gymnosperms are the less advanced type; about 700 species of gymnosperms are known. The largest group of gymnosperms is the conifers ("cone-bearers"), which include such commercially important forest trees as pine, fir, spruce, and redwood.

Two types of cones are present: male cones, which produce pollen, and female cones, which bear ovules. The ovules are located on the surfaces of specialized structures called cone scales. After wind-mediated pollination, the sperm reaches the egg via a pollen tube, and the fertilized egg develops into an embryo. Upon maturation, the cone scales, which are appressed during early development, separate from each other, allowing the naked seeds to fall to the ground.

Angiosperms, the more advanced type of seed plant, first became abundant during the Cretaceous period, about 100 million years ago. Today, they dominate the landscape, easily outcompeting their cousins, the gymnosperms. About 250,000 species are known, but many more remain to be characterized. A typical angiosperm life cycle, that of *Zea mays* (corn), is shown in Figure. The major innovation of the angiosperms is the flower; hence they are referred to as *flowering plants.* There are other anatomical differences between angiosperms and gymnosperms, but none so crucial and far-reaching as the mode of reproduction.

The vegetative plant represents the diploid sporophyte generation. Meiosis occurs in the male and female flowers, represented by the tassels and ears, respectively. The haploid microspores (male spores) develop into pollen grains, and the single surviving haploid megaspore (female spore) divides mitotically to form the embryo sac (megagametophyte). The egg forms in the embryo sac.

Pollination leads to the formation of a pollen tube containing two sperm cells (the microgametophyte). Finally, double fertilization results in the formation of the diploid zygote, the first stage of the new sporophyte generation, and the triploid endosperm cell.

FLOWER STRUCTURE AND THE ANGIOSPERM LIFE CYCLE

The flower consists of several leaflike structures attached to a specialized region of the stem called the receptacle. Sepals and petals are the most leaflike. Petals have the primary function of attracting insects to serve as pollinators, accounting for their often showy and brightly colored appearance.

The stamen is the male sexual structure, and the pistil is the female sexual structure. The pistil is composed of one or more united carpels; the pistil, or in some flowers a whorl of pistils, is sometimes referred to as the gynoecium.

The stamen consists of a narrow stalk called the filament and a chambered structure called the anther. The anther contains tissue that gives rise to pollen grains. The pistil consists of the stigma (the tip where pollen lands during pollination), the style (an elongated structure), and the ovary. The ovary, the hollow basal portion of the pistil, completely encloses one or more ovules. Each ovule, in turn, contains an embryo sac, the structure that gives rise to the female gamete, the egg.

After landing on the stigma, the pollen grain germinates to form a long pollen tube, which penetrates the tissues of the style and ultimately enters the cavity of the ovary, which houses the ovule. Within the ovary, the pollen tube enters the ovule and deposits two haploid sperm cells in the embryo sac.

One sperm cell fuses with the egg to produce the zygote; the other typically fuses with the two polar nuclei to produce a specialized storage tissue termed the endosperm, which provides nutrients to the growing embryo. Endosperm tissue also provides the bulk of the worldrquote s food supply in the form of cereal grains. As in conifers, in angiosperms the outer tissues of the ovule harden into a protective seed coat. Angiosperm seeds have a second layer of protective tissues, the fruit. The fruit consists of the ovary wall and, in some cases, receptacle tissue.

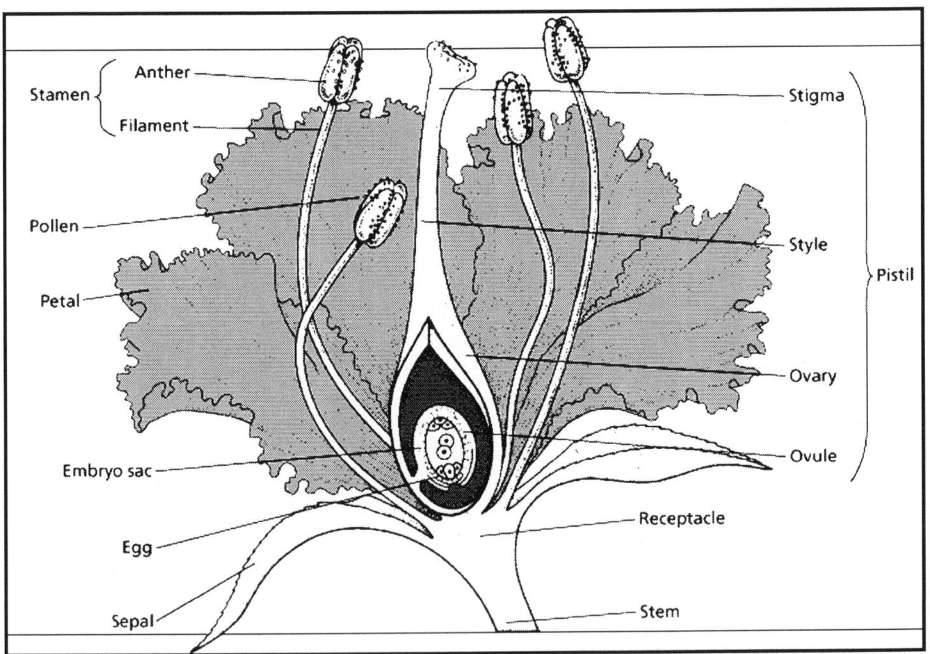

Fig. Schematic Representation of an Idealized flower of the Angiosperms.

Angiosperms are divided into two major groups, dicotyledons (dicots) and monocotyledons (monocots). This distinction is based primarily on the number of cotyledons, or seed leaves. In addition, the two groups differ with respect to other anatomical features, such as the arrangement of their vascular tissues, and their floral structure. As the dominant plant group on Earth, and because of their great economic and agricultural importance, angiosperms have been studied much more intensively than other types of plants, and they are discussed extensively in this book. Plant physiologists have focused on a relatively small number of species that represent convenient experimental systems for the study of specific phenomena. Therefore, while we focus on these famous few, it is important to keep in mind the tremendous diversity of form and function that exists within the angiosperms, and the even greater diversity of form and function that is found within the plant kingdom as a whole.

PLANT TISSUE SYSTEMS: DERMAL, GROUND, AND VASCULAR

Dermal tissue. The epidermis is the dermal tissue of young plants undergoing primary growth. It is generally composed of specialized, flattened polygonal cells that occur on all plant surfaces. Shoot surfaces are usually coated with a waxy cuticle to prevent water loss and are often covered with hairs, or trichomes, which are epidermal cell extensions. Pairs of specialized epidermal cells, the guard cells, are found surrounding microscopic pores in all leaves. The guard cells and pores are called stomata (singular stoma), and they permit gas exchange (water loss, CO_2 uptake, and O_2 release or uptake) between the atmosphere and the interior of the leaf.

The root epidermis is adapted for absorption of water and minerals, and its outer wall surface typically does not have a waxy cuticle. Extensions from the root epidermal cells, the root hairs, increase the surface area over which absorption can take place. Ground tissue. Making up the bulk of the plant are cells termed the ground tissue. There are three types of ground tissue: parenchyma, collenchyma, and sclerenchyma.

- Parenchyma, the most abundant ground tissue, consists of thin-walled, metabolically active cells that carry out a variety of functions in the plant including photosynthesis and storage
- Collenchyma tissue is composed of narrow, elongated cells with thick primary walls. Collenchyma cells provide structural support to the growing plant body, particularly shoots, and their thickened walls are nonlignified so they can stretch as the organ elongates. Collenchyma cells are typically arranged in bundles or layers near the periphery of stems or leaf petioles.
- Sclerenchyma consists of two types of cells, sclereids and fibers. Both have thick secondary walls and are frequently dead at maturity. Sclereids occur in a variety of shapes, ranging from roughly spherical to branched, and are widely distributed throughout the plant. In contrast, fibers are narrow, elongated cells that are commonly associated with vascular tissues. The main function of sclerenchyma is to provide mechanical support particularly to parts of the plant that are no longer elongating.

In the stem, the pith and the cortex make up the ground tissue. The pith is located within the cylinder of vascular tissue, where it often exhibits a spongy texture because of the presence of large intercellular air spaces. If the growth of the pith fails to keep up with that of the surrounding tissues, the pith may degenerate, producing a hollow stem. In general, roots lack piths, although there are exceptions to this rule. In contrast, the cortex, which is located between the epidermis and the vascular cylinder, is present in both stems and roots.

At the boundary between the ground tissue and the vascular tissue in roots, and occasionally in stems, is a specialized layer of co rtex known as the endodermis. This single layer of cells originates from cortical tissue at the innermost layer of the root cortex and forms a cylinder that surrounds the central vascular tissue, or stele. Early in root development, a narrow band composed of the waxy substance suberin is formed in the cell walls circumscribing each endodermal cell. These suberin deposits, called Casparian strips, form a barrier in the endodermal walls to the intercellular movement of water, ions, and other water-soluble solutes to the vascular cells. Leaves have two interior layers of ground tissue that are collectively known as the mesophyll. The palisade parenchyma consists of closely spaced, columnar cells located beneath the upper epidermis. There is usually one layer of palisade parenchyma in the leaf. Palisade parenchyma cells are rich in chloroplasts and are a primary site of photosynthesis in the leaf. Below the palisade parenchyma are i rregularly shaped, widely spaced spongy mesophyll cells. The spongy mesophyll cells are also photosynthetic, and the large spaces between these cells allow diffusion of carbon dioxide. The spongy mesophyll also contributes to leaf flexibility in the wind, and this flexibility facilitates the movement of gases within the leaf. Vascular tissues: xylem and phloem. The vascular tissue is composed of two major conducting systems: the xylem and the phloem. The xylem transports water and mineral ions from the root to the rest of the plant. The phloem distributes the products of photosynthesis and a variety of other solutes throughout the plant. The tracheids and vessel elements are the conducting cells of the xylem. Both of these cell types have elaborate secondary-wall thickenings and lose their cytoplasm at maturity; that is, they are dead when functional. Tracheids overlap each other, whereas vessel elements have open end walls and are arranged end to end to form a larger unit called a vessel. Other cell types present in the xylem include parenchyma cells, which are important for the storage of energy-rich molecules and phenolic compounds, and sclerenchyma fibers.

The sieve elements and sieve cells are responsible for sugar translocation in the phloem. The former are found in angiosperms; the latter perform the same function in gymnosperms. Like vessel elements, sieve elements are often stacked in vertical rows, forming larger units called sieve tubes, whereas sieve cells form overlapping arrays. Both types of conducting cells are living when functional, but they lack nuclei and central vacuoles and have relatively few cytoplasmic organelles. Substances are translocated from sieve cell to sieve cell laterally through circular or oval zones containing enlarged pores, called sieve areas. In contrast, sieve tubes translocate substances through large pores in the end walls of the sieve elements, called sieve plates. Sugar movement through sieve tubes is more efficient and rapid than through sieve cells and represents a more evolutionarily advanced mechanism. Sieve elements are associated with, and depend on, densely cytoplasmic parenchyma cells called companion cells. The analogous cells adjacent to the sieve cells of gymnosperms are called albuminous cells. Companion cells provide proteins and metabolites necessary for the functions of the sieve tube elements. In addition, the phloem frequently contains storage parenchyma and fibers that provide mechanical support.

THE STRUCTURES OF CHLOROPLAST GLYCOSYLGLYCERIDES

The structures of the three plant glycosylglycerides, monogalactosyldiacylglycerol (MGDG), digalactosyldiacylglycerol (DGDG) and sulfoquinovosyldiacylglycerol (SQDG), are shown in Figure. The two galactolipids, MGDG and DGDG, are uncharged but polar; the sulfolipid, SQDG, is negatively charged as well as being polar (Harwood 1997).

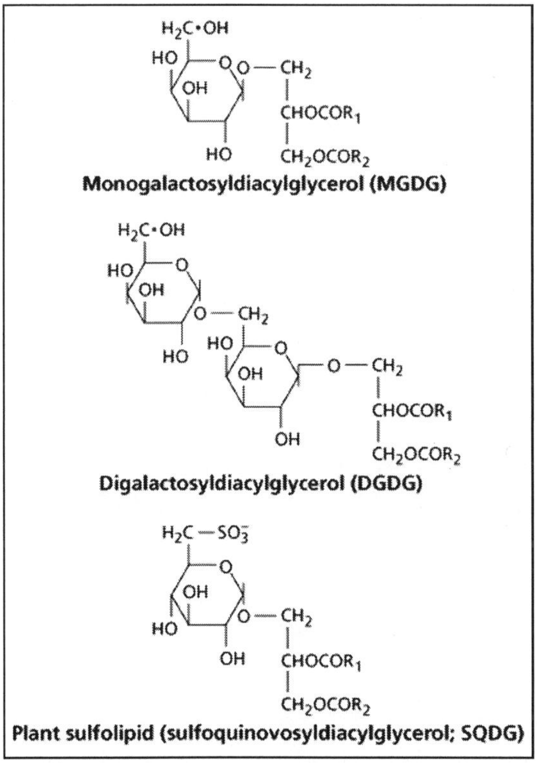

Monogalactosyldiacylglycerol (MGDG)

Digalactosyldiacylglycerol (DGDG)

Plant sulfolipid (sulfoquinovosyldiacylglycerol; SQDG)

Fig. The Structures of Chloroplast Glycosylglycerides.

MGDG and DGDG are galactolipids; SQDG is a sulfolipid. Note that the two galactolipids are uncharged, while the sulfolipid is charged.

THE MULTIPLE STEPS IN CONSTRUCTION OF THE CELL PLATE FOLLOWING MITOSIS

During Mitosis in plants, the daughter chromosomes separate during anaphase and migrate toward their respective poles. In contrast to animal cells, the spindle poles of plant cells is quite diffuse, so that the chromosomes move in nearly parallel fashion.

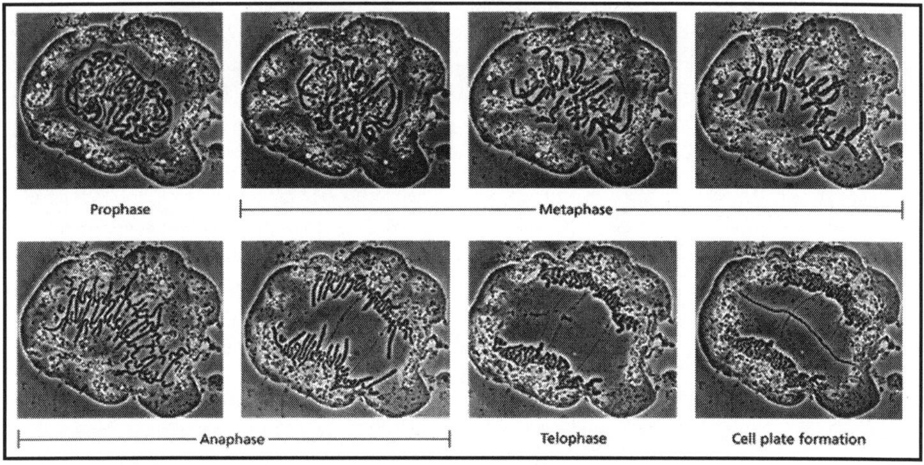

Fig. Mitosis in a living Endosperm cell of the African blood lily, *Haemanthus Katharinae*, Recorded with time-lapse Microscopy.

At the end of teleophase, cytokinesis begins with the formation of the cell plate. This process involves fusion of many small secretory vesicles, and attachment of the resulting structure to the plasma membrane. A diagram of the steps involved in cell plate formation is shown in Figure.

This structure is called the fusion tube network (FTN). The transition from the first to the second stage of cell plate formation can be inhibited by caffeine. (B) Formation of a tubulo-vesicular network (TVN). The contents of the vesicles, mainly pectins, represent the precursors from which the new middle lamella is assembled outside the cell. In the next stage, vesicle fusion increases, forming a tubulo-vesicular network (TVN), and the membranes become coated with either clathrin or other proteins. (C) In the third stage, the central region of the growing cell plate forms a tubular network (TN), with vesicle fusion occurring at the growing edges where the remaining microtubules are located. (D) In the final stage, the cell plate contacts and adheres to the plasma membrane of the parent cell. At the same time the tubular network expands to form a fenestrated sheet. (E) At the end of mitosis, the phragmoplast disappears, the cell enters interphase, and microtubules reappear in the cytosol near the plasma membrane, where they play a role in the deposition of cellulose microfibrils during cell wall growth. (F) Electron micrograph of a cell plate forming in a root tip of a beet, (*Beta vulgaris*) (10,000×) MT, microtubule; VE, vesicles; N, nucleus; NE, nuclear envelope; P, cell plate.

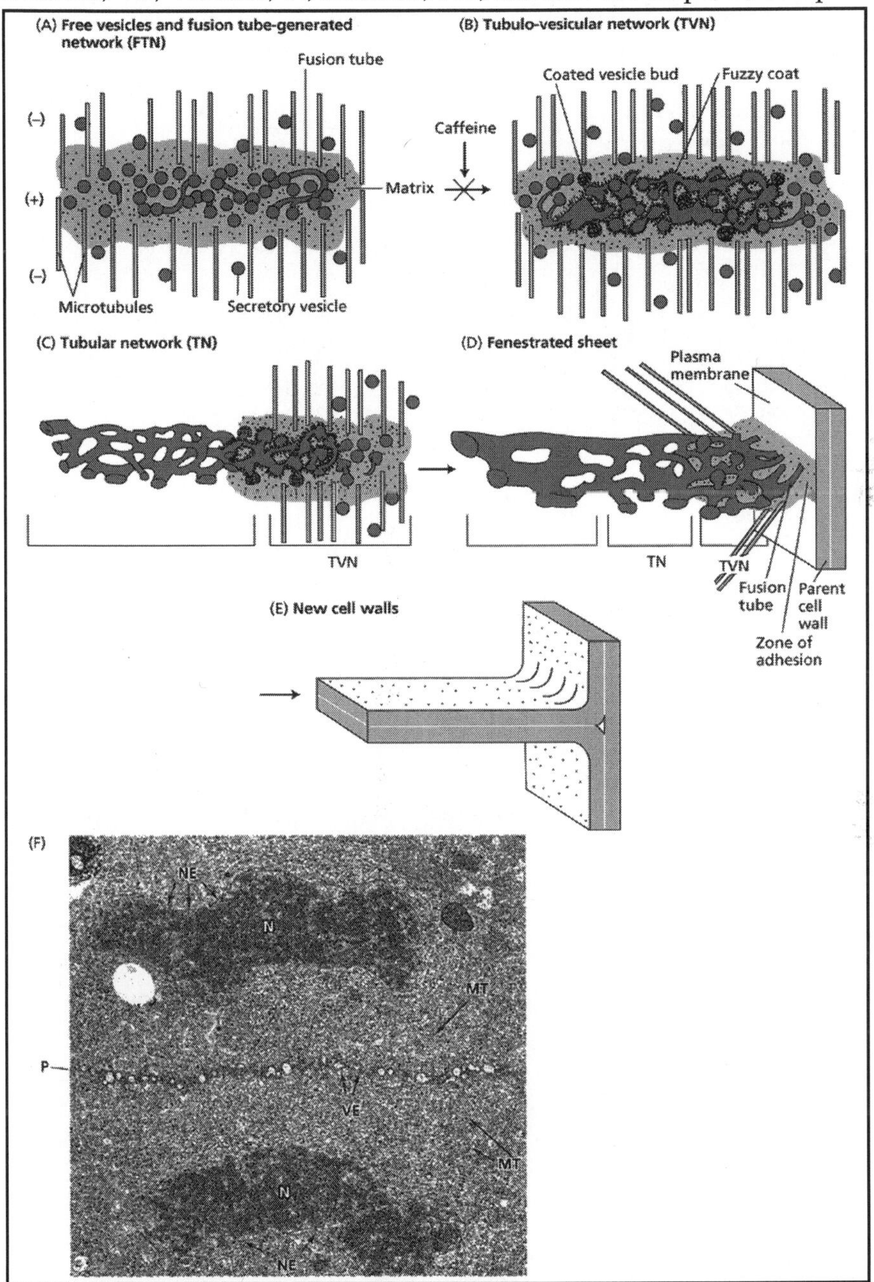

Fig. (A) In the first step, Golgi Vesicles, some of which are Interconnected via *fusion tubes,* Aggregate in the Spindle Midzone Area.

PLANT CELL MODEL AND PARTS

Plant cells are the cells found in plants comprising sub cellular organelles. Following is a plant cell model for kids that will give your child a visual of the plant cell and a better understanding of its various parts.

The cells are the basic units of life that work together to perform life sustaining functions in both animal and plant worlds. Plant cells are eukaryotic cells having thick and rigid cell walls.

Eukaryotic cells are cells that contain complex structures enclosed within membranes called nucleus or nucleus envelop, within which the genetic material is present. These cells are present in almost all living organisms including animals, plants and fungi.

Plant cells differ from animal cells in that they have three different structures known as cell wall, vacuoles and plastids. However, they lack centrioles and intermediate filaments which are present in animal cells. Read more on plant cell structure and parts.

Cell Membrane

It permits the waste material to exit the cell and regulates the movement of materials in and out of the cell. It also acts as a barrier between the inside of the cell and the outside, so that the chemical conditions on both sides can be different.

Cell Wall

It provides rigid structural support to the cell from which structures, like, leaves and stems are produced. It allows the circulation and distribution of water, minerals and nutrients in and outside the cell. The wall also controls the presence of pathogen microbes and development of tissues in the cell.

Golgi Body

It stores, packages, and circulates the lipids and proteins made in the endoplasmic reticulum. It stores the proteins into packages called vesicles.

Rough Endoplasmic Reticulum

It synthesizes and exports proteins and glycoproteins throughout the cell.

Lysosomes

These organelles digest food by breaking down larger molecules into smaller ones, using special proteins.

Cytoplasm

It helps in maintaining the cell shape. Along with this, it plays an important role in the internal movement of cell organelles, cell mobility and muscle fibre contraction. It also distributes oxygen and nutrients to different parts of the cell.

Nucleolus

It is the most prominent structure in the nucleus wherein ribosomes are made.

Vacuole

The plant cell contains a large, single vacuole (an enclosed compartment) which is used to store water, compounds and minerals that help in plant growth.

Ribosomes

These are packets of RNA (Ribonucleic Acid) and are called as protein builders or synthesizers of the cell.

Chloroplasts

They are the food producers of the cell, containing chlorophyll, the green pigment essential for photosynthesis. Their main function is to generate sugars and starches.

Nucleus

It's the most important part of the cell, comprising chromosomes, i.e. structures made up of genetic information that helps in cell growth and reproduction. Read more on cell nucleus: structure and functions.

Nuclear Envelope

It's an enclosure that surrounds nucleus and its contents. Unlike cell membrane which has pores and spaces for RNA and proteins to pass through, it keeps the chromatin and nucleolus inside the nucleus.

Smooth Endoplasmic Reticulum

It's main function is to package proteins for transport, synthesize membrane phosolipids, and secrete calcium. It also performs transformation of bile pigments, glycogenolysis (the breakdown of glycogen), and detoxification of different drugs and chemical agents.

Mitochondria

It provides energy to the cell by combining sugar molecules with oxygen to generate carbon dioxide and water.

Amylosplast

It's an organelle present in some plant cells that stores starch.

Druse Crystal

It's a granular type of crystal found in plant vacuoles. It is composed of calcium oxalate and is considered to deter herbivory.

Centrosome

Also known as Microtubule Organizing Centre, it's a region in the cell where microtubules are produced that perform a variety of functions ranging from transportation to structural support. Though both plant and animal cell centrosomes play similar roles during cell division, plant cell centrosomes are simpler and do not contain centrioles.

Peroxisomes

These are membrane bound packets of oxidative enzymes that convert fatty acids into sugar and assist chloroplasts in photo-respiration.

Golgi Vesicles

These vesicles transfer proteins and lipids to different parts of the cells.

PLANT CELL MODEL

Following are some simple instructions or plant cell model ideas, to construct a model of a plant cell:

- Enumerate and collect all possible materials required for the model and look out for the different cell parts in a biology textbook.
- Use a heavy sturdy wooden base to construct the base of the model so that it can hold the weight and facilitate presentation of the model.
- Construct the cell nucleus and ensure that it's round in shape. Paint the nucleus in one colour so that it can be identified as a distinct cell part.
- With the help of a dowel rod or thick, straight stick, connect the nucleus to the base of the model. For this, drill a hole into the wooden base, stick the dowel rod into the hole and connect the nucleus to the rod.
- The cell wall and cell membrane of the plant cell model should be rectangular in shape. The outer wall is the cell wall which should be made of a strong material like hard plastic or wood and the cell membrane can be made of cellophane or thin plastic. Now attach both the walls to the nucleus and the model base.
- Add various other key components of the plant cell to the model. Like, for chlorophyll, use a piece of green fabric stuffed with cotton. Sew it and attach it to the model.
- Finally, prepare a report on the various plant cell parts and their corresponding functions.

CELLULAR TOTIPOTENCY

Totipotency is the ability of all living cells potentially to regenerate whole new individuals - the trick is how to trigger them to do it and so to demonstrate it. If I remember correctly, Dolly the sheep was cloned from a cell from another sheep's udder - a differentiated cell, which in normal circumstances would have remained as such. Similarly in plants, it is possible to trigger single epidermal cells in oilseed rape, sunflower, and I am sure there are others, to develop into embryos and subsequently to whole plants. Even immature pollen grains have this potential - these immature cells can be switched to develop into whole plants, even though they only have one set of chromosomes, and give rise to haploid plants, although sometimes these may diploidise during culture, to generate homozygous diploid plants. One of the hardest things in plant tissue culture is to get tissues from mature trees to illustrate totipotency, but by forcing a rejuvenation step, it is even possible to do this with some species - I believe quite a bit of work has been carried out on pines. Simple cuttings, and the SAPS tissue culture system depends on this ability of cells to switch their development - hence root formation from hypocotyls, and root formation from woody cuttings, which allow new plants to be obtained for propagation.

We described the different media used for cell/tissue culture, and also discussed culture of individual differentiated cells. However, in plant tissue culture experiments, unless single cell cultures are absolutely necessary, more often we use explants (pieces of differentiated tissues) to initiate their growth in culture. This is also necessary to obtain callus from which single cell cultures are derived. During the last few decades, even for development of new crops or for improvement in the characteristics of existing crops (e.g. to obtain disease free crop, etc.), the techniques of tissue culture in combination with genetic manipulation have become important tools.

For instance, as an alternative to vegetative propagation, shoot tip propagation was used commercially for rapid and consistent reproduction of elite or difficult to propagate genotypes., When an explant from differentiated tissue is used for culture on a nutrient medium, the non-dividing, quiescent cells first undergo certain changes to achieve a meristematic state. The phenomenon of the reversion of mature cells to the meristematic state leading to the formation of callus is called dedifferentiation. The component cells of callus have the ability to form a whole plant, a phenomenon described as redifferentiation.

These two phenomenons of dedifferentiation and redifferentiation are inherent in the capacity described as cellular totipotency, a property found only in plant cells and not in animal cells.In other words, while a differentiated plant cell retains its capacity to give rise to a whole plant, an animal cell loses this capacity of regeneration after differentiation.

Although, generally a callus phase is involved before the cells can undergo redifferentiation leading to regeneration of a whole plant, but rarely the dedifferentiated cells can give rise to whole plants directly without an intermediate callus phase. These aspects of tissue culture and its application in clonal propagation and in the general on of hereditary variation called somaclonal variation.

Some plants have cells that are able to convert from a structured, or differentiated, state to a totipotent state. One example is the ZZ plant (Zamioculcas zamiifolia). This plant is so cellularly precocious that a mere piece of leaf or stem can be used to grow a new plant. Other plants and trees can produce spontaneous buds on the stem or trunk. These emerge from the activity of the meristematic tissue of the cambium layer, and since they are not buds that are normally produced in the axils of leaves, they are called adventitious buds. This can happen even when the tree in question is not the kind that will root from cuttings. The formation of adventitious buds begins with the production of a callus, which is essentially an undifferentiated lump of cells resembling a small tumor. Through genetic and hormonal processes, cells differentiate into growing points which then produce new stems with leaves. A similar process can result in adventitious roots as well.

Advances in the understanding of totipotency and how cells can be stimulated to become totipotent has resulted in an entire industry, the business of micropropagation, or what is commonly referred to as "tissue culture". While micropropagation labs do culture tissue (meristematic tissue), this work is not the same as what is done in a scientific tissue culture laboratory. The work done in scientific labs is much more esoteric and amazing. I studied the science of tissue culture when I attended graduate school, and we did things like isolating single cells from a callus and removing the cell wall using enzymes to yield a naked plant cell called a protoplast.

These are important in scientific research because genes are more easily inserted into plant cells when they don't have a cell wall around them. We grew pure callus cultures and cell lines that were, essentially, all totipotent cells. Using the right combinations of plant hormones, we could then cause the cells to turn into plant shoots with roots. These and other experiments done in scientific tissue culture labs are not the kinds of activities engaged in by micropropagation operations, but they do demonstrate vividly the power of totipotency in plant growth. By contrast, commercial tissue culture is the business of producing many thousands of clones of particular plants using meristematic tissue. Different plants require different formulations of media, as the formula that is ideal for one plant will produce no results with some other plants. Conventional micropropagation requires a sterile or aseptic environment and the operating costs of such a facility are substantial. However, without totipotency, none of this would be possible at any cost.

WATER AND PLANT CELLS

CALCULATING CAPILLARY RISE

The water properties of cohesion, adhesion and surface tension, give rise to the phenomenon of capillarity, the movement of water for small distances up a capillary tube. The smaller the tube radius, the higher the capillary rise. How far the water will move may be calculated using the following formula:

$$\text{Capillary rise} = \frac{14.9 \times 10^{-6}\text{m}^2}{\text{radius}}$$

where both capillary rise and radius are expressed in meters.

For a xylem vessel with 25 µm radius, the capillary rise is about 0.6 m. This distance is much too small to be significant for water transport up tall trees.

Fibrous materials such as cell walls can act like wicks to draw water by capillarity from nearby xylem. This capillary action ensures that cell wall surfaces that are directly exposed to the air, such as those in leaf mesophyll, remain wetted and do not dry out. Because the cell wall capillaries have a tiny radius, about 10 m^{-8}, very large physical forces can be generated in the water just below the evaporative surfaces of cell walls.

CALCULATING HALF-TIMES OF DIFFUSION

Diffusion Is Rapid over Short Distances but Extremely Slow over Long Distances

From Fick's law, one can derive an expression for the time it takes for a substance to diffuse a particular distance. If one defines conditions such that all the solute molecules are concentrated at the starting position, then the concentration front moves away from the starting position, as shown for a later time point in Figure. As the substance diffuses away from the starting point, the concentration gradient becomes less steep (Δc_s decreases) and thus net movement becomes slower. The time it takes for the substance at any given distance from the starting point to reach one-half of the concentration at the starting point ($t_c = ½$) is given by the following equation:

$$t_{c=1/2} = \frac{(\text{distance})^2}{D_s} K$$

where K is a constant and D_s is the diffusion coefficient. The above equation shows that the time required for a substance to diffuse a given distance increases in proportion to the *square* of that distance. Let us consider two numerical examples. First, how long it would take a small molecule to diffuse across a typical cell? The diffusion coefficient for a small molecule like glucose is about 10^{-9} m^2 s^{-1}, and the cell length may be 50 µm. Thus, for this example:

$$t_{c=1/2} = \frac{\left(50 \times 10^{-6}\text{m}\right)^2}{10^{-9}\text{m}^2\text{s}^{-1}} = 2.5 \text{ s}$$

This calculation shows that small molecules diffuse over cellular dimensions rapidly. What about diffusion over longer distances? Calculating the time needed for the same substance to move a distance of 1 m (e.g., the length of a corn leaf), we find:

$$t_{c=1/2} = \frac{(1\text{m})^2}{10^{-9}\text{m}^2\text{s}^{-1}} = 10^9 \text{ s} \approx 32 \text{ years}$$

a value that exceeds by orders of magnitude the life span of a corn plant, which lives only a few months. This shows that diffusion in solutions can be effective within cellular dimensions but is far too slow for mass transport over long distances. Diffusion is of great importance as a driving force for the water vapour lost from leaves, because the diffusion coefficient for a molecule in air is much greater than in aqueous solutions.

Alternative Conventions for Components of Water Potential

Students planning further study of plant water relations should note that the components of water potential defined in the text are sometimes given different names and symbols. In particular, the equation,

$$\Psi_w = \Psi_s = \Psi_p$$

is often replaced by the following equivalent equation:

$$\Psi_w = -\pi + P$$

In this alternative convention, P is the same as Ψ_p. It is the hydrostatic pressure of the solution, and may be positive, as in turgid cells, or negative, as in xylem water. The symbol π is called osmotic pressure and is the negative of Ψ_s. That is, π has positive values, and Ψ_s has negative values. "Osmotic pressure" is the term that physical chemists, zoologists, and many others use to denote the effect of dissolved solutes on the free energy of water. Most handbooks of physics and chemistry use the term "osmotic pressure" and the symbol π. The negative sign in front of π in the equation above accounts for the reduction in water potential (Ψ_w) by dissolved solutes. Thus $\Psi_s = -\pi$. A very interesting, if somewhat unconventional, account of the history and physical meaning of osmotic pressure is given by Hammel and Scholander.

Unfortunately, some authors have mixed the conventions for Ψ_s and π, leading to unnecessary confusion about what is meant by the symbol π. Thus, π is sometimes incorrectly called osmotic potential instead of osmotic pressure, and it may be used either as a positive quantity or as a negative quantity.

Regarding matric potential, it is usually designated by the symbol *t*.

CAN NEGATIVE TURGOR PRESSURES EXIST IN LIVING CELLS?

One of the challenging aspects of understanding plant water relations is the remarkable range of pressures, both positive and negative, that occur within the bodies of plants. Negative pressures (or tensions), which depend upon the cohesive strength of water coupled with the strength of lignified cell walls to resist deformation, play an important role in water transport through the xylem. Positive pressures, which depend upon the semipermeable nature of the plasma membrane and the elastic nature of primary cell walls, occur in all hydrated living plants cells but can be especially large in sieve tubes and guard cells. Living plant cells are typically assumed to have only positive pressures. However, there appears to be no reason that negative pressures could not also occur within the cytoplasm of living plant cells. This topic explores the existence and potential role of negative turgor pressures in plants.

When a cell loses water in air, turgor declines and solute concentrations increase. At the turgor loss point ($\Psi_p = 0$), the hydrostatic pressure in the cell sap is equal to atmospheric pressure, meaning that no net force is exerted on the cell wall. If water continues to be lost from the cell, the pressure within the cytoplasm drops below atmospheric pressure, resulting in a force imbalance that collapses the cell wall. The deformation of living cells upon desiccation is called cytorrhysis. Note that the plasma membrane is pressed against the cell wall throughout desiccation (i.e., plasmolysis does not occur) because the hydrostatic pressure in cytoplasm remains greater than the hydrostatic pressure in apoplast.

In the living cells described above, a decrease in cell water potential below the turgor loss point is balanced by an increase in solute concentration as the volume of the cell decreases. Thus, true "negative" pressures do not develop. In contrast, xylem conduits have rigid cell walls that resist deformation, allowing them to sustain negative pressures without imploding. This raises the question of what happens when water is lost from living cells that have thick walls or are embedded in a rigid matrix of cells (e.g., xylem parenchyma). Might these cells resist deformation and thus develop negative turgor pressures?

It is important to emphasize that this is a controversial area, due in large part to the absence of direct measurements of negative turgor pressures within cells.

However, before reviewing the evidence for negative turgor pressures, it is worthwhile to consider what physiological effects might result from the development of such negative pressures in living cells. The major outcome of negative turgor is that it allows stiff-walled cells to decrease in water potential without undergoing major changes in cell volume or osmotic concentration. Because cytorrhysis might cause physical damage to the wall and/or cell membranes, while the high concentrations of solutes resulting from the reduction in cell volume might adversely affect the conformation of membranes and proteins, it is possible to imagine a physiological role or benefit for the development of negative turgor pressures in plant cells. We now turn this around and ask if there are any downsides to the generation of negative turgor pressures in living cells? In the xylem, the primary issue associated with negative pressures is cavitation; could this also be a concern for living cells? The primary mechanism for cavitation in plants is air seeding, reflecting the fact that the probability of the de novo formation of gas voids in water (either by homogeneous or heterogeneous nucleation) is extremely low.

In air seeding, air is sucked in through the cell wall, where it then "nucleates" the transition to the vapour phase. For air seeding to occur in a living cell experiencing negative pressures, air would have to be pulled through the very small pores of the cell wall. While one can imagine this happening, the movement of air across the cell wall would result in plasmolysis and thus the immediate release of any tension in the cytoplasm because the plasma membrane is not capable of withstanding any significant pressure.

Other costs associated with the ability of living cells to generate negative pressures include the metabolic costs of producing rigid cell walls, as well as any limitations lignified walls might place on physiological function. In addition, a strategy for avoiding cell damage due to desiccation via the generation of negative pressures might impose limitations on cell size. The strength of a cell to withstand collapse (and thus generate negative pressures) is inversely proportional to cell size.

Evidence for negative turgor pressure in plants is limited, in part reflecting the fact that few researchers have devoted much attention to this issue. Living cells with flexible (unlignified) cell walls deform relatively easily upon desiccation and thus do not appear to support negative pressures. However, measurements of the forces needed to collapse cell walls suggest that living cells with thick walls can withstand forces in the range of 1.0 MPa. Visual examination of tissues adapted to withstand very cold temperatures also provides indirect evidence for negative turgor.

For example, frozen ray parenchyma cells often do not exhibit significant deformation, despite the very strong desiccatory effects (low water potential) of extracellular ice. One can hypothesize that the existence of negative pressures within these cells acts to balance the water potential gradient across the cell membrane (i.e., between the cytoplasm and ice formed within the apoplast). In summary, negative turgor pressure remains an intriguing but little-studied area of plant water relations. While it does not appear to form in cells such as the leaf mesophyll, that can deform easily, its existence in living cells with either lignified cell walls or where the cell is embedded in a matrix of lignified tissues (e.g., living cells within wood) cannot be ruled out. The potential benefits of negative turgor pressures in terms of preventing mechanical and osmotic damage associated with severe desiccation makes this topic worthy of further study.

THE MATRIC POTENTIAL

In discussions of soils, seeds, and cell walls, one often finds reference to yet another component of Ψ_w: the matric potential (Ψ_m). The matric potential is used to account for the reduction in free energy of water when it exists as a thin surface layer, one or two molecules thick, adsorbed onto the surface of relatively dry soil particles, cell walls, and other materials.

The matric potential does not represent a new force acting on water, because the effect of surface interactions can theoretically be accounted for by an effect on Ψ_s and Ψ_p. In dry materials, however, this surface interaction effect often cannot easily be separated into Ψ_p and Ψ_s components in dry materials, so they are frequently bulked together and designated as the matric potential.

It is generally not valid to add Ψ_m to independent measurements of Ψ_s and Ψ_p to arrive at a total water potential. This is particularly true for water inside hydrated cells and cell walls, where matric effects are either negligible or they are accounted for by a reduction in Ψ_p. For instance, the negative pressure in water held by cell wall microcapillaries at the evaporative surfaces of leaves, discussed on page 45 of the textbook, is sometimes described as a wall matric potential. Care is needed to avoid inconsistencies when accounting for this physical effect in definitions of Ψ_p, Ψ_s, and Ψ_m.

MEASURING WATER POTENTIAL

Plant scientists have expended considerable effort in devising accurate and reliable methods for evaluating the water status of a plant. Four instruments that have been used extensively to measure Ψ_w, Ψ_s, and Ψ_p are described here: psychrometer, pressure chamber, cryoscopic osmometer, and pressure probe.

Psychrometer (Ψ_w Measurement)

Psychrometry (the prefix "psychro-" comes from the Greek word *psychein*, "to cool") is based on the fact that the vapour pressure of water is lowered as its water potential is reduced. Psychrometers measure the water vapour pressure of a solution or plant sample, on the basis of the principle that evaporation of water from a surface cools the surface.

One psychrometric technique, known as *isopiestic psychrometry*, has been used extensively by John Boyer and coworkers and is illustrated in Figure. Investigators make a measurement by placing a piece of tissue sealed inside a small chamber that contains a temperature sensor (in this case, a thermocouple) in contact with a small droplet of a standard solution of known solute concentration (known Ψ_s and thus known Ψ_w).

If the tissue has a lower water potential than that of the droplet, water evaporates from the droplet, diffuses through the air, and is absorbed by the tissue.

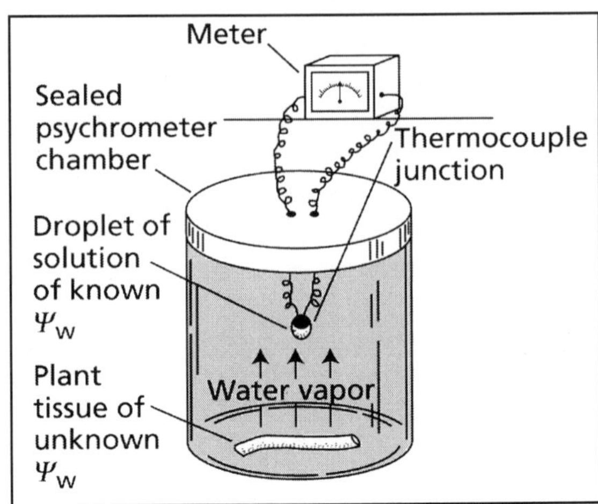

Fig. Diagram illustrating the use of Isopiestic Psychrometry to Measure the Water Potential of a Plant Tissue.

This slight evaporation of water cools the drop. The larger the difference in water potential between the tissue and the droplet, the higher the rate of water transfer and hence the cooler the droplet. If the standard solution has a lower water potential than that of the sample to be measured, water will diffuse from the tissue to the droplet, causing warming of the droplet.

Measuring the change in temperature of the droplet for several s olutions of known Ψ_w makes it possible to calculate the water potential of a solution for which the net movement of water between the droplet and the tissue would be zero signifying that the droplet and the tissue have the same water potential.

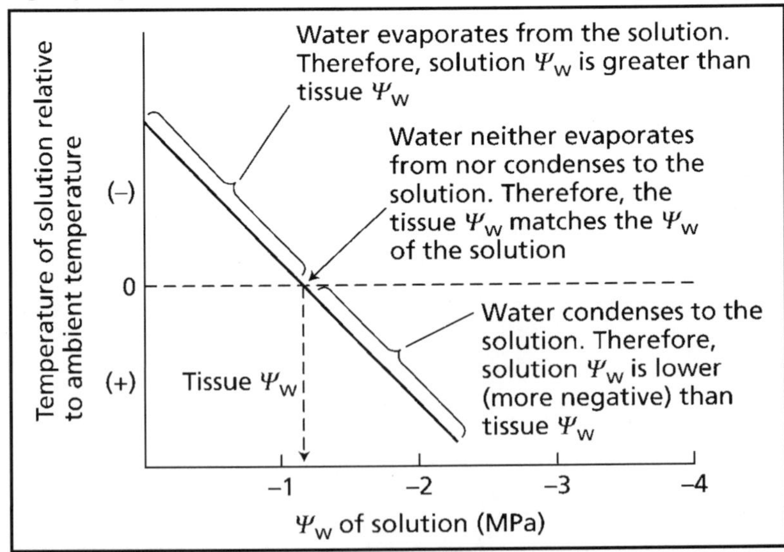

Psychrometers can be used to measure the water potentials of both excised and intact plant tissue. Moreover, the method can be used to measure the Ψ_s

of solutions. This can be particularly useful with plant tissues. For example, the Ψ_w of a tissue is measured with a psychrometer, and then the tissue is crushed and the Ψ_s value of the expressed cell sap is measured with the same instrument. By combining the two measurements, researchers can estimate the turgor pressure that existed in the cells before the tissue was crushed ($\Psi_p = \Psi_w - \Psi_s$).

A major difficulty with this approach is the extreme sensitivity of the measurement to temperature fluctuations. For example, a change in temperature of 0.01°C corresponds to a change in water potential of about 0.1 MPa. Thus, psychrometers must be operated under constant temperature conditions. For this reason, the method is used primarily in laboratory settings.

Pressure Chamber (Ψ_w Measurement)

A relatively quick method for estimating the water potential of large pieces of tissues, such as leaves and small shoots, is by use of the pressure chamber. This method was pioneered by Henry Dixon at Trinity College, Dublin, at the beginning of the twentieth century, but it did not come into widespread use until P. Scholander and coworkers at the Scripps Institution of Oceanography improved the instrument design and showed its practical use.

In this technique, the organ to be measured is excised from the plant and is partly sealed in a pressure chamber. Before excision, the water column in the xylem is under tension. When the water column is broken by excision of the organ (i.e., its tension is relieved allowing its Ψ_p to rise to zero), water is pulled rapidly from the xylem into the surrounding living cells by osmosis. The cut surface consequently appears dull and dry. To make a measurement, the investigator pressurizes the chamber with compressed gas until the distribution of water between the living cells and the xylem conduits is returned to its initial, pre-excision, state.

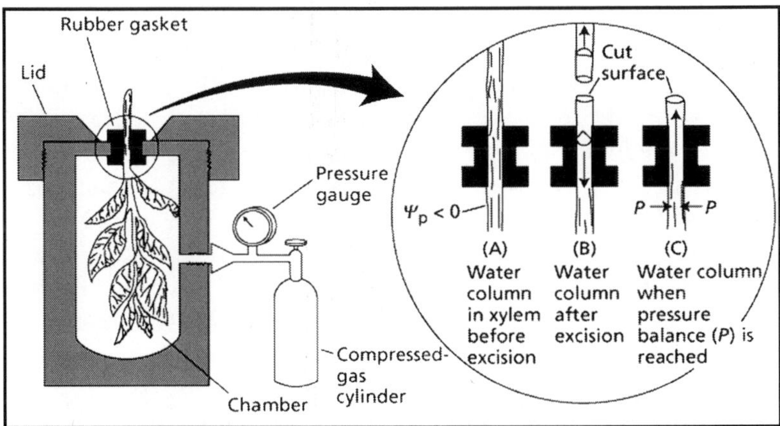

Fig. The Pressure Chamber Method for Measuring plant water Potential.

This can be detected visually by observing when the water returns to the open ends of the xylem conduits that can be seen in the cut surface. The pressure needed to bring the water back to its initial distribution is called the *balance pressure* and is readily detected by the change in the appearance of the cut surface, which becomes wet and shiny when this pressure is attained.

The diagram at left shows a shoot sealed into a chamber, which may be pressurized with compressed gas. The diagrams at right show the state of the water columns within the xylem at three points in time: (A) The xylem is uncut and under a negative pressure, or tension. (B) The shoot is cut, causing the water to pull back into the tissue, away from the cut surface, in response to the tension in the xylem. (C) The chamber is pressurized, bringing the xylem sap back to the cut surface.

The pressure chamber is often described as a tool to measure the tension in the xylem. However, this is only strictly true for measurements made on a non-transpiring leaf or shoot (for example, one that has been previously enclosed in a plastic bag). When there is no transpiration, the water potential of the leaf cells and the water potential in the xylem will come into equilibrium.

The balancing pressure measured on such a non-transpiring shoot is equal in magnitude but opposite in sign to the pressure in the xylem (Ψ_p). Because the water potential of our non-transpiring leaf is equal to the water potential of the xylem, one can calculate the water potential of the leaf by adding together Ψ_p and Ψ_s of the xylem, provided one collects a sample of xylem sap for determination of Ψ_s. Luckily Ψ_s of the xylem is usually small (> -0.1 MPa) compared to typical midday tensions in the xylem (Ψ_p of –1 to –2 MPa). Thus, correction for the Ψ_s of the xylem sap is frequently omitted.

Balancing pressure measurements of transpiring leaves are more difficult to interpret. The fact that water is flowing from the xylem to the leaf means that differences in water potential must exist. When the transpiring leaf or shoot is cut off, the tension in the xylem is instantly relieved and water is drawn into the leaf cells until the water potentials of the xylem and the leaf cells come into equilibrium.

Because the total volume of the leaf cells is much larger than the volume of sap in the xylem, this equilibrium water potential will be heavily weighted towards that of the leaf. Thus, any measurement of the balancing pressure on such a leaf or shoot will result in a value that is approximately the water potential of the leaf, rather than the tension of the xylem. (To be exact, one would have to add the Ψ_s of the xylem sap to the negative of the balancing pressure to get the leaf water potential.) One can explore the differences between the water potential of the xylem and the water potential of a transpiring leaf by comparing balancing pressures measured on covered (i.e., non-transpiring) versus uncovered (transpiring) leaves.

Pressure chamber measurements provide a quick and accurate way of measuring leaf water potential. Because the pressure chamber method does not require delicate instrumentation or temperature control, it has been used extensively under field conditions. For a more complete description of the theory and operation of the pressure chamber.

Cryoscopic Osmometer (Ψ_s Measurement)

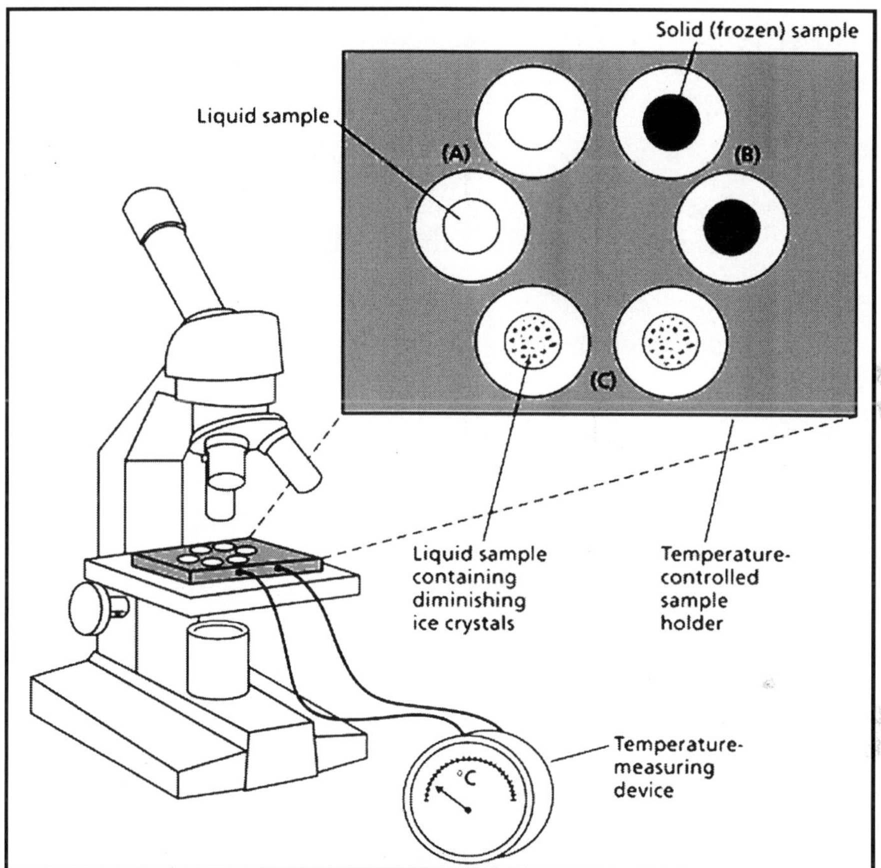

Fig. A Cryoscopic Osmometer Measures the Concentration of total dissolved solutes by Measuring the Freezing-point Depression of a Solution.

(A) Very small liquid samples are loaded onto the temperature-controlled stage of a microscope. (B) When the temperature is quickly reduced, the samples supercool and freeze. (C) Slowly warming the stage causes the samples to thaw. The temperature at which the last ice crystal melts provides a measure of the melting point of the sample.

The cryoscopic osmometer measures the osmotic potential of a solution by measuring its freezing point. Solutions have *colligative* properties that collectively depend on the number of dissolved particles and not on the nature of the solute. For example, solutes reduce the vapour pressure of a solution, raise its boiling point, and lower its freezing point. The specific nature of the solute does not matter. One of the colligative *properties* of solutions is the decrease in the freezing point as the solute concentration increases. For example, a solution containing 1 mol of solutes per kilogram of water has a freezing point of –1.86°C, compared with 0°C for pure water.

Various instruments can be used to measure the freezing-point depression of solutions. With a cryoscopic osmometer, solution samples as small as 1 nanoliter (10^{-9} L) are placed in an oil medium located on the temperature-controlled stage of a microscope.

The very small sample size allows sap from single cells to be measured and permits rapid thermal equilibration with the stage. To prevent evaporation, the investigator suspends the samples in oil-filled wells in a silver plate (silver has high thermal conductivity). The temperature of the stage is rapidly decreased to about –30° C, which causes the sample to freeze.

The temperature is then raised very slowly, and the melting process in the sample is observed through the microscope. When the last ice crystal in the sample melts, the temperature of the stage is recorded (note that the melting and freezing points are the same). It is straightforward to calculate the solute concentration from the freezing-point depression; and from the solute concentration (c_s), Ψ_s is calculated as $-RTc_s$.This technique has been used to measure droplets extracted from single cells

Pressure Probe (Ψ_p Measurement)

If a cell were as large as a watermelon or even a grape, measuring its hydrostatic pressure would be a relatively easy task. Because of the small size of plant cells, however, the development of methods for direct measurement of turgor pressure has been slow.

Using a micromanometer, Paul Green at the University of Pennsylvania developed one of the first direct methods for measuring turgor pressure in plant cells. In this technique, an air-filled glass tube sealed at one end is inserted into a cell.

The high pressure in the cell compresses the trapped gas, and from the change in volume one can readily calculate the pressure of the cell from the ideal gas law (pressure × volume = constant). This method works only for cells of relatively large volume, such as the giant cell of the filamentous green alga *Nitella.* For smaller cells, the loss of cell sap into the glass tube is sufficient to deflate the cell and this yields artifactually low pressures.

For higher plant cells, which are several orders of magnitude smaller in volume than *Nitella,* a more sophisticated device, the pressure probe, was developed by Ernest Steudle, Ulrich Zimmermann, and their colleagues in Germany. This instrument is similar to a miniature syringe. A glass microcapillary tube is pulled to a fine point and is inserted into a cell.

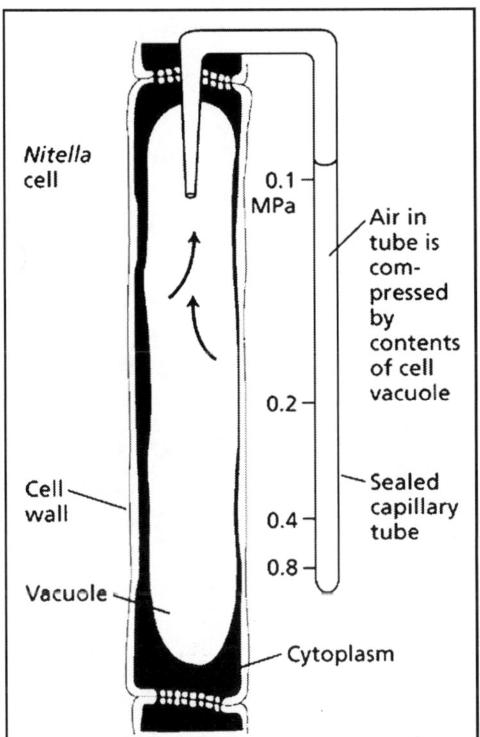

Fig. Use of the micromanometer, a pressure probe, to measure cell turgor pressure. *Nitella* cells (which are particularly large—about 100 mm in diameter and many centimeters long) were used for these measurements.

The microcapillary is filled with silicone oil, a relatively incompressible fluid that can be readily distinguished from cell sap under a microscope.

When the tip of the microcapillary is first inserted into the cell, cell sap begins to flow into the capillary because of the initial low pressure of that region. Investigators can observe such movement of sap under the microscope and counteract it by pushing on the plunger of the device, thus building up a pressure. In such fashion the boundary between the oil and the cell sap can be pushed back to the tip of the microcapillary. When the boundary is returned to the tip and is held in a constant position, the initial volume of the cell is restored and the pressure inside the cell is exactly balanced by the pressure in the capillary. This pressure is measured by a pressure sensor in the device. Thus the hydrostatic pressure of individual cells may be measured directly.

The primary advantage of this method over the one shown in Figure is that cell volume is minimally disturbed. Minimal disturbance is of great importance for the tiny cells that are typical of higher plants, in which loss of even a few picoliters (10^{-12} L) of fluid can substantially reduce turgor pressure.

This method has been used to measure Ψ_p and other parameters of water relations in cells of both excised and intact tissues of a variety of plant species.

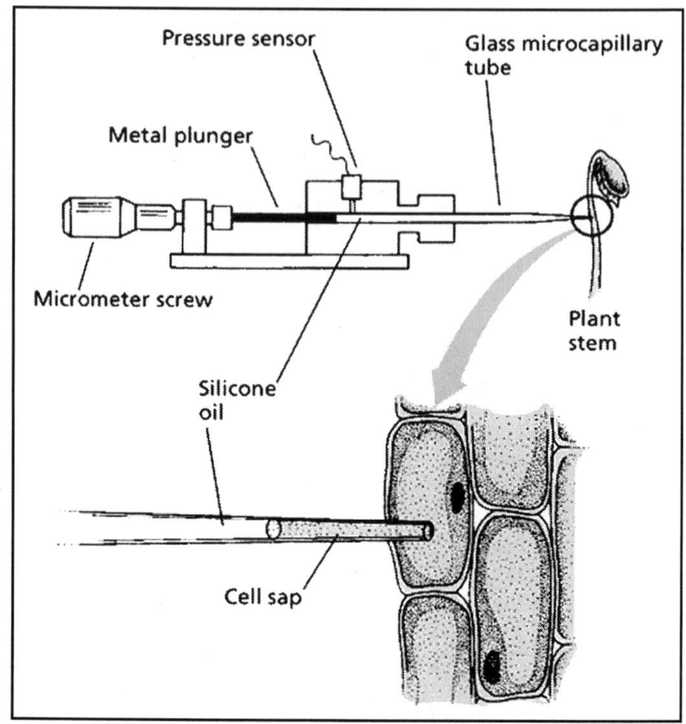

Fig. Diagram of the Simplest Pressure Probe (not to scale).

The primary limitation of this method is that some cells are too small to measure. Furthermore, some cells tend to leak after being stabbed with the capillary, and others plug up the tip of the capillary, thereby preventing valid measurements. The pressure probe has also been adapted to measure positive and negative values of Ψ_p in the xylem. However, technical problems with cavitation limit the measurement of negative Ψ_p by this technique.

UNDERSTANDING HYDRAULIC CONDUCTIVITY

Consider a cell with an initial water potential of –0.2 MPa, submerged in pure water. From this information we know that water will flow into the cell and that the driving force is $\Delta\Psi_w$ = 0.2 MPa, but what is the initial rate of movement? The rate depends on permeability of the membrane to water, a property usually called the hydraulic conductivity (*Lp*) of the membrane.

Driving force, membrane permeability, and flow rate are related by the following equation:

Flow rate = driving force × hydraulic conductivity

Hydraulic conductivity expresses how readily water can move across a membrane and has units of volume of water per unit area of membrane per unit time per unit driving force (for instance, $m^3\ m^{-2}\ s^{-1}\ MPa^{-1}$ or $m\ s^{-1}\ MPa^{-1}$). The larger the hydraulic conductivity, the larger the flow rate. In textbook

Figure, the hydraulic conductivity of the membrane is 10^{-6} m s^{-1} MPa^{-1}. The transport (flow) rate (J_v) can then be calculated from the following equation:

$$J_v = Lp(\Delta\Psi_w)$$

where J_v is the volume of water crossing the membrane per unit area of membrane and per unit time (m^3 m^{-2} s^{-1} or, equivalently, m s^{-1}). Please note that this equation assumes that the membrane is ideal—that is, that solute transport is negligible and water transport is equally sensitive to $\Delta\Psi_s$ and $\Delta\Psi_p$ across the membrane. Nonideal membranes require a more complicated equation that separately accounts for water flow induced by $\Delta\Psi_s$ and by $\Delta\Psi_p$.

In our example, J_v has a value of 0.2×10^{-6} m s^{-1}. Note that J_v has the physical meaning of a velocity. We can calculate the flow rate in volumetric terms (m^3 s^{-1}) by multiplying J_v by the surface area of the cell.

The resulting value is the *initial* rate of water transport. As water is taken up, cell Ψ_w increases and the driving force ($\Delta\Psi_w$) decreases. As a result, water transport slows with time.

WILTING AND PLASMOLYSIS

Plasmolysis is the separation of plant cell cytoplasm from the cell wall as a result of water loss. It is unlikely to occur in nature, except in severe conditions. Plasmolysis is induced in the laboratory by immersing a plant cell in a strongly saline or sugary solution, so that water is lost by osmosis.

If onion epidermal tissue is immersed in a solution of calcium nitrate, cells rapidly lose water by osmosis and the protoplasm of the cells shrinks. This occurs because the calcium and nitrate ions freely permeate the cell wall and encounter the selectively permeable plasma membrane. The large vacuole in the centre of the cell originally contains a dilute solution with much lower osmotic pressure than that of the calcium nitrate solution on the other side of the membrane.

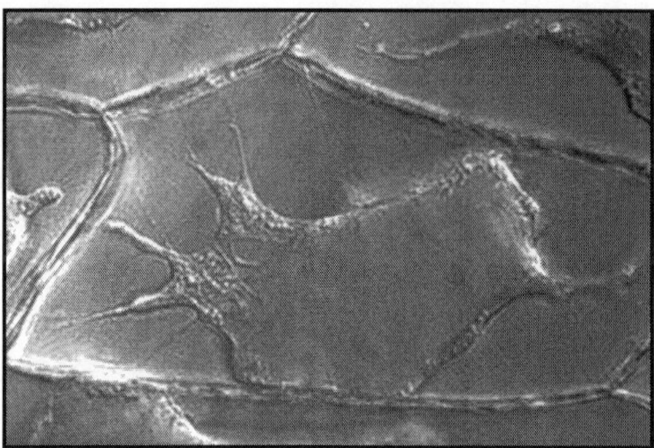

Fig. Plasmolysis of an Epidermis Cell of *Allium cepa* after Addition of Calcium Nitrate.

The vacuole thus loses water and becomes smaller. The space between the cell membrane and the cell wall enlarges and the plasma membrane and the protoplasm within it contract to the centre of the cell. Strands of cytoplasm extend to the cell wall because of plasma membrane-cell wall attachment points. Plasmolysed cells die unless they are transferred quickly from the salt or sugar solution to water.

Both chloroplast and vacuole become unevenly shaped. Strands of cytoplasm remain between the shrinking protoplasm and the neighboring cells.

6

Biotechnology in Plant Cell

Clonal propagation refers to the process of asexual reproduction by multiplication of genetically identical copies of individual plants. The vegetative propagation of plants is labour-intensive, low in productivity and seasonal. The tissue culture methods of plant propagation, known as 'micropropagation' utilizes the culture of apical shoots, axillary buds and meristems on suitable nutrient medium.The regeneration of plantlets in cultured tissue was described by Murashige in 1974. Fossard gave a detailed account of stages of micropropagation.

The micropropagation is rapid and has been adopted for commercialization of important plants such as banana, apple, pears, strawberry, cardamom, many ornamentals (*e.g.* Orchids) and other plants.The micropropagation techniques are preferred over the conventional asexual propagation methods because of the following reasons: (a) In the micropropagation method, only a small amount of tissue is required to regenerate millions of clonal plants in a year., (b) micropropagation is also used as a method to develop resistance in many species., (c) in vitro stock can be quickly proliferated as it is season independent. (d) long term storage of valuable germplasm possible.

The steps in micropropagation method are:

- *Initiation of culture*: From an explant like shoot tip on a suitable nutrient medium,
- Multiple shoots formation from the cultured explant,
- Rrooting of in vitro developed shoots and,
- Transplantation - transplantation to the field following acclimatization.

The factors that affect micropropagation are:

- Genotype and the physiological status of the plant *e.g.* plants with vigorous germination are more suitable for micropropagation.,
- The culture medium and the culture environment like light, temperature etc. For example an illumination of 16 hours a day and 8 hours night is satisfactory for shoot proliferation and a temperature of 25°C is optimal for the growth.

The benefits of micropropagation this method are:

- Rapid multiplication of superior clones can be carried out through out the year, irrespective of seasonal variations.
- Multiplication of disease free plants *e.g.* virus free plants of sweet potato (Ipomea batatus), cassava (Manihot esculenta)
- Multiplication of sexually derived sterile hybrids
- It is a cost effective process as it requires minimum growing space.

SOMACLONAL VARIATION

The genetic variations found in the in vitro cultured cells are collectively referred to as somaclonal variation and the plants derived from such cells are called as 'somaclones'. It has been observed that the long-term callus and cell suspension culture and plants regenerated from such cultures are often associated with chromosomal variations. It is this property of cultured cells that finds potential application in the crop improvement and in the production of mutants and variants (*e.g.* disease resistance in potato).

Larkin and Scowcroft working at the division of Plant Industry, C.S.I.R.O., Australia gave the term 'somaclones' for plant variants obtained from tissue cultures of somatic tissues. Similarly, if the tissue from which the variants have been obtained is having gametophytic origin such as pollen or egg cell, it is known as 'gametoclonal' variation.

They explained that it may be due to:

- Reflection of heterogeneity between the cells and explant tissue,
- A simple representation of spontaneous mutation rate, and
- Activation by culture environment of transposition of genetic materials.

Shepard *et al.* also contributed by screening about 100 somaclones produced from leaf protoplasts of Russet Burbank. They found that there was a significant amount of stable variation in compactness of growth habit, maturity, date, tuber uniformity, tuber skin colour and photoperiodic requirements. Somaclonal Variations has been used in plant breeding programmes where the genetic variations with desired or improved characters are introduced into the plants and new varieties are created that can exhibit disease resistance, improved quality and yield in plants like cereals, legumes, oil seeds tuber crops etc. Somaclonal variation is applicable for seed

APPLICATIONS OF SOMACLONAL VARIATIONS

- Methodology of introducing somaclonal variations is simpler and easier as compared to recombinant DNA technology.
- Development and production of plants with disease resistance *e.g.* rice, wheat, apple, tomato etc.
- Develop biochemical mutants with abiotic stress resistance *e.g.* aluminium tolerance in carrot, salt tolerance in tobacco and maize.

- Development of somaclonal variants with herbicide resistance *e.g.* tobacco resistant to sulfonylurea
- Development of seeds with improved quality *e.g.* a new variety of Lathyrus sativa seeds (Lathyrus Bio L 212) with low content of neurotoxin.
- Bio-13 – A somaclonal variant of Citronella java (with 37% more oil and 39% more citronellon), a medicinal plant has been released as Bio-13 for commercial cultivation by Central Institute for Medicinal and Aromatic Plants.
- Supertomatoes- Heinz Co. and DNA plant Technology Laboratories (USA) developed Supertomatoes with high solid component by screening somaclones which helped in reducing the shipping and processing costs.

Production of Virus Free Plants

The viral diseases in plants transfer easily and lower the quality and yield of the plants. It is very difficult to treat and cure the virus infected plants therefore te plant breeders are always interested in developing and growing virus free plants. In some crops like ornamental plants, it has become possible to produce virus free plants through tissue culture at the commercial level.

This is done by regenerating plants from cultured tissues derived from:

- Virus free plants,
- Meristems which are generally free of infection - In the elimination of the virus, the size of the meristem used in cultures play a very critical role because most of the viruses exist by establishing a gradient in plant tissues. The regeneration of virus-free plants through cultures is inversely proportional to the size of the meristem used.,
- Meristems treated with heat shock (34-36°C) to inactivate the virus,
- Callus, which is usually virus free like meristems.
- Chemical treatment of the media- attempts have been made to eradicate the viruses from infected plants by treating the culture medium with chemicals *e.g.* addition of cytokinins suppressed the multiplication of certain viruses.

Among the culture techniques, meristem-tip culture is the most reliable method for virus and other pathogen elimination. Viruses have been eliminated from a number of economically important plant species which has resulted in a significant increase in the yield and production *e.g.* potato virus X from potato, mosaic virus from cassava etc. These virus free plants are not disease resistant so there is a need to maintain stock plants to multiply virus free plants whenever required.

Production of Synthetic Seeds

In synthetic seeds, the somatic embryos are encapsulated in a suitable matrix (*e.g.* sodium alginate), along with substances like mycorrhizae, insecticides, fungicides and herbicides. These artificial seeds can be utilized for the rapid and mass propagation of desired plant species as well as hybrid varieties.

The major benefits of synthetic seeds are:

- They can be stored up to a year with out loss of viability
- Easy to handle and useful as units of delivery
- Can be directly sown in the soil like natural seeds and do not need acclimatization in green house.

Mutant Selection

An important use of cell cultures is in mutant selection in relation to crop improvement. The frequency of mutations can be increased several fold through mutagenic treatments and millions of cells can be screened. A large number of reports are available where mutants have been selected at cellular level. The cells are often selected directly by adding the toxic substance against which resistance is sought in the mutant cells. Using this method, cell lines resistant to amino acid analogues, antibiotics, herbicides, fungal toxins etc have actually been isolated.

Production of Secondary Metabolites

The most important chemicals produced using cell culture are secondary metabolites, which are defined as' those cell constituents which are not essential for survival'. These secondary metabolites include alkaloids, glycosides (steroids and phenolics), terpenoids, latex, tannins etc. It has been observed that as the cells undergo morphological differentiation and maturation during plant growth, some of the cells specialize to produce secondary metabolites. The in vitro production of secondary metabolites is much higher from differentiated tissues when compared to non-differentiated tissues. The cell cultures contribute in several ways to the production of natural products.

These are:

- A new route of synthesis to establish products *e.g.* codeine, quinine, pyrethroids,
- A route of synthesis to a novel product from plants difficult to grow or establish *e.g.* thebain from Papaver bracteatum,
- A source of novel chemicals in their own right *e.g.* rutacultin from culture of Ruta,
- As biotransformation systems either on their own or as part of a larger chemical process *e.g.* digoxin synthesis.

The Advantages of In Vitro Production of Secondary Metabolites

- The cell cultures and cell growth are easily controlled in order to facilitate improved product formation.

- The recovery of the product is easy.
- As the cell culture systems are independent of environmental factors, seasonal variations, pest and microbial diseases, geographical location constraints, it is easy to increase the production of the required metabolite.
- Mutant cell lines can be developed for the production of novel and commercially useful compounds.
- Compounds are produced under controlled conditions as per the market demands.
- The production time is less and cost effective due to minimal labour involved.

INVESTMENTS IN PLANT BIOTECHNOLOGY

The implicit rationale for IPR protection is that they promote investments in plant breeding and bio-engineering. This hypothesis is contested by section of analysts. Nevertheless it is a fact that the structure of IPRs do influence investments. IPRs induce their own pattern of innovations. The time taken for R&D to fructify as innovations is a crucial aspect guiding investments.

Often IPR regimes play a key role in influencing the time pattern of innovations. It is a well-known fact that conventional plant breeding methods are dilatory and time consuming. It requires 7 to 8 generations of repeated breeding to reduce heterozygosity of new genotypes. As Vossen observes Arabica Coffee subject to three to four cycles of breeding still display heterozygosity and it is not until 20 to 25 years (spanning 4 to 5 generations) that one arrives at true breeding varieties. This time lag is even more pronounced for the Arabusta coffee variety (inter-specific hybrid of Arabica and Robusta Coffee) where segregation of unfavourable characteristics could still occur even after several generations of backcrossing and selfing.

Another interesting example is the one cited by Evenson about the long gestation period in regard to rice varieties in India. According to Evenson, India released 306 rice varieties for planting during the period 1965-1986 after performing 20,000 crosses over a period of over 15 years since 1950.

It must be understood that this long process of backcrossing was preceded by considerable time spent on location and identification of individuals plants with requisite traits. The speed or pace at which plant varieties qualify for being released for planting have economic implications on investment behaviour in seed/plant breeding industry. Asexual modes of plant breeding which include plant genetic 'engineering' and transgenic techniques could obviate the long gestation period associated with pre-release preparations encountered conventional breeding.

It is a different matter than biosafety concerns may delay releases of these varieties. The time lag in release of plant varieties fulfilling the criteria of novelty, distinctness, uniformity and stability (or alternatively the requirements of novelty, inventiveness and commercial application) raise fundamental issues from the view-point of capital investments. The time lag in varietal release is potentially least in the case of transgenic and non-sexually produced varieties while for traditional or classical methods the time lag can be greater. The UPOV 1991 based plant IPR regimes could induce longer time lags as compared to the situation when an IPR regime is based on the 1978 version of the UPOV. This is due to the fact that the minimum genetic distance concept implicit in UPOV 1991 precludes 'close cousin' varieties from acquiring protection rights, unless the techniques of genetic distance determination by themselves are inadequate.

UPOV 1978 would facilitate capital investments to some extent as the release of plant varieties is periodically more frequent. Even under the plant patent regimes, investments could bring about short-term returns, though the volume of investments will have to be larger on account of capital intensity of advanced plant genetic engineering technologies.

By comparison, UPOV 1991 regime is least conducive to capital investments as plant variety releases can be painfully slow under this regime. Therefore sustenance of conventional plant breeding under UPOV 1991 regimes, can only be ensured by infusion of low discount capital. Paradoxically it is the reduced availability of low discount public funds for plant breeding that has induced countries like India to go in for Plant Varieties Protection Legislations modelled on UPOV lines. The implicit assumption in undertaking this step is that private funds for R&D in plant breeding will 'crowd in', if protection plant varieties rights are accorded. But the reality is that in liberalised deregulated developing economies, private investments have 'crowded in', in short return promising avenues.

In other words, private investment funds for R&D are not likely to be channellised into medium to long return yielding areas. Then the relative rates of discount will be r1 > rb < rc and relative present worth will be pa < pb > pc.

Consequently classical plant breeding activities will suffer for want of funds in the UPOV 1991. In other words, UPOV 1991 modelled national legislations will not necessarily hold attraction for private sector investors in R&D projects involving classical plant breeding as compared to say, UPOV 1978 or the Plant Patent regimes. Rather private investments would be catalysed better by *sui-generis* plant protection legislations, which partake of plant patent elements and focus in the main on hybrids and genetically engineered varieties.

Given the linkages between plant breeding and the seed industry these perverse investment behaviour trends can produce major economic consequences. The Seed Laws in various countries, both advanced and developing, are getting to be progressively oriented towards restricting market circulation to varieties that are conferred with plant breeder or patent rights. The UPOV guidelines on DUS are used world wide not only as the basis for establishing varietal distinctiveness and descriptions but also for seed certification purposes. The DUS guidelines are employed for recognising and registering not only the 'basic' seeds used for multiplication of 'pedigree' seeds but also for certification of the pedigree seeds themselves.

Evenson explains how absence of IPR rights will reduce Marginal Variable Costs (MVC) of seed multiplication. Tougher controls by certification of seeds could ensure that seed supply is kept in a 'constricted mode' and MVC of seed multiplication are raised. This in turn, will produce its own supply-demand dynamics. The implications of IPR regimes in terms of the supply curves for seeds. The situation of 'liberal' plant varieties protection laws (which do not provide for genetic distance) and equally liberal 'seed laws' which permit non-registered varieties (including landraces) to be sold in open markets, the macro-supply curve of seeds will be elastic (S1). When these regimes shift towards 'genetic dependency' PBR regimes and tighter seed laws that prohibit marketing of non-registered varieties, the supply curve will tend to be relatively inelastic (S2).

Analogously, demand curves could shift from D1 (wherein demand for protected varieties will be elastic on account of availability of landraces and non-protected varieties in the markets) to D2 (where the demand for the protected variety will be inelastic on account of ban on sale of landraces and other non-protected varieties). In the D1-S1 situation, supply constraints and demand flexibility could induce low price regimes while in D2-S2 the reverse situation would occur. D2-S2 situation is not desirable for countries like India for other reasons as well.

The experience of transitional economies of East Europe, which had re-modelled their Seed Laws in conformity with European Community Seed Laws, has been sad. Farmers of the East European bloc cultivating and conserving landraces have been adversely affected by new seed laws, which by prohibiting trade in non-registered varieties not fulfilling the DUS criteria have contributed to the erosion of landraces in these countries. It is thus clear that agro-biodiversity can be adversely affected by the uniformity criteria of DUS, which brings about resultant changes in demand and supply position of seeds.

IPR REGIMES FOR PLANT BIOTECHNOLOGY

Article 27(3) of the TRIPs Agreement lays down that members shall provide for the protection of plant varieties either by patents or by an effective *sui-generis* system or by any combination thereof. Article 27(3) in particular is critical to India as this requires us to provide protection of plant varieties by conferring plant breeders rights either by way of patents or through *sui-generis* plant variety protection legislations. Countries with strong R&D base in plant genetic engineering such as USA have robust Plant Utility Patent Legislations. India is certainly not inclined to adopt patent protection regimes for its plant varieties. India is rather inclined to adopt a *sui-generis* legislation which is non-'patent' based. There are many reasons for this proclivity. India is one of the ten mega-diversity countries of the world and a rich storehouse of landraces of principal agricultural crops. India has a strong R&D base in conventional plant breeding methods. Its strength in plant genetic engineering is impressive but not an overwhelming factor by comparison. The first two strengths explain India's disinclination towards plant patent regimes or towards a *sui-generis* legislation which is 'patent'-driven.

While Plant Utility Patents Act provide for broad patents over plant varieties, traits and genes and even the physical parts of the plants, plant breeders rights provide for IPR only over varieties. As is well known, since 1990s the UPOV (Union for the Protection of Varieties) has been largely viewed by developing countries as offering the best regime for positioning their national legislations.

The central feature of the UPOV is the protection it affords to plant breeders who produce plant varieties that fulfil the criteria of distinctiveness uniformity and stability (DUS). The current version of the UPOV, *viz.* UPOV 1991 had added additional criteria of 'new' to DUS thus rendering DUS as NDUS. Contrary to the popular notion, the NDUS criteria of UPOV 1991 is not substantively different from the principles of 'novelty', 'inventiveness' and 'industrial application' (NII) which applies for patents. The criteria of 'novelty' and 'inventiveness' in Patent Laws are covered by the criteria of 'new' and 'distinct' in UPOV 1991. Thus by distinctness, the UPOV means a variety of plant which is 'clearly distinguishable from other varieties whose existence is a matter of common knowledge'. It is apparent that this term captures the attributes of 'novelty' and 'inventiveness' implicit in Patent Laws. Even in respect of 'uniformity' and 'stability' criteria the UPOV does not offer different recipes.

True, by 'stability', the UPOV conveys that 'relevant characteristics of protected plant variety remain unchanged either for a specified period or after repeated propagations or cycles of propagations'. It is also true that 'stability' is a difficult criteria for a plant breeder to fulfil. Attainment of 'stability' criterion is problematic for cross-pollinated plants and non-single homozygous lines of autogamous plant varieties. This, in turn, reduces the commercialisation potential of the plant variety since inconsistency of genetic quality jeopardises commercial application of the plant variety. The same holds true of the criterion of "uniformity". Therefore, the NDUS criteria of UPOV 1991 are homologous to the NII criteria implicit in Plant Utility Patent Laws.

The UPOV has undergone two major amendments since its inception in 1961, *i.e.* in 1978 and 1991. The main concern with regard to currently operational UPOV 1991 has been its stringent provisions, which include the additional attribute of 'new' to the DUS criteria. The scope of UPOV 1991 in terms of coverage is larger. While plant varieties of all taxa are covered by UPOV 1991 only plant varieties of nationally designated species were covered by UPOV 1961 and 1978. Further by defining clearly that 'variety' represents a 'group of plants', or 'single' or 'several plants' or 'one' or 'several parts' of a plant(s) the UPOV 1991 has considerably enlarged the scope of its coverage. Going by this definition even sexual and vegetatively propagated varieties are covered by UPOV 1991.

The US PVPA legislation has been quick to adopt the UPOV 1991 formula. The amendments of 1994 to the US, PVPA broadens the term 'plant varieties' to include all materials harvested from the protected varieties. The other drastic feature of UPOV 1991 has been abridgement of the 'plant breeders exemptions' and 'farmers privileges' conferred by the earlier versions. Earlier versions of UPOV allowed farmers to re-use seeds generated from their farms for self-use, and allowed breeders to freely use protected varieties for further improvements. These concessions gave UPOV a 'less stringent' face in comparison to 'patent' laws. UPOV 1991 gives an option to National Governments to disallow the farmers privileges of retaining or re-using seeds for self cultivation. UPOV 1991 also restricts the breeders rights by enlarging the right of the breeder to close variations of his cultivar or 'essentially derived varieties'. In other words, the breeder of a cosmetically bred variety would have to buy genetic dependency rights from the derived variety prior to commercialising the derived version. Further the duration of protection have also been extended under UPOV 1991 to 20 years for crops and to 25 years for trees and vines. Thus in terms of these rigourous features, UPOV 1991 eliminates the arbitrage between plant varieties protection and patent regimes.

POST R&D REGULATIONS AND INVESTMENTS

Post-R&D regulations play a critical role in deciding on the pace of investments in the biotechnology sector. Even if a start-up enterprise enjoys a favourable IPR regime, that rewards its inventions, it is unlikely that it will invest in product development, in case post-R&D regulations are rigid and dilatory in nature. This then forms the backdrop for strategic alliances between start-up ventures and downstream industrial complexes. Multi-layered regulations in the post-R&D phase involve complexities which are expressed in the delays in commercial application/utilization of invented products or technologies.

In the case of third generation plant biotechnologies, such as transgenics, the problem is compounded by the fact that regulations by themselves are complex and less understood by the regulators themselves. A transgenic plant would have to undergo multi-stage regulatory checks and clearances before finding commercial application.

India is uniquely situated as far as biotechnology investments and infrastructure are concerned. In the area of plant biotechnology, advances have been made in transgenics of rice and wheat carrying stress tolerant genes such as CodA, COR47 and HVA1. The larger challenge is to take these products through the regulatory processes. More formidable is the issue of market development and appropriate pricing of transgenic seeds.

There are many imponderables here. Market shares for seeds in India have varied from crop to crop and from region to region. The 'demand' for transgenic plants or seeds may be even more uncertain, given the 'safety' dimensions and risk aversion propensity of Indian farmers.

In the dry agricultural pockets of Central and South India, farmers tend to be attached to traditional drought-hardy varieties of plants, which in turn contributes to their initial 'inertia' to shift to transgenics. Thus the task of marketing and pricing of novel biotechnology plants gets to be complex in such environment of 'demand incoherence'. A high margin pricing strategy for seeds may choke demand for transgenics.

Thus the degree of risks in commercializing transgenic plants could be of a high order. Given these realities, the idea of strategic alliances between start-ups and downstream entities looks to be realistic. The fact that premier R&D institutions such as the Indian Institute of Science, has set up its own start-up biotechnology venture, renders a strategy of tie-ups, a logical option. Much will depend upon the complexities inherent in product development and the willingness of competent downstream entities to signup agreements. To a large extent, this will depend upon enabling policies at the upper end of the value chain.

To sum up, India's agricultural biotechnology industry has the necessary policy and regulatory support to promote active R&D. However, in the critical area of product development and commercialization much more attention needs to be paid. The multi-tiered regulatory framework that exists could seriously affect the performance of start-up companies that would like to run through the entire life-cycle of biotechnology product development. Consequently, partnerships and strategic alliances should be encouraged between start-up ventures and established companies that have high product development capabilities. The experience of other countries in the Asia Pacific region could be a key input in framing development strategies for biotechnology products in India.

BIOTECHNOLOGY AND PLANTS

Today, biotechnology is being used as a tool to give plants new traits that benefit agricultural production, the environment, and human nutrition and health. The purpose of this publication is to provide basic information about plant biotechnology and to give examples of its uses.

The goal of plant breeding is to combine desirable traits from different varieties of plants to produce plants of superior quality. This approach to improving crop production has been very successful over the years. For example, it would be beneficial to cross a tomato plant that bears sweeter fruit with one that exhibits increased disease resistance. To do this, it takes many years of crossing and backcrossing generations of plants to obtain the desired trait. Along the way, undesirable traits may be manifested in the plants because there is no way to select for one trait without affecting others. Another limitation of traditional plant selection is that breeding is restricted to plants that can sexually mate.

Advances in scientific discovery and labouratory techniques during the last half of the twentieth century led to the ability to manipulate the deoxyribonucleic acid (DNA) of organisms, which accelerated the process of plant improvement through the use of biotechnology.

USE OF PLANT ORGANISMS IN AGRICULTURAL AND INDUSTRIAL PROCESSES

Plant biotechnology, in the broadest sense, employs a range of biological, chemical, and engineering disciplines to develop the use of plant organisms in agricultural and industrial processes.

The most obvious use of plant products in traditional biotechnology was for food. Over the centuries, farmers have learned to select edible crops for yield and quality. New and even better varieties were produced when early humans learned to cross-pollinate plants and became plant breeders. Early humans also exploited plants in other ways. In the beginning it was for fuel and clothing, but as time went on plants provided vegetable dyes for colouring, herbs, spices, drugs, perfumes, rubber, flavourings, oils, waxes, and a host of other products.

Although biotechnology has been with us for a long time, it is only now receiving the interest of the general public. This new interest in biotechnology is due to the revolution in molecular biology that has occurred over the last 30 to 40 years. Advances in DNA technology have made it possible to move genes between widely divergent organisms, creating new combinations of trits never before possible to move genes between widely divergent organisms, creating new combinations of traits never before possible in nature. These advances were originally developed for use in the laboratory where they revolutionized our ability to ask questions – and obtain meaningful answers - about fundamental biological problems. However, this new technology is now being exploited by industry, where the economic potential is immense.

PROMISING TECHNIQUE FOR CROP IMPROVEMENT

Biotechnology is a new and promising technique for crop improvement and novel uses of plants, animals and micro-organisms. Its potential to reduce or eliminate major constraints to crop productivity and to enhance food quality are widely recognized world wide. Plant biotechnology is a set of biological techniques developed through basic research and now applied to research and product development through the use of plants. In a Darwinian view of life, the fitness of an organism is measured only by its ability to replace itself with healthy, fertile offspring. Sexual reproduction is not the sole means by which flowering plants reproduce. Many species can also reproduce asexually, creating offspring that are genetically identical to themselves. Plant propagation is the art and science of multiplication of plant material involving the application of the principles of plant growth and development. Successful propagation of plants requires an understanding of plant morphology, physiology and genetic principles to assist in devising methods and to cope with unexpected difficulties. Therefore, in recent years, biotechnology has provided a powerful means to supplement traditional method through the use of molecular genetics in cloning and sequencing of genes leading to the analysis of the genome structure, evolution and expression.

Plant biotechnology encompasses broad fields of study, which includes plant tissue culture, genetic engineering and molecular markers. These disciplines are gradually becoming a functional aspect of classical breeding programmes. One of the areas of major impact is the control of plant growth and development, specifically vegetative and generative reproduction and propagation. Tissue culture manipulations play important role in plant breeding. Their routine application is determined by the possibility of recovering a great number of plants from cultures of desired cultivants

GENETIC MODIFICATION

The term "genetic modification" in general would mean any change in hereditary traits, but by legal documents it is understood as a synonym for transgenosis called also recombinant DNA technique or genetic engineering. It consists in direct adjustment of hereditary record carrier, mostly DNA but in certain viruses might be also RNA. The part of organism´s DNA chain is removed or in most cases new piece is added. By the removal the function of the particular gene is abolished. The introduced part, called insert, carries one or several genes usually isolated from other organisms and is prepared in such a way that the genes are expressed, *i.e.,* their products are synthesised by the modified organism and new trait occurs. Monogenic traits are introduced most easily. Polygenic traits depends on exact tuning of several gene cooperation.

From practical reasons helper genes are used for easy selection of modified cells and their identification. The new genes are included in the genome and transferred to the progeny. Helper genes may but need not be transferred with them. The product of genetic modification is by most documents called Genetically Modified Organism (GMO), in the protocol to CBD (Convention on Biological Diversity) known as the Cartagena Protocol Living Modified Organism (LMO). Genetic modification is a new technique, therefore it is regulated in most countries to avoid unexpected consequences. Based on accumulating experience rules for effective and safe use has been developed.

MOLECULAR GENETICS AND RECOMBINANT DNA TECHNOLOGY

Although modern biotechnology is based on advances in molecular genetics and recombinant DNA technology, exploitation of these advances in plants has been facilitated by two other established techniques in plant physiology. They are the culture of tissue, cells, and protoplasts and the technique of protoplast fusion.

Tissue culture is the technique of maintaining plant tissues indefinitely on an artificial medium. It has long been exploited as a system for the study of cell division, enlargement, and differentiation in laboratories around the world. Although plant tissue culture is most often carried out on solid medium, containing agar, tissues can also be cultured in liquid medium (for example somatic embryo culture). With a relatively small investment in space, technical support, and materials, tissue culture has made it possible to produce literally millions of high-quality, genetically uniform plants.

Tissue culture is very useful for:

- *In vitro* production of secondary metabolites,
- *In vitro* propagation (cryopreservation, culture systems, elicitors, growth regulators)
- *In vitro*, somaclonal and gametoclonal selection,
- Bioreactors (cell suspension cultures, hairy root cultures, stirred tank bioreactors),
- Cell, tissue and embryomanipulation (embryo rescue, haploid production, somatic embryogenesis),
- Microspore and macrospore culture (anther and pollen culture).

Protoplasts are plant cells that have had their cell walls removed by digestion with the enzyme cellulase and other cell wall-degrading enzymes. Protoplasts can be isolated from virtually any plant tissue and, in many cases, remain capable of sustained cell division.

As with tissue culture, cells regenerated from protoplasts can be induced to divide and form callus and, eventually, to regenerate fertile plants. Species that have been fully regenerated from protoplasts include potato, tobacco, pepper and tomato. Removal of the constraints imposed by the cell wall has stimulated many fundamental studies on plant cell biology. Perhaps most significant is the capacity of protoplast to fuse, giving rise to somatic hybrids. This allows hybrids to be formed between species that are genetically incompatible. Moreover, plants have cytoplasmic genomes that are maternally inherited. Some of these genes have practical as well as theoretical importance. Protoplast fusion makes it possible to produce new cytoplasmic-nuclear combinations without the time-consuming and expensive backcrossing required by conventional breeding. Finally, and perhaps most importantly, the absence of a cell wall facilitates the genetic transformation of plants by the direct transfer of genes into protoplasts. Plants that are subsequently regenerated will be transformed and express the new gene or genes.

PLANT PROTECTION IN MODERN AGRICULTURE

High productivity levels associated with modern agriculture depend heavily on protection of the crop species against competition from weeds and attack by viruses, fungi and predatory insects. Even a few weeds per hectare, or a minor insect or fungus infestation, can markedly reduce agricultural yields and, consequently, farmers incomes. Since the late 1940s, weeds, pathogens, and predatory insects have been controlled by extensive use of herbicides, fungicides, and insecticides.

Serious questions are being raised about the carryover of these agricultural chemicals into the food chain and their potentially harmful effects on consumers and the ecosystem. The heavy use of herbicides also had some undesirable side effects, not least of which is a dramatic rise in the number of species exhibiting herbicide resistance. The advent of genetic engineering presents opportunities for novel methods of plant protection against such pests with decreased reliance on potentially dangerous chemical controls. One of them is plant biotechnology.

The development of herbicide-resistant cultivars of crop species has the potential to be commercially as well as environmentally significant, because they allow farmers to control herbicide-sensitive weeds with a single application of herbicide after the resistant crop plants have germinated and begun to grow. Another strategy is to achieve the herbicide resistance by overexpression of tolerant enzymes. Insects and plant diseases are responsible for annual crop losses in excess of 10 to 15 per cent worldwide. To date, the most successful approach to engineering insect resistance exploits naturally occurring bacterial enterotoxins. Strategies based on the natural defence system are synthesis of pathogenesis related proteins (for example chitinase) or a form of "vaccination" (cultured plant cells have been transformed with a cDNA clone for a coat protein of the virus).

PLANT PATHOGEN DIAGNOSIS

The first and most important step in managing a plant disease is to correctly identify it. Although some diseases can be diagnosed quickly by visual examination, others require laboratory testing for diagnosis. These laboratory procedures may take days or even weeks to complete and are, in some cases, relatively insensitive.

Fortunately, as the result of advances in biotechnology, new products and techniques are becoming available that will complement or replace time-consuming laboratory procedures. Many products already are available for use, while others are still in developmental stages. The economical and vital importance of healthy crops makes from the plant pathogen diagnostics a subject of a continuous intense research with numerous challenges for further development.

MOLECULAR IMPROVEMENT OF PLANTS

The molecular improvement of plants requires a thorough understanding of the molecular basis of important plant processes and the identification and characterization of genes conferring agronomically important traits. During the past fifteen years, great development of molecular and cellular studies in plants including the molecular mapping of plant genomes has been performed. The concept of DNA markers based on polymerase chain reaction has revolutionized the study of chromosome segments, including minute regions, and has led to new opportunities such as map-based cloning and plant breeding. Linkage maps composed of DNA markers for a wide range of plant species, both models and crops, have been constructing and help to solve various research tasks.

Several major types of DNA markers have been widely developed and used - Restriction Fragment Length Polymorphisms (RFLP), Random Amplified Polymorphic DNAs (RAPD), Simple Sequence Repeats (SSR), Amplified Fragment Length Polymorphisms (AFLP), more recently Single Nucleotide Polymorphisms (SNP) and Insertions/Deletions (In/Del). The next generation maps are going to be those based on SNPs. DNA sequences of a plant organism can be learned from a physical map which consists of overlapping DNA fragments spanning the genome. Such complete knowledge is available in Arabidopsis thaliana since 2000, nearly in rice and is increasing in maize. Genetic mapping helps us to understand the structure, function and evolution of plant genomes. It can be an important tool for agricultural crop improvement by means of location of genes for both qualitative and quantitative characters, provides a basis for positional cloning of genes and, consequently, for genetic modification of plant genomes with genes of plant origin. Recent work has shown that the genetic maps of many closely related species are quite similar with respect to the content and location of genes, and scientists are trying to determine how the genetic map of one species may be applied to others by comparative mapping. Moreover, the finding that also distantly related species have co-linear maps for segments of their genomes leads to the synteny.

Potentially the most exciting new technologies for genetic analysis are oligonucleotide chips, capillary array electrophoresis and matrix laser desorption ionization, time of flight mass spectrometry (MALDI-TOF MS). It will be interesting to observe the impact of these all promising technologies on understanding of plant genome organization.

BIODIVERSITY OF PLANTS IN AGRICULTURE/HORTICULTURE

From about 250 of vascular species of plants occured on the Earth as low as 30,000 can be classified as edible plants and only 7,000 is included among crop plants. Nevertheless the world food security depends of more than 95 per cent on 30 of main crop plants. It shows evidence of close diversity of human sustenance. An interspecific genetic diversity (*i.e.* genetic diversity of plant crops) consists as a matter of fact in origin of the species. In terms of UNEP in Rio de Janeiro in 1992 The "Convention on Biological Diversity" was adopted and came into operation in 1993.

The Czech Republic subscribed this document in June 1993; The "Convention on Biological Diversity" came into operation in March 1994 and the statutory text is published in Collection of Laws No. 134/1999. The main point of The Convention consists in preservation of biological diversity of plants and animals, their genetic base and variability of ecosystems. Thus the object of the preservation is biodiversity in all the three components: specific, genetic and ecosystematic diversity and their contexture. The Convention distinguishes the preservation of biodiversity in a locality of natural occurence ("in situ") and outside a natural occurence ("ex situ"). The "in situ" conservation requires the conservation of all parts of ecosystem.

The special case of the "in situ" conservation is "on farm" conservation asserting for a rescue of local and traditional species or strains endangered by extinction – it deals with fodder crop and medicinal plants in the region of the Czech Republic. The best way of protecting wild plants related to crop plants are national parks, protection zones of water sources and pasture bands, where the "in situ" conservation of genetic diversity is associated with conservation of nature.

The "ex situ" conservation of genetic diversity is realised especially in gene banks (the number of gene banks all over the wold is cca 1300) in which is stored about 90 per cent from 6.1 milion of samples, however cca half of which can be duplicate. The preservation of genetic diversity in The Czech Republic is realised by The "National programme on Plant Genetic Resources Conservation and Utilisation". There is collected cca 46,000 of plant samples in the following 14 workplaces in The Czech Republic:

Department of Gene Bank, workplace Praha – Ruzyne, Research Institute of Crop Production Department of Gene Bank, workplace Olomouc, Research Institute of Crop Production Research Station for Viticulture in Karlstejn, Research Institute of Crop Production Agricultural Research Institute Kromeriz Ltd.

Agritec, Research, Breeding and Services Ltd. Potato Research Institute Ltd. Hop Research Institute Ltd. Research and Breeding Institute of Pomology Holovousy Ltd. Silva Tarouca Research Institute for Landscape and Ornamental Gardening Research Institute for Fodder Crops Troubsko Ltd. Grass Research Station Roznov – Zubri, Oseva PRO Ltd. Oil Plants Research Institute in Opava, Oseva PRO Ltd. Faculty of Horticulture, Mendel University of Agriculture and Forestry in Brno Research Station of Viticulture Znojmo Ltd. 80 per cent of samples is generatively and 20 per cent vegetatively reproduced. Another way of genetic conservation (mainly specific) are botanical gardens.

GENE TRANSFER METHODS IN PLANTS

To achieve genetic transformation in plants, we need the construction of a vector (genetic vehicle) which transports the genes of interest, flanked by the necessary controlling sequences *i.e.* promoter and terminator, and deliver the genes into the host plant. The two kinds of gene transfer methods in plants are:

VECTOR-MEDIATED OR INDIRECT GENE TRANSFER

Among the various vectors used in plant transformation, the Ti plasmid of Agrobacterium tumefaciens has been widely used. This bacteria is known as "natural genetic engineer" of plants because these bacteria have natural ability to transfer T-DNA of their plasmids into plant genome upon infection of cells at the wound site and cause an unorganized growth of a cell mass known as crown gall. Ti plasmids are used as gene vectors for delivering useful foreign genes into target plant cells and tissues. The foreign gene is cloned in the T-DNA region of Ti-plasmid in place of unwanted sequences.

To transform plants, leaf discs (in case of dicots) or embryogenic callus (in case of monocots) are collected and infected with Agrobacterium carrying recombinant disarmed Ti-plasmid vector. The infected tissue is then cultured (co-cultivation) on shoot regeneration medium for 2-3 days during which time the transfer of T-DNA along with foreign genes takes place. After this, the transformed tissues (leaf discs/calli) are transferred onto selection cum plant regeneration medium supplemented with usually lethal concentration of an antibiotic to selectively eliminate non-transformed tissues. After 3-5 weeks, the regenerated shoots are transferred to root-inducing medium, and after another 3-4 weeks, complete plants are transferred to soil following the hardening (acclimatization) of regenerated plants. The molecular techniques like PCR and southern hybridization are used to detect the presence of foreign genes in the transgenic plants.

VECTORLESS OR DIRECT GENE TRANSFER

In the direct gene transfer methods, the foreign gene of interest is delivered into the host plant cell without the help of a vector. The methods used for direct gene transfer in plants are:

Chemical mediated gene transfer *e.g.* chemicals like polyethylene glycol (PEG) and dextran sulphate induce DNA uptake into plant protoplasts. Calcium phosphate is also used to transfer DNA into cultured cells.

Microinjection where the DNA is directly injected into plant protoplasts or cells (specifically into the nucleus or cytoplasm) using fine tipped (0.5 - 1.0 micrometer diameter) glass needle or micropipette. This method of gene transfer is used to introduce DNA into large cells such as oocytes, eggs, and the cells of early embryo. Electroporation involves a pulse of high voltage applied to protoplasts/cells/tissues to make transient (temporary) pores in the plasma membrane which facilitates the uptake of foreign DNA. The cells are placed in a solution containing DNA and subjected to electrical shocks to cause holes in the membranes. The foreign DNA fragments enter through the holes into the cytoplasm and then to nucleus.

Particle gun/Particle bombardment - In this method, the foreign DNA containing the genes to be transferred is coated onto the surface of minute gold or tungsten particles (1-3 micrometers) and bombarded onto the target tissue or cells using a particle gun (also called as gene gun/shot gun/ microprojectile gun).The microprojectile bombardment method was initially named as biolistics by its inventor Sanford. Two types of plant tissue are commonly used for particle bombardment- Primary explants and the proliferating embryonic tissues.

Transformation - This method is used for introducing foreign DNA into bacterial cells *e.g.* E. Coli. The transformation frequency (the fraction of cell population that can be transferred) is very good in this method. *E.g.* the uptake of plasmid DNA by E. coli is carried out in ice cold $CaCl_2$ (0-50C) followed by heat shock treatment at 37-450C for about 90 sec. The transformation efficiency refers to the number of transformants per microgram of added DNA. The $CaCl_2$ breaks the cell wall at certain regions and binds the DNA to the cell surface. Conjuction - It is a natural microbial recombination process and is used as a method for gene transfer. In conjuction, two live bacteria come together and the single stranded DNA is transferred via cytoplasmic bridges from the donor bacteria to the recipient bacteria.

Liposome mediated gene transfer or Lipofection - Liposomes are circular lipid molecules with an aqueous interior that can carry nucleic acids. Liposomes encapsulate the DNA fragments and then adher to the cell membranes and fuse with them to transfer DNA fragments. Thus, the DNA enters the cell and then to the nucleus. Lipofection is a very efficient technique used to transfer genes in bacterial, animal and plant cells.

SELECTION OF TRANSFORMED CELLS FROM UNTRANSFORMED CELLS

The selection of transformed plant cells from untransformed cells is an important step in the plant genetic engineering. For this, a marker gene (*e.g.* for antibiotic resistance) is introduced into the plant along with the transgene followed by the selection of an appropriate selection medium (containing the antibiotic). The segregation and stability of the transgene integration and expression in the subsequent generations can be studied by genetic and molecular analyses.

PLANT GERMPLASM CONSERVATION

Germplasm refers to the sum total of all the genes present in a crop and its related species.

The conservation of germplasm involves the preservation of the genetic diversity of a particular plant or genetic stock for it's use at any time in future. It is important to conserve the endangered plants or else some of the valuable genetic traits present in the existing and primitive plants will be lost. A global organization- International Board of Plant Genetic Resources (IBPGR) has been established for germplasm conservation and provides necessary support for collection, conservation and utilization of plant geneic resources through out the world.

The germplasm is preserved by the following two ways:

1. *In-situ conservation*: The germplasm is conserved in natural environment by establishing biosphere reserves such as national parks, sanctuaries. This is used in the preservation of land plants in a near natural habitat along with several wild types.
2. *Ex-situ conservation*: This method is used for the preservation of germplasm obtained from cultivated and wild plant materials. The genetic material in the form of seeds or in vitro cultures are preserved and stored as gene banks for long term use.

In vivo gene banks have been made to preserve the genetic resources by conventional methods *e.g.* seeds, vegetative propagules, etc. In vitro gene banks have been made to preserve the genetic resources by non - conventional methods such as cell and tissue culture methods. This will ensure the availability of valuable germplasm to breeder to develop new and improved varieties.

The methods involved in the in vitro conservation of germplasm are:

- *Cryopreservation*: In cryopreservation, the cells are preserved in the frozen state. The germplasm is stored at a very low temperature using solid carbon dioxide (at -79°C), using low temperature deep freezers (at -80°C), using vapour nitrogen (at -150°C) and liquid nitrogen (at-196°C). The cells stay in completely inactive state and thus can be conserved for long periods. Any tissue from a plant can be used for cryopreservation *e.g.* meristems, embryos, endosperms, ovules, seeds, cultured plant cells, protoplasts, calluses. Certain compounds like- DMSO (dimethyl sulfoxide), glycerol, ethylene, propylene, sucrose, mannose, glucose, praline, acetamide etc are added during the cryopreservation. These are called cryoprotectants and prevent the damage caused to cells (by freezing or thawing) by reducing the freezing point and super cooling point of water.
- *Cold Storage*: Cold storage is a slow growth germplasm conservation method and conserves the germplasm at a low and non-freezing temperature (1-9°C). The growth of the plant material is slowed down in cold storage in contrast to complete stoppage in cryopreservation and thus prevents cryogenic injuries. Long term cold storage is simple, cost effective and yields germplasm with good survival rate. Virus free

strawberry plants could be preserved at 10°C for about 6 years. Several grape plants have been stored for over 15 years by using a cold storage at temperature around 9°C and transferring them in the fresh medium every year.

- *Low pressure and low oxygen storage*: In low- pressure storage, the atmospheric pressure surrounding the plant material is reduced and in the low oxygen storage, the oxygen concentration is reduced. The lowered partial pressure reduces the in vitro growth of plants. In the low-oxygen storage, the oxygen concentration is reduced and the partial pressure of oxygen below 50 mmHg reduces plant tissue growth. Due to the reduced availability of O_2, and reduced production of CO_2, the photosynthetic activity is reduced which inhibits the plant tissue growth and dimension. This method has also helped in increasing the shelf life of many fruits, vegetables and flowers.

APPLICATION OF IMMOBILIZED CELLS

Immobilization of plant cells is considered to be of importance in research and development in plant cell cultures, because of the potential benefits that could be provided:

- The extended viability of cells in the stationary (and producing) stage, enabling maintenance of biomass over a prolonged time period;
- Simplified downstream processing (if products are secreted);
- The (putative) promotion of differentiation, linked with enhanced secondary metabolism;
- Higher cell density enabling a reduced bioreactor size, thereby reducing costs and the risk of contamination;
- Reduced shear sensitivity (especially with entrapped cells);
- Promotion of secondary metabolite secretion, in some cases;
- Flow-through reactors can be used enabling greater flow rates;
- Minimization of fluid viscosity increase, which in cell suspension causes mixing and aeration problems.

An immobilization system which could maintain viable cells over an extended period of time and release the bulk of the product into the extracellular medium in a stable form, could dramatically reduce the costs of phytochemicals production in plant cell culture. However, an immobilized system also has the problems described below:

- Immobilization is normally limited to cases where production is decoupled from cell growth;
- The initial biomass must be grown in suspension;
- Secretion of product into the extracellularly medium is imperative;
- Where secretion occurs there may be problems of extracellular degradation of the products;

- When gel entrapment is used, the gel matrix introduces an additional diffusion barrier.

Due to these problems, a system with commercial potential has not yet been developed in plant tissue cultures. However, various immobilization methods have been developed, i.e., entrapment, adsorption and covalent coupling.

Some preliminary results have been obtained with immobilized cells. Early work with C. roseus, showed that agar, agarose and carageenan were all suitable immobilization matrices suitable for maintenance of cell viability; but alginate was superior in terms of ajmalicine production. The accumulation of serpentine by C. roseus and anthraquinones by Morinda citrifolia were both enhanced in the immobilized state when compared with freely suspended cells. It should be noted however, that the possibility that alginate acts as an elicitor of secondary metabolism cannot be ruled out. Agar has been shown to stimulate shikonin accumulation in L. erythrorhizon cultures. Lambe and Rosevear have successfully immobilized C. roseus cells in polyacrylamide with alginate and observed prolonged viability and increased productivity.

Adsorption immobilization has been successfully used with a number of plant species. Capsicum frutescens cells immobilized on polyurethane foam produced 50 times as much capsaicin as suspension cells. Similarly, Solanum nigrum cells accumulated glycoalkaloids to levels exceeding those found in suspensions. Datura innoxia cells accumulated tropane alkaloids with a profile similar to that of the intact plant, whilst in free suspensions productivity was markedly suppressed. In general it appears that mild immobilization either through gel entrapment or surface adsorption enhances productivity and prolongs the viability of cultured cells. As described, immobilized cells can also be used as biocatalysts for biotransformations. Such a system compares favourably with the use of freely suspended cells since, in the case of immobilization, the catalyst is theoretically reusable and the product is easily separated from the biomass. The most appropriate example is that of the 12-hydroxylation of ß-methyldigitoxin to ß-methyldigoxin with alginate-entrapped Digitalis lanata cells. The enzyme activity was maintained by the immobilized cultures for a period of 61 days. Furthermore, the product was located in the extracellular medium. Mild permeabilization of the cells may enable biotransformation rates to be increased.

Polyurethane-immobilized C. frutescens cells fed capsaicin precursors produced this metabolite at levels of up to 10 times those of non-fed cultures. DiCosmo *et al.* found that glass fibres can be used as a carrier of plant cells to produce useful plant metabolites. Papaver somniforum cells were immobilized on fabric of loosely woven polyester fibres arranged in a spiral configuration on stainless steel support frame by Kurz *et al.* to produce sanguinarine, an antibiotic in oral hygiene. The yield was 3.6 mg/g-fw. by immobilized cells and was more than twice as much as by suspension cells.

PRODUCT SECRETION

Many plant products produced by cell cultures have been reported to be accumulated intracellularly. However, it may be possible to produce much higher level of product if it was secreted into the medium. This is because the product intracellularly accumulated sometimes inhibits its own synthesis by regulation mechanisms such as product inhibition and repression. For many immobilized plant cell systems to work it is essential that a significant amount of product is released into the medium. Two types of immobilization system with C. roseus *i.e.* gel entrapment in polysaccharide beads and in polyacrylamide sheets, have both exhibited alkaloid release by mechanisms that do not appear to be associated with losses in viability. Capsicum frutescens cells immobilized on polyurethane released capsaicin entirely into the medium, although other species immobilized by the same method retained the product intracellularly.

To enhance release, permeabilization of cell membranes has been attempted, but with only limited success. Brodelius tested five permeabilizing agents on three different species, and although product release was achieved, cell viability dropped in most cases. The exceptions were DMSO and Triton X-100, applied to C. roseus cells. Other attempts to permealize cells have also resulted in non-viable populations and the use of electroporation has also resulted in a viability decrease. The release of betanin by Beta vulgaris cells ultrasonicated for 20 to 60 seconds has been reported, with no apparent effect on cell viability. Chenopodium rubrum cells, immobilized in alginate beads, secreted the red betacyanin pigment amaranthin into the medium. However, the pigment was subsequently degraded; chitosan and DMSO permitted further product release into the extracellular medium, but this was also accompanied by product degradation. Low concentrations of chitosan (0.01%) and DMSO (5.7%) incubated with the cultures for 96 hr did not appear to affect viability significantly, but a longer incubation period had a deterlerious effect.

The lack of an appropriate means of product release is a serious problem in terms of an industrial approach particularly based on immobilization of plant cells. It is perhaps necessary to examine in vivo physiological mechanisms of release. Thus, some researchers have been investigating the factors involved in vacuolar phytochemical storage. The accumulation of indole alkaloids in C. roseus vacuoles has been attributed to an ion-trap mechanism whereby the basic indole alkaloids are trapped in the acidic vacuole due to their positive charge at low pH, preventing diffusion across the tonoplast. This mechanism has also been demonstrated for quinoline alkaloid accumulation in Cinchona species.

Active uptake mechanisms have also been reported for indole alkaloids in C. roseus vacuoles and isoquinoline alkaloids in Fumaria capreolata L. The physiological significance of these two possible mechanisms is not yet clear, since there is obviously strong evidence for both. In terms of product release, it is pertinent to note that in cell cultures, an efflux of alkaloids was observed under certain conditions, indicating an equilibrium between the intracellular and extracellular compartments that could be perturbed by medium acidification with subsequent product release. The release of serpentine by C. roseus cells was observed when the cells were filtered and resuspended in fresh or conditioned medium and it was suggested that temporary membrane uncoupling was responsible. Although most plant cell culture products were shown to be accumulated intracellularly, Shuler *et al.* recently reported that almost all of taxol produced by Taxus brevifolia cell cultures was detected in the culture filtrate. This is an exceptional case in plant tissue culture research.

MUTAGENESIS

In the fermentation industry, induction of genetic mutant strains of microorganisms is ubiquitous, and auxotrophic and/or regulatory mutants are used extensively to produce a variety of products including amino acids, nucleotides, antibiotics, etc. However, mutagenesis has limited applicability to plant cell cultures, because of their diploid genetic make-up: the chance of obtaining a double mutation in a target gene is less than 1 in 106. Although, in principle, haploid plants can be produced from anther cultures; in practice haploid cell cultures tend to revert to the diploid state. This makes the chance of isolating over-producing cells from mutagen treatment of haploid cells very low. Furthermore, biosynthetic pathways of many secondary metabolites and their regulation mechanisms in higher plants are not always understood precisely, therefore, it is also difficult to know what kind of mutants should be induced in order to increase product synthesis.

However, Berlin induced p-fluorophenylalanine resistant cell lines of tobacco cell cultures and found that five lines of N. tabacum and five lines of N. glauca out of 31 resistant cell lines accumulated higher levels of phenolics. He and his colleagues also reported that a resistant strain of N. tabacum produced 6 to 10 times higher levels of cinnamoyl putrescine than that of the parent strain.

Fig. shows the comparison of enzyme activities involved in the biosynthesis of cinnamoyl putrescines in a parent strain, TX1 and a resistant strain, TX4. Enzyme activities in the p-fluorophenylalanine resistant strain were much higher than those of the parent. Berlin *et al.* also found that a parent strain of C. roseus produced catharanthine only in the production medium but its tryptophan analog resistant mutant accumulated the same alkaloid even in the growth medium. Among several types of analogues tested, 4-fluorotryptophan was found to be the most efficient of them.

Increase of metabolite levels using regulatory mutants is theoretically possible and selection of suitable analogues for this purpose could be an important factor in order to produce a variety of products. Several research groups have used mutagens such as N-methyl-N'-nitro-N-nitrosoguanidine, ethylmethane sulfonate, N-ethyl-N-nitrosourea or X-rays to induce high yielding strains. In China, a fine callus strain of Anisodus acutangulus was derived through irradiation with 4000 R of X-ray, and the level of scopolamine in the cells was 0.177 mg/g d.wt., which was about 30% higher than of the parent. The productivity was stable. Deus treated C. roseus with X-rays and obtained a cell line having a high producing activity of serpentine whose level was 2%.

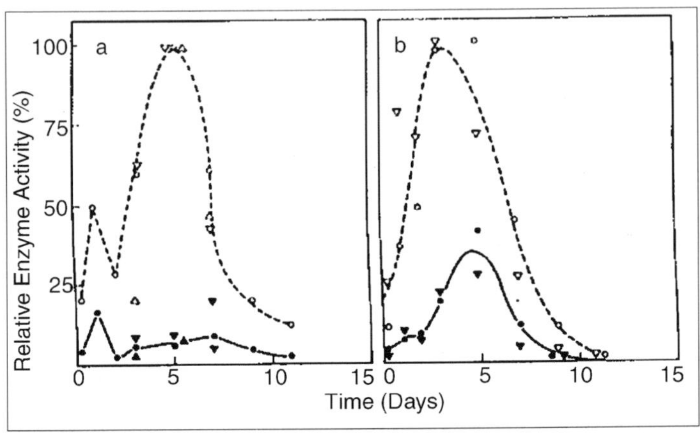

Fig. Comparison of Enzyme Activities

A cell line of Lavendula vera producing a high level of free biotin was obtained by gamma ray irradiation for 1 hr. with 60_{Co} (dose 10 KR h) by Watanabe *et al*. The line was found to contain 7 times the amount, 0.425 μg/g-fresh wt., of free biotin compared with the original unselected cells, 0.061 μg/g fresh wt., and 4.5 times that found in the leaves.

In spite of the difficulty of obtaining auxotrophic mutants in plant cells, several groups have been actively working with this approach using haploid cells or protoplasts although their aim is not always for plant metabolite production. Induction of a mutant having altered permeability could also be important, because plant cells generally accumulate their metabolites intracellularly, which is disadvantageous in commercial production because the amount of compounds produced is usually low. If the cells excreted products in large amounts, the product cost would be reduced. According to Berlin's results, Thuja occidentalis excreted monoterpenoids, but the levels in the medium were only 5% of those in the mother plant.

The same group reported that Macleya microcarpa cells excreted nearly all the alkaloids detectable in the culture flask. Yamakawa *et al*. cultivated Tinospora rumphii cells in 14 ml medium and found 0.57 (5.3% in dry weight cells) of isoquinoline alkaloids in the cells and 0.5 mg in the culture filtrate after 7 days cultivation. By replacement of the medium with fresh medium after 3 days of cultivation, 0.50 mg of the alkaloids in the cells, and 1.02 mg in the filtrate were accumulated. It seems then that the products must be excreted to some extent. Unfortunately, the secretion mechanisms for secondary products in higher plant cells have not yet been elucidated and extensive fundamental studies are required before meaningful manipulation can take place.

THE ANALYSIS OF PLANT GROWTH

New cells form continually in the apical meristems. Cells enlarge slowly in the apical meristem and more rapidly in the subapical regions. The resulting increase in cell volume can range from severalfold to one hundredfold, depending on the species and environmental conditions. Classically, plant growth has been analyzed in terms of cell number or overall size (i.e., mass). However, these measures tell only part of the story.

Tissue growth is neither uniform nor random. The derivatives of the apical meristems expand in predictable and site-specific ways, and the expansion patterns in these subapical regions largely determine the size and shape of the primary plant body. The total growth of the plant can be thought of as the sum of the local patterns of cell expansion.

The analysis of the motions of cells or "tissue elements" (and the related problem of cell expansion) is called *kinematics*. In this section we will discuss both the classical definitions of growth and the more modern, kinematic approach. As we will see, the advantage of the kinematic approach is that it allows one to describe the growth patterns of organs mathematically in terms of the expansion patterns of their component cells.

Plant Growth Can Be Measured in Different Ways

Growth in plants is defined as an irreversible increase in volume. The largest component of plant growth is cell expansion driven by turgor pressure. During this process, cells increase in volume manyfold and become highly vacuolate. However, size is only one criterion that may be used to measure growth. Growth also can be measured in terms of change in fresh weight—that is, the weight of the living tissue—over a particular period of time. However, the fresh weight of plants growing in soil fluctuates in response to changes in water status, so this criterion may be a poor indicator of actual growth. In these situations, measurements of dry weight are often more appropriate.

Cell number is a common and convenient parameter by which to measure the growth of unicellular organisms, such as the green alga *Chlamydomonas*. In multicellular plants, however, cell number can be a misleading growth measurement because cells can divide without increasing in volume.

Growth is assessed by a count of the number of cells per milliliter at increasing times after the cells are placed in fresh growth medium. Temperature, light, and nutrients provided are optimal for growth. An initial lag period during which cells may synthesize enzymes required for rapid growth is followed by a period in which cell number increases exponentially.

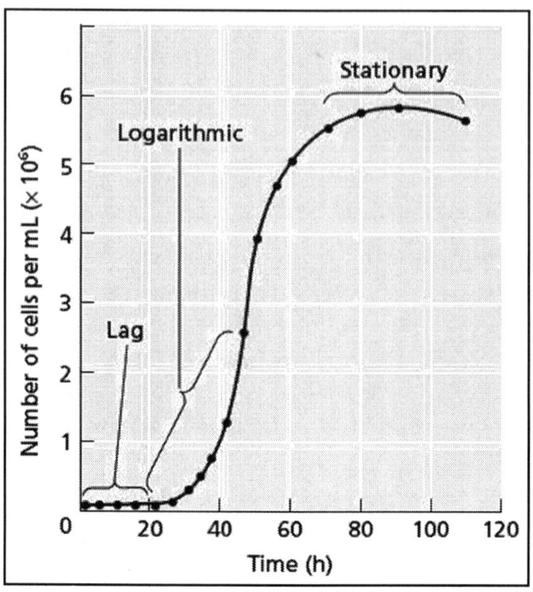

Fig. Growth of the unicellular green alga *Chlamydomonas*.

This period of rapid growth is followed by a period of slowing growth in which the cell number increases linearly. Then comes the stationary phase, in which the cell number remains constant or even declines as nutrients are exhausted from the medium.

For example, during the early stages of embryogenesis, the zygote subdivides into progressively smaller cells with no net increase in the size of the embryo. Only after it reaches the eight-cell stage does the increase in volume begin to mirror the increase in cell number. Because the zygote is an especially large cell, this lack of correspondence between an increase in cell number and growth may be unusual, but it points out the potential problem in equating an increase in cell number with growth.

Although cell number may not always be a reliable measure of plant growth, under most circumstances dividing cells, particularly in meristems, double in volume during their cell cycle. Therefore, an increase in cell number, such as the increase brought about by the activity of the apical meristems, does contribute to plant growth. However, the largest component of plant growth is the rapid cell expansion that occurs in the subapical region after cell division ceases. Because all the cells of the plant axis elongate under normal conditions, the greater the number of cells produced by the apical meristem, the longer the axis will be. For example, when Arabidopsis plants are transformed with a gene that encodes cyclin, a key component of the cell cycle regulatory machinery, the cells of the apical meristem progress through their cell cycles more rapidly, so more cells form per unit time. As a result, the roots of these transgenic plants have more cells and are substantially longer than the roots of wild-type plants grown under similar conditions. New cells form continually in the apical meristems. With each new round of cell division and associated cell expansion, the older derivatives are displaced a small distance from the apex. As the cells recede farther from the apex, the rate of displacement is greatly accelerated. By viewing plant growth as a process of cell displacement from the apex, we can apply the principles of kinematics.

The Production of Cells by the Meristem Is Comparable to a Fountain

Moving fluids such as waterfalls, fountains, and the wakes of boats can generate specific forms. The study of the motion of fluid particles and the shape changes that the fluids undergo is called kinematics. The ideas and numerical methods used to study these fluid forms are useful for characterizing meristematic growth. In both cases, an unchanging form is produced even though it is composed of moving and changing elements.

An example of an unchanging form composed of changing and displaced elements in plants is the hypocotyl hook of a dicot, such as the common bean. As the bean seedling emerges from the seed coat, the apical end of the hypocotyl bends back on itself to form a hook. The hook is thought to protect the seedling apex from damage during growth through the soil. During seedling growth (in soil or dim light) the hook migrates up the stem, from the hypocotyl into the epicotyl and then to the first and second internodes, but the form of the hook remains constant.

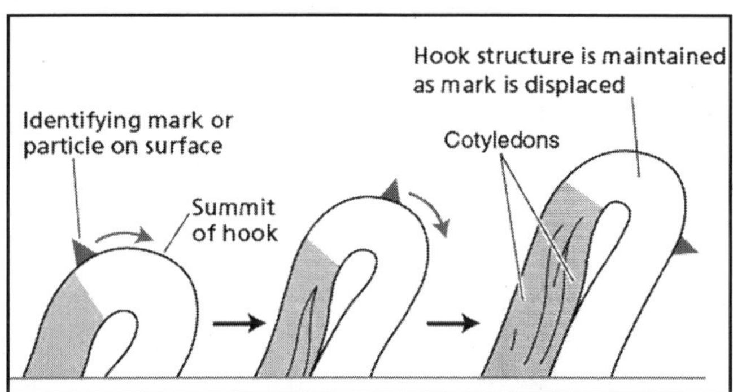

Fig. The dicot Hypocotyl hook is an example of a Constant form Composed of changing Elements.

The hooked form is maintained over time, while different tissues first curve and then straighten as they are displaced from the seedling apex during growth. If a mark is placed at a fixed point on the surface, it will be displaced (indicated by the arrow), appearing to flow through the hook over time. If we mark a specific epidermal cell on the seedling stem located close to the seedling apex, we can watch it as it flows into the hook summit, then down into the straight region below the hook. The mark is not crawling over the plant surface, of course; plant cells are cemented together and do not experience much relative motion during development. The change in position of the mark relative to the hook implies that the hook is composed of a procession of tissue elements, each of which first curves and then straightens as it is displaced from the plant apex during growth. The steady form is produced by a parade of changing cells. A root tip is another example of a steady form composed of changing tissue elements. Here, too, the form is observed to be steady only when distance is measured from the root tip. A region of cell division occupies perhaps 2 mm of the root tip. The elongation zone extends for about 10 mm behind the root tip. Phloem differentiation is first observed beginning at 3 mm from the tip, and functional xylem elements may be seen at about 12 mm from the tip. A marked cell near the tip will seem to flow first through the region of cell division, then through the elongation zone and into the region of xylem differentiation, and so on. This shifting implies that developing tissue elements first divide and elongate, and then differentiate. In an analogous fashion, the shoot bears a succession of leaves of different developmental stages. During a period of 24 hours, a leaf may grow to the same size, shape, and biochemical composition that its neighbour had a day earlier. Thus, shoot form is also produced by a parade of changing elements that can be analyzed with kinematics. Such an analysis is not merely descriptive; it permits calculations of the growth and biosynthetic rates of individual tissue elements (cells) within a dynamic structure.

Tissue Elements Are Displaced during Expansion

As we have seen, growth in shoots and roots is localized in regions at the tips of these organs. Regions with expanding tissue are called growth zones. With time, meristems move away from the plant base by the growth of the cells in the growth zone. If successive marks are placed on the stem or root, the distance between the marks will change, depending on where they are within the growth zone. In addition, all of these marks will move away from the tip of the root or shoot, but their rate of movement will differ depending on their distance from the tip. From another perspective, if you were to stand at the tip of a root that had marks placed at intervals along the axis, you would see that all marks would move farther away from you with time. The reason is that discrete regions on the plant axis experience displacement as well as expansion during growth and development.

As Regions Move Away from the Apex, Their Growth Rate Increases

As a given region of the plant axis moves away from the apex, its growth velocity increases (the rate of elongation accelerates) until a constant limiting velocity is reached equal to the overall organ extension rate. The reason for this increase in growth velocity is that with time, progressively more tissue is located between the moving particle and the apex, and progressively more cells are expanding, so the particle is displaced more and more rapidly. In a rapidly growing maize root, a tissue element takes about 8 hours to move from 2 mm (the end of the meristematic zone) to 12 mm (the end of the elongation zone). Beyond the growth zone, elements do not separate; neighboring elements have the same velocity (expressed as the change in distance from the tip per unit of time), and the rate at which particles are displaced from the tip is the same as the rate at which the tip moves through the soil. The root tip of maize is pushed through the soil at 3 mm h^{-1}. This is also the rate at which the nongrowing region recedes from the apex, and it is equal to the final slope of the growth trajectory.

The Growth Velocity Profile Is a Spatial Description of Growth

The velocities of different tissue elements are plotted against their distance from the apex to give the spatial pattern of growth velocity, or growth velocity profile. Growth velocity increases with position in the growth zone. A constant value is obtained at the base of the growth zone. The final growth velocity is the final, constant slope of the growth trajectory equal to the elongation rate of the organ, as discussed in the previous section. In the rapidly growing maize root, the growth velocity is 1 mm h^{-1} at 4 mm, and it reaches its final value of nearly 3 mm h^{-1} at 12 mm.

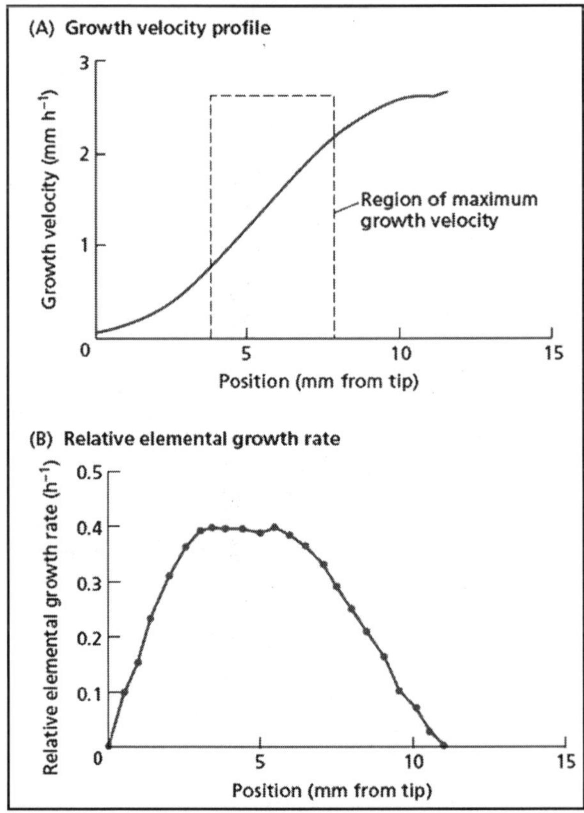

Fig. The Growth of the Primary root of Zea mays (maize) can be represented Kinematically by two related growth curves.

(A) The growth velocity profile plots the velocity of movement away from the tip of points at different distances from the tip. This tells us that growth velocity increases with distance from the tip until it reaches a uniform velocity equal to the rate of elongation of the root. (B) The relative elemental growth rate tells us the rate of expansion of any particular point on the root. It is the most useful measure for the physiologist because it tells us where the most rapidly expanding regions are located.

If the growth velocity is known, the relative elemental growth rate, which represents the fractional change in length per unit time, can be calculated. The relative elemental growth rate shows the location and magnitude of the extension rate and can be used to quantify the effects of environmental variation on the growth pattern.

The Relative Elemental Growth Rate

One way to characterize the growth pattern along an axis mathematically is to plot the distance of a tissue element from the apex versus time. The resulting curve is called the growth trajectory. The slope of the curve at any given point is equivalent to the growth velocity at that region of the axis. With the apex as the reference point, the growth trajectory is concave. Marked tissue elements at different locations may be followed simultaneously to generate a family of growth trajectories.

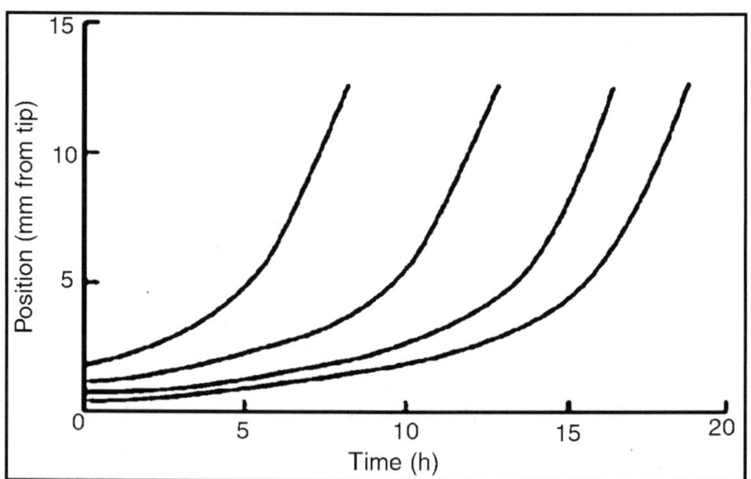

Fig. Growth Trajectories of *Zea mays* (maize):Growth Trajectories plot the positions of points spaced along the root axis as a Function of time.

Note that distance is measured from the root tip. This is termed a material specification of growth because real or material particles are followed.

As the regions of the plant axis move away from the apex, their growth velocity increases (i.e., the cells accelerate) until a constant limiting velocity is reached equal to the overall organ extension rate. The reason for this increase in growth velocity is that with time, progressively more tissue is located between the moving particle and the apex, and progressively more cells are expanding, so the particle is displaced more and more rapidly.

In a rapidly growing maize root, a tissue element takes about 8 hours to move from 2 mm (the end of the meristem) to 12 mm (the end of the elongation zone). The root tissue element accelerates through this region.

Beyond the growth zone, elements do not separate; neighboring elements have the same velocity, and the rate at which particles are displaced from the tip is the same as the rate at which the tip is pushed away from the soil surface. The root tip of maize is pushed through the soil at 3 mm h^{-1}. This is also the rate at which the nongrowing region recedes from the apex, and it is equal to the final slope of the growth trajectory. The growth trajectory is a material description of growth because an individual element is followed through time. The growth trajectory provides a way to convert spatial information into a developmental time course. For example, if the amount of suberin in the root increases with position, the growth trajectory will give information on the time required for the observed accumulation of suberin in cells moving between any two spatial points.

To characterize the local expansion rate, we must consider the growth velocities at the apical and basal ends of a small segment of tissue. If the apical end of the segment is moving faster than the basal end, the segment is elongating. The velocity difference, divided by the segment length, gives the relative growth rate of the segment. Now, imagine that the segment shrinks to a point x. To find the relative growth rate, r, at point x, we can use calculus: The value of r is given by the velocity gradient, dv/dx, where v is the growth velocity. This measure of the local growth rate is called the relative elemental growth rate. It represents the fractional change in length per unit time and has units of h^{-1}. If the growth velocity is known, the relative elemental growth rate can be evaluated by differentiation of the growth velocity with respect to position. Or we can approximate r by measuring the relative growth rates of closely spaced segments of initial length L. For each segment, $r = (1/L)(dL/dt)$.

The relative elemental growth rate tells us the rate of expansion of any particular point on the root. It is the most useful measure of root growth rate for the physiologist because it tells where the most rapidly expanding regions are located.

The relative elemental growth rate profile, $r(x)$, is a spatial description of the expansion pattern in the growth zone. In the maize root, the relative elemental growth rate has a maximum of 0.40 h^{-1} at 4 mm and drops to zero by 12 mm. That is, the tissue element at 4 mm is expanding 40% h^{-1}, while tissue beyond 12 mm has ceased expansion. Again, to understand the profile in physiological terms, we must consider that individual cells expand more and more rapidly as they are displaced to 4 mm and then less and less rapidly as they are displaced through the basal part of the growth zone.

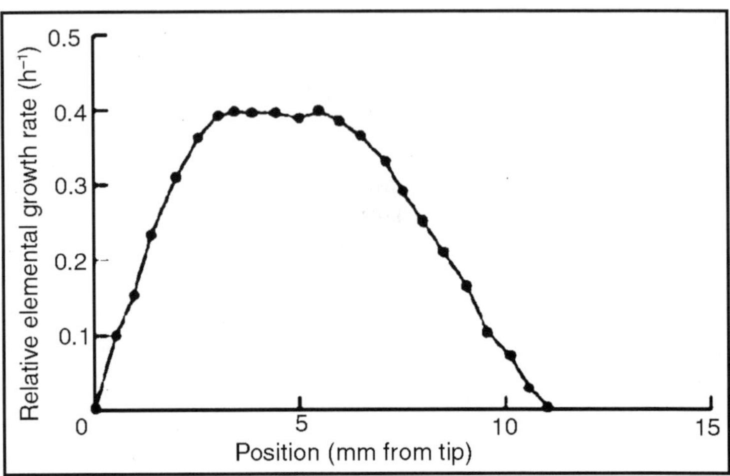

Fig. Relative Elemental Growth rate of *Zea mays* (maize).

The relative elemental growth rate profile shows the location and magnitude of the extension rate and can be used to quantify the effects of environmental variation on the growth pattern. It is also the basis for studying the physiology of growth. The graph of $r(x)$ can be compared to the distribution pattern of any substance thought to be regulating the growth rate. A close correlation between the two curves would provide circumstantial evidence for the role of the substance as a limiting factor for growth.

Embryonic Dormancy

Embryo Maturation Requires Specific Gene Expression

The Arabidopsis embryo enters dormancy after it has generated about 20,000 cells. Dormancy is brought about by the loss of water and a general shutting down of gene transcription and protein synthesis—not only in the embryo, but also throughout the seed. To adapt the cell to the special conditions of dormancy, specific gene expression is required.

For example, the Abscisic Acid Insensitive3 (*ABI3*) and *FUSCA3* genes are necessary for the initiation of dormancy and are sensitive to the hormone abscisic acid, which is the signaling molecule that initiates seed and embryo dormancy.

ABI3 also controls the expression of genes encoding the storage proteins that are deposited in the cotyledons during the maturation phase of embryogenesis.The *LE*AFY *C*OTYLEDON*1* (*LEC1*) gene also is active in late embryogenesis. Because *lec1* mutants cannot survive desiccation and do not enter dormancy, the embryos die unless they are rescued through isolation before desiccation occurs.

The rescued embryos will germinate in culture and produce fertile plants, which are like wild-type plants except that they lack the 7S storage protein and they have leaflike cotyledons with trichomes on their upper surface. The normal appearance and development of the mature *lec1* mutants indicates that the *LEC1* gene is required only during embryogenesis. Although the most obvious defects of the *lec1* mutants are seen only in the maturation phase embryo, mRNA from *LEC1* gene expression can be detected throughout embryogenesis. It has been proposed that *LEC1* is a general repressor of vegetative development and its expression is necessary throughout embryogenesis.

POLARITY OF FUCUS ZYGOTES

Plant cells perceive their position within a tissue and assume the polarity appropriate for that tissue. In addition, cells can perceive and act on information coming from their environment to alter their polarity and redirect growth when necessary.

Cellular polarity must be established to orient subsequent expansion. Cells that are initially apolar, such as zygotes, spores, and pollen grains, later become polarized and initiate tip growth and have been especially valuable in the analysis of the induction and establishment of polarity. A wide variety of external gradients, including temperature, light intensity, pH, ions, and electric-potential gradients, can polarize and orient the axis of growth of these cells. Thus, polarity can be experimentally manipulated without the complexities associated with multicellular tissues. Much of our understanding of the establishment of polarity in tip-growing cells is derived from studies of rhizoid emergence in zygotes of the brown alga *Fucus*.

Polarity of the *Fucus* Zygote is Initially Unstable but Later Fixed

Unlike higher plants, and even unlike most other algae, certain brown algae, particularly *Fucus* and *Pelvetia*, shed their eggs and sperm into the sea, where fertilization occurs. The zygote initially lacks a cell wall and is completely apolar. However, a polar axis is established within a few hours of fertilization in response to unilateral light. The axis is not stable at first and can be changed many times simply by a change in the direction of the light.

The polar axis becomes fixed within 10 to 14 hours after fertilization, after which it cannot be altered by a change in the direction of the light, and a structure that will become the rhizoid begins to protrude from the darkened side. Even though the zygote is still a single cell, it is polarized, with two structurally different poles, the thallus and rhizoid poles, which have radically different fates. The first zygotic division is asymmetric and occurs at right angles to the polar axis. It establishes two cells, each having the identity of the pole from which it emerges. The smaller *rhizoid cell* produces a filament of cells and the holdfast that attaches the plant to rocks. The larger *thallus cell* divides to produce a globular mass of cells, from which the main body of the plant, the thallus, forms.

Calcium Channel Redistribution Leads to a Transcellular Current

A dye known as fluorescent dihydropyridine (FL-DHP), which binds to calcium channels in mammalian cells and can interfere with calcium uptake, has been used as a probe to examine the polarization of the *Fucus* zygote cell membrane.

During the first 6 to 8 hours after fertilization, and in response to unilateral light, FL-DHP receptors, presumably calcium channels, are redistributed and become localized on the dark side of the *Fucus* zygote at what will become the rhizoid pole. If the direction of the light is changed, the location of the FL-DHP–binding proteins also changes, again becoming concentrated on the dark side of the zygote. One of the earliest events in the establishment of polarity after exposure of the zygotes to unilateral light is the induction of a positive current that flows into the cell on the shaded side and loops back through the external medium. This transcellular current, which is due largely to calcium ions, is on the order of 100 pA (1 picoampere = 10^{-12} ampere) and develops in advance of any morphological sign of polarity.

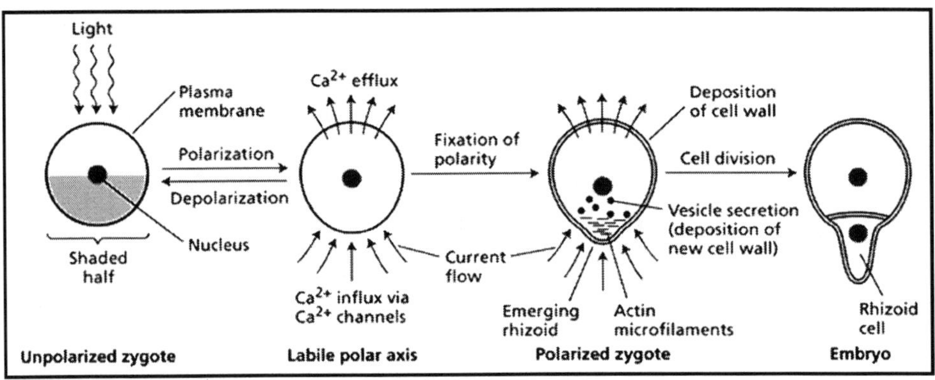

Fig. The Events Involved in the Establishment of Polarity in Zygotes of the brown alga *Fucus*.

(A) The zygote is polarized by an asymmetric stimulus from the environment, such as unilateral light. (B) A current flows (charged calcium ions move) through the polarized but still spherical zygote at the site at which the rhizoid will emerge, driven by Ca^{2+} uptake in the shaded half of the cell, from which the rhizoid will emerge. The current (Ca^{2+}) flows out on the opposite side.

(C) Cell polarity becomes fixed when actin microfilaments assemble at the site of rhizoid emergence and a cell wall is assembled around the zygote. The cell wall completely surrounds the zygote, but its composition differs in the rhizoid and thallus halves. Vesicles containing sulfated polysaccharides are transported by actin filaments to the plasma membrane and deposited in the cell wall only at the site of rhizoid emergence. (D) Finally, the zygote divides, and the rhizoid cell grows at its tip. The relevance of this phenomenon to the establishment of cell polarity is indicated by the fact that the site at which positive charge flows into the cell is also the site of subsequent rhizoid outgrowth. Transcellular ionic currents also have been observed in some tip-growing cells of higher plants, including root hairs and germinating pollen grains, but they have not yet been detected in plant cells that expand by diffuse growth. The ionic current in polarizing *Fucus* zygotes likely arises from the redistribution of ion channels in the plasma membrane. This current generates an electric-potential difference across the membrane that could alter other membrane properties, such as the distribution of membrane proteins.

The current also alters the ionic composition of the cytosol; the magnitude of the effect depends on the ions that form the current, which vary in different cell types. Because the Ca^{2+} concentration in the cytosol is usually very low (between 10^{-6} and 10^{-8} μM), even a small increase in intracellular Ca^{2+} concentration could have a marked effect.

Changes in Cytosolic Calcium Have Been Measured

The fluorescent calcium-indicator dye *fura*-2 has been used to detect intracellular cytosolic calcium gradients in both plant and animal cells. Since the excitation spectrum of fura-2 shifts when it binds calcium, we can determine the intracellular concentrations of both forms of the dye (with and without bound calcium) by exciting them with the appropriate two wavelengths.

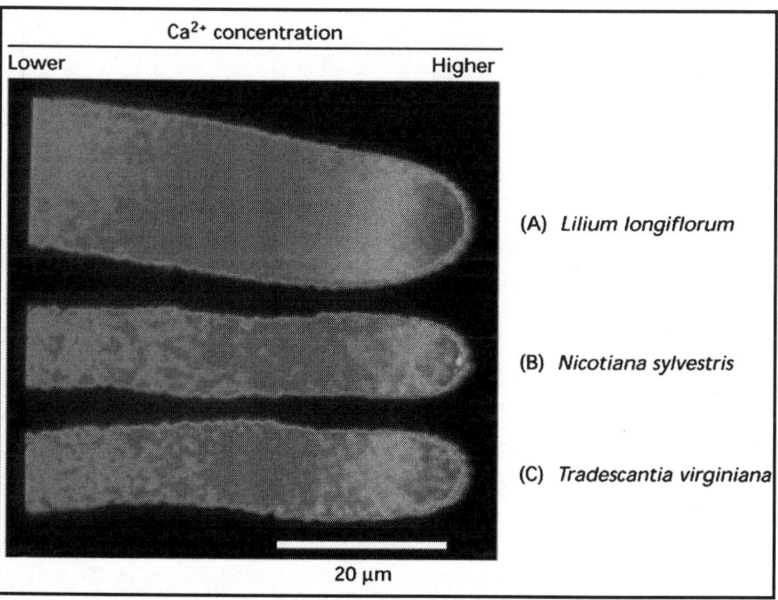

Fig. Tip-growing pollen tubes exhibit a steep intracellular gradient of calcium ion concentration, with the highest levels at the growing tip.

The ratio of the two emissions provides a measure of the calcium concentration that is independent of the dye concentration. Using this technique to measure the cytosolic Ca^{2+} gradient in growing *Fucus* rhizoids, researchers observed a gradient from 450 nM at the tip to 100 nM at the base.

This calcium ion gradient may be important for the polarized assembly of actin filaments and vectoral discharge of Golgi-derived vesicles at the rhizoid pole. Calcium ion gradients also have been shown to be established in other tip-growing cells, such as germinating pollen grains, where the direction of growth and the region in which Golgi vesicles carrying cell wall components are deposited are predicted by the calcium gradient.

Pollen tubes from three species of plants were microinjected with a fluorescent calcium indicator dye to demonstrate this Ca^{2+} gradient, ranging from about 1 μM at the extreme tip to about 0.2 μM at the base.

The Actin Cytoskeleton Becomes Polarized

Inhibition of protein synthesis by treatment with the drug *cycloheximide* delays rhizoid formation but does not stop the establishment of polarity. This result suggests that although rhizoid formation requires the synthesis of new proteins, the establishment of polarity does not, so polarity development must involve a redistribution of existing components. Actin filaments become preferentially localized in the rhizoid half of the zygote and extend from the vicinity of the nucleus to the plasma membrane at the site marked by localized FL-DHP–binding proteins as polarity develops.

Actin filaments are dynamic, and their assembly from a pool of monomeric actin subunits is highly regulated. Actin filaments are localized in the rhizoid half of the cell presumably in response to a change in the assembly or disassembly dynamics, and this localization involves a redistribution of the cellular actin. The fixation of polarity can be blocked by *cytochalasin B*, a drug that destroys actin filaments. Treatment with cytochalasin B also disrupts the ionic current that develops during polarization. Actin filaments participate in cell secretion by transporting secretory vesicles to the plasma membrane in many types of cells.

Secretion of Golgi-Derived Vesicles Is Localized

Cytoplasmic organelles become asymmetrically distributed before the polarity of the cell is fixed. Golgi-derived vesicles accumulate at the site of the future outgrowth in the rhizoid half of the *Fucus* zygote as polarization is induced by light. For several hours after the induction of polarity, the asymmetric distribution of vesicles can be changed by subsequent treatment with light from a different direction. During this phase when changes are possible, cytoplasmic vesicles accumulate at the prospective site of rhizoid formation on the shaded side of the cell. Subsequently, the vesicles fuse with the plasma membrane, secreting their contents into the cell wall, leading to rhizoid emergence. Evidence suggests that calcium gradients play a role in triggering vesicle discharge at the pole. The fixation of light-induced polarity in *Fucus* zygotes also requires the presence of the cell wall. Removal of the wall with cell wall–digesting enzymes does not prevent the induction of polarity by unilateral light, but it does prevent fixation of the axis. The orientation of the embryonic axis remains unstable until a new cell wall is synthesized.

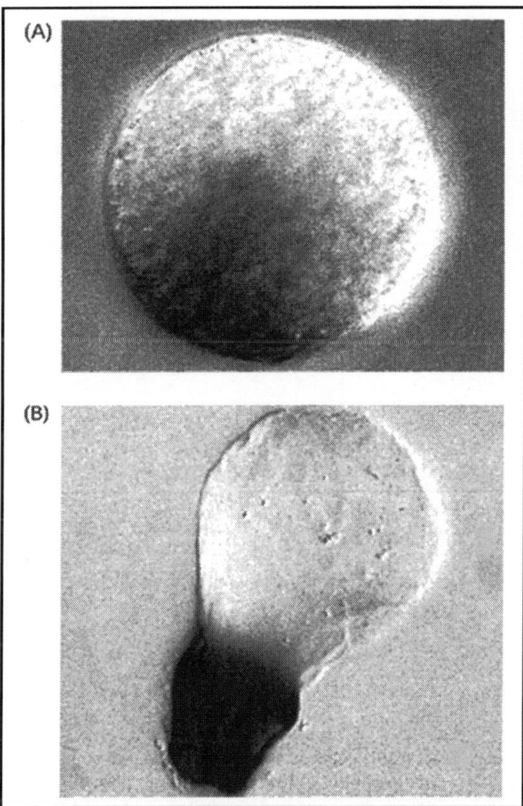

Fig. Cytoplasmic Vesicles that contain Specific cell wall Components become Asymmetrically Distributed early in the Development of *Fucus* zygote polarity.

(A) A toluidine blue stained *Fucus* zygote before fixation of the polar axis. The toluidine blue dye stains a wall component in vesicles or in walls. (B) The embryo after establishment of rhizoid–thallus polarity, with toluidine blue staining confined to the cell wall in the rhizoid portion of the embryo.

From these observations it is possible to envisage a chain of events that might occur during the establishment of cell polarity in tip-growing cells.

- The ionic currents that arise early in the process may initiate the formation of an intracellular Ca^{2+} gradient.
- The increase in intracellular calcium at the tip may help organize and stabilize the microfilament components of the cytoskeleton.

- The cytoskeleton directs vesicle traffic to the growing tip.
- A calcium gradient, perhaps immediately adjacent to the plasma membrane, may promote vesicle fusion with the plasma membrane, although this remains to be demonstrated.
- Finally, cell wall deposition fixes the polarity of the cell, possibly by interacting with transmembrane proteins.

THE PREPROPHASE BAND OF MICROTUBULES

As we saw, microtubules in the cortical cytoplasm of growing plant cells are important in determining the orientation of cellulose microfibrils in the cell wall. When dividing cells enter the G_2 phase of the cell cycle, the interphase cortical microtubule array is replaced by a narrow band of interconnected microtubules and microfilaments encircling the nucleus. This circular array of microtubules and microfilaments is called the preprophase band. The preprophase band was first observed in 1966 by Jeremy Pickett-Heaps and Donald Northcote at Cambridge University. It appears as a transient ring of cytoskeletal elements in the cortical cytoplasm, near the plasma membrane. Generally the preprophase band is 2 to 3 μm wide and contains from 10 to more than 100 microtubules, interspersed and interconnected with actin microfilaments.

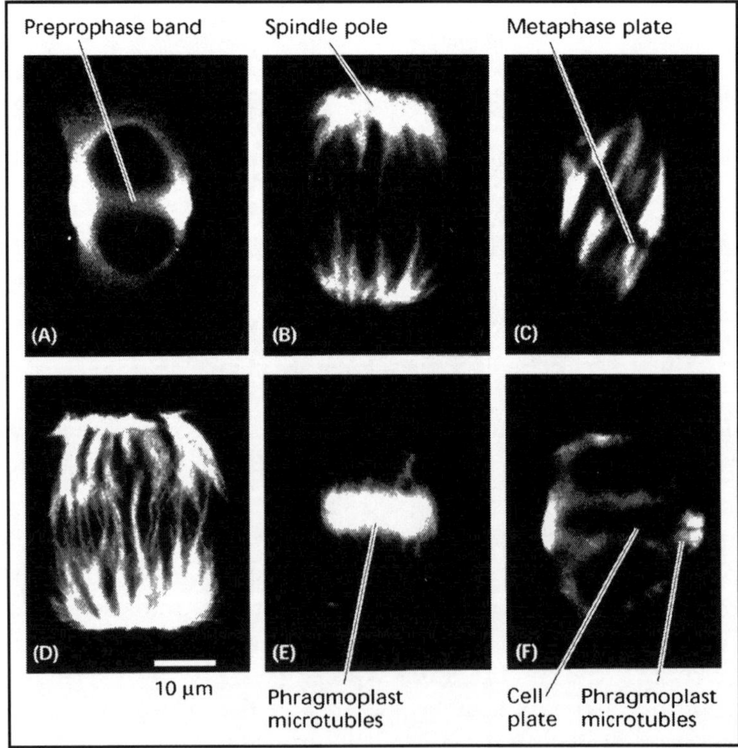

Fig. Immunofluorescent Staining of the Cytoskeleton in onion (*Allium cepa*) root tip cells with an Antibody to the Microtubule protein, tubulin.

As cells progress through their cell cycle, different cytoskeletal arrays that contain microtubules are formed. (A) A preprophase band appears after DNA replication is complete, but before the cell enters mitosis. It can be seen as two bright spots on either side of the unstained nucleus; the image of the band is fainter as it passes across the nucleus. (B) The mitotic spindle forms in late prophase, with spindle microtubules emanating from the polar regions to form the spindle fibers. (C) The spindle fibers hold the chromosomes (dark line) in the metaphase plate. In this cell the alignment of the plate is oblique, which is relatively unusual. (D) The chromosomes move toward their poles within the array of spindle microtubules. (E) The phragmoplast forms in telophase and is seen here as a bright band in the centre of the cell that consists of microtubules, microfilaments, and vesicles in the early stage of cell plate formation. (F) Phragmoplast microtubules form an annular ring around the perimeter of the cell plate and appear only at the edges of the cell. Bar = 10 μm.

As the cell enters mitosis, microtubules begin to assemble at the surface of the nuclear envelope, and as the nuclear envelope breaks down, they invade the nuclear area to form the mitotic spindle. The preprophase band disappears as the spindle forms, leaving behind an actin-depleted zone. The plane in which cytokinesis occurs is determined before the cell initiates mitosis and is accurately predicted by the position of the preprophase band. It has been suggested that the preprophase band determines the orientation of the mitotic spindle and that it marks the site at which the new cell wall will form during cytokinesis, but this has not been established. The position of the preprophase band and the plane of cytokinesis both may be determined by another mechanism, possibly sensitive positional information.

GENES REQUIRED FOR SHOOT APICAL MERISTEM FORMATION: CLASS III HD-ZIP GENES

Another class of transcription factors, encoded by CLASS III HD-ZIP class genes, has recently been shown to play essential roles in establishing the Shoot Apical Meristem (SAM). Mutants which lack activity of the previously discussed *REV* gene, as well as the related *PHABULOSA* and *PHAVOLUTA* genes, are unable to form an SAM. As will be discussed later, these genes also promote establishment of adaxial tissues in the leaves. As meristems often form in close association with these tissues (e.g., in the axils of leaves), models have been proposed in which the expression of CLASS III HD-ZIP genes promote what could be termed "central" cell identities. In the stem, the "central" identity would include regions that contain apical initials, as opposed to the organogenic regions of the flanking peripheral zone. In the context of leaf development, this "central identity" would encompass the adaxial tissues, including the asymmetrically positioned xylem elements of the leaf as well as closely associated domains from which axillary meristems arise. This type of model receives further support from studies in which the overexpression of CLASS III HD-ZIP class genes causes SAMs to form in abnormal positions.

The regulation of CLASS III HD-ZIP genes is complex. Their polarized patterns of expression, and the mutant phenotypes that result from perturbing this expression, suggests that the normal patterning of the apical domain depends on radial patterning processes. As discussed earlier, the proteins encoded by these genes contain START domains, suggesting their activity is regulated by as yet undefined sterol type signaling molecules. The region that encodes the START domain is further distinguished by characteristics that suggest these genes are targets of microRNA (miRNA) mediated repression of transcript levels.

ORGANELLES OF PLANT CELL

Plants are highly evolved eukaryotic organisms that comprise of membrane bound cell organelles. Even though plants and animals belong to eukaryotic groups, they differ in certain characteristic features.

For example, a plant cell possesses a well-developed cell wall and large vacuoles, while an animal cell lacks such structural parts. In addition to these, a plant cell lacks centrioles and intermediate filaments, which are present in an animal cell.

A typical plant cell is made up of cytoplasm and organelles. Scientific studies have been done regarding plant cell organelles and their functions. Each of the organelles of a plant cell has specific functions, without which the cell cannot operate properly.

LIST OF PLANT CELL ORGANELLES

When it comes to plant cell organelles, they are more or less similar to animal cells, except that the latter lacks chloroplast organelles, that are responsible for photosynthesis. Following is a list of organelles found in plant cell:

Nucleus

Nucleus (plural nuclei) is a highly specialized cell organelle, which stores the genetic component (chromosomes) of the particular cell. It serves as the main administrative center of the cell, by coordinating the metabolic processes like cell growth, cell division and protein synthesis.

Plastids

Plastids are collective terms for organelles that carry pigments. In a plant cell, chloroplasts are the most prominent forms of plastids, which contain the green chlorophyll pigment. Because of these chloroplast plastids, a plant cell has the ability to undergo photosynthesis in the presence of sunlight and synthesize its own food.

Ribosomes

Ribosomes are plant cell organelles that comprise of proteins (40 per cent) and ribonucleic acid or RNA (60 per cent). They are important organelles responsible for the synthesis of proteins. Each ribosome consists of two parts, a larger subunit and a smaller subunit.

Mitochondria

Mitochondria (singular mitochondrion) are spherical to rod-shaped organelles present in the cytoplasm of the plant cell. They break down the complex carbohydrates and sugars into usable forms, for the plant. As mitochondria indirectly supply energy for the plant cell, they are also called as the powerhouse of the cell.

Golgi Body

Golgi body is also referred to as golgi complex or golgi apparatus. It plays a major role in transporting chemical substances in and out of the cell. After the endoplasmic reticulum synthesizes lipids and proteins, golgi body alters and prepares them for exporting outside the cell.

Endoplasmic Reticulum

Endoplasmic reticulum (ER) is the connecting link between the nucleus and cytoplasm of the plant cell. Basically, it is a network of interconnected and convoluted sacs that are located in the cytoplasm.

Based on the presence or absence of ribosomes, ER can be of smooth or rough types. The former type lacks ribosomes, while the latter is covered with ribosomes. Overall, endoplasmic reticulum serves as a manufacturing, storing and transporting structure for glycogen, proteins, steroids and other compounds.

Vacuoles

Vacuoles are the storage organelles that help in regulating turgor pressure of the plant cell. In a plant cell, there can be more than one vacuole. However, the centrally located vacuole is larger than others, which stores all sorts of chemical compounds. Vacuoles also assist in intracellular digestion of complex molecules and excretion of waste products.

Peroxisomes

Peroxisomes are cytoplasmic organelles of the plant cell, which contains certain oxidative enzymes. These enzymes are used for the metabolic breakdown of fatty acids into simple sugar forms. Another important function of peroxisomes is to help chloroplasts in undergoing photorespiration process.

Well! This was brief information regarding plant cell organelles, their structure and specific functions. As we have seen, coordination of the functions of plant cell organelles is crucial for carrying out the physiological and biochemical functionalities of the plant.

Cell Nucleus: Structure

The structure of a cell nucleus consists of nuclear membrane (nuclear envelope), nucleoplasm, nucleolus and chromosomes. Nucleoplasm, also known as karyoplasm, is the matrix present inside the nucleus. Let's discuss in brief about the several parts of a cell nucleus.

Nuclear Membrane

The nuclear membrane is a double-layered structure that encloses the contents of the nucleus. The outer layer of the nuclear membrane is connected to the endoplasmic reticulum. A fluid-filled space or perinuclear space is present between the two layers of a nuclear membrane. The nucleus communicates with the remaining of the cell or cytoplasm through several openings called nuclear pores. Nuclear pores are the sites for the exchange of large molecules (proteins and RNA) between the nucleus and cytoplasm.

Chromosomes

Chromosomes are present in the form of strings of DNA and histones (protein molecules) called chromatin. Chromatin is further classified into heterochromatin and euchromatin based on the function. The former type is a highly condensed trancriptionally inactive form, mostly present in adjacent to the nuclear membrane. Euchromatin is a delicate, less condensed organization of chromatin, which is found abundantly in a transcribing cell.

Nucleolus

The nucleolus is a dense, spherical-shaped structure present inside the nucleus. Some of the eukaryotic organisms have nucleus that contains up to four nucleoli. The nucleolus plays an indirect role in protein synthesis by producing ribosomes. Ribosomes are cell organelles made up of RNA and proteins; they are transported to the cytoplasm, which are then attached to the endoplasmic reticulum. Ribosomes are the protein-producing structures of a cell. Nucleolus disappears when a cell undergoes division and is reformed after the completion of cell-division.

Cell Nucleus: Functions

Speaking about the functions of a cell nucleus, it controls the hereditary characteristics of an organism and is responsible for the protein synthesis, cell division, growth and differentiation.

Here is a list of the functions carried out by a cell nucleus:

- Storage of hereditary material, the genes in the form of long and thin DNA (deoxyribonucleic acid) strands, referred to as chromatins.

- Storage of proteins and RNA (ribonucleic acid) in the nucleolus.
- Nucleus is a site for transcription in which messenger RNA (MRNA) are produced for the protein synthesis.
- Exchange of hereditary molecules (DNA and RNA) between the nucleus and rest of the cell.
- During the cell division, chromatins are arranged into chromosomes.
- Production of ribosomes (protein factories) in the nucleolus.
- Selective transportation of regulatory factors and energy molecules through nuclear pores.

As the nucleus regulates the integrity of genes and gene expression, it is also referred to as the control center of a cell. Overall, the cell nucleus stores all the chromosomal DNA of an organism.

The environment which we inhabit and live our daily lives, is an incorporation of several kinds of components. These components, although may not share any resemblance, yet are in a mutual dependency with each other. These components not only refer to the living things, but other elements which are classified as the non-living things on the Earth. Both these elements are important in their own domain in the environment, as they work to maintain the delicate balance in the ecosystem. And this very balance helps to regulate all those things which makes life possible on the Earth. Ergo, with any of these things missing in this system, the biological organization will suffer nothing less than chaotic events. So let's proceed further to understand these types of components and what role do they play in the levels of organization of life.

LEVELS OF ORGANIZATION

In an ascending manner, the levels of biological organization are as follows:

Subatomic Particles

Any substance is composed mainly of three subatomic particles. They include the protons, neutrons and electrons. As you must be aware, protons refer to positively charged particles, while neutrons are ones which posses no charge. These two kinds reside inside the nucleus of the atom. Now, about the electrons, they posses negative charge and rotate around the nucleus, in different energy shells. Another subatomic particle is what is known as the photon - possessing zero mass and rest energy and traverses with the speed of light. It is actually defined as a quantum of electromagnetic energy.

Atoms

After the subatomic particles, now we come to the next point in the levels of organization, know as atoms. All objects are made up of matter in this universe. And the basic building blocks of these matter are known as atoms.

These components consist of equal number of protons and neutrons. However, a single element might have two different atoms, wherein, there might be a difference in the number of neutrons. The center of the atom is known as the nucleus, and it comprises of protons and neutrons. This structure of the atomic nucleus is orbited by negatively charged electrons.

Small Molecules

By saying small molecules, I am referring to those levels of organization, which comprises of amino acids, fatty acids, glucose, etc.

Macromolecules

Now these are the large molecules which are contained in a cell, and are required for carrying out the vital processes of life. Examples include carbohydrates, lipids, proteins, nucleic acids, etc.

Molecular Assemblies

A level higher than macromolecules, these assemblies are defined as sets which comprise of one or more molecular entity.

Organelle

Every cell has a specialized part which serve as organs. These are known as organelle. Examples are nucleus, mitochondria, etc.

Cells

This is one of the important levels of organization, wherein, cells are known to play vital roles. Every living creature on this Earth consists of cells, which are primarily, their structural and functional units. For instance, microorganisms like bacteria are known to be single-celled organisms. While, we humans and other organisms, are known to have as many as 100,000,000,000,000 cells!

Tissues

Tissue is nothing but a term that is used to refer to a group of cells. There are several functions in the body of living organisms. And different function is assigned to different tissues. For example, muscle tissue, nervous tissue, etc., are found in animals, while plants, have meristematic tissues and permanent tissues.

Organs

Again, when tissues are grouped together, they form what is known as an organ, which also has an important place in the levels of organization. As you are aware, heart, lungs, kidneys, stomach are organs. And these organs are enabled by the different tissues to carry out their specific duties.

Organ System

As you must have guessed already, this system is a formed by different kinds of organs. The system is assigned to perform higher levels of work in the body. This is more important in multicellular organisms like us and animals. A very simple example of an organ system is the circulatory system. It is a collaboration of heart, blood and blood vessels.

Organisms

Now when all the above levels of organization are taken in one picture, what is derived is known as organism like plants, animals, humans, bacterium, etc.

Population

It is here that we speak about a broader concept in the levels of organization of living things. A group of organisms, belonging to the same species may have the capability to inter-breed with each other. And when they inhabit a particular area, they are known as population. Its example can be the population of frogs in a pond.

Species

Species is a term that is used to mark the distinction between different kinds of organisms. For example, human species are different from animal species and species of plants.

Community

Taking it down to the last three of the levels of organization, we are speaking of community. It is nothing but a reference to a group of organism of different species, which populate a given area and stay in mutual interaction with each other.

Ecosystem

To define it in a layman language, interaction of a group of organism with the environment that they live in, is an ecosystem. For example, a pond ecosystem may consists of organisms such as fish, frogs, insects, etc., who depend on the water, mud, plants, etc., for survival.

Biosphere

Finally, we have reached the last block of the pyramid of the levels of organization, with what is known as biosphere. It encompasses all. Meaning, it consists of the regions which lie on, above and below the surface of the Earth, including the atmosphere where the living organisms thrive.

So there you are with a quick explanation on the different levels of organization, which we, along with several other components are living. As we can see, each level serves as a building block for the other, and this is what lays emphasis on the importance of all the components.

MODERN PLANT BIOTECHNOLOGY

Plants are made of millions of cells all working together. Every cell of a plant has a complete "instruction manual" or genome that is inherited from the parents of the plant as a combination of their genomes.

Genes are found within the genome and serve as the "words" of the instruction manual. When a cell reads a word, or in scientific terms "expresses a gene," a specific protein is produced. Proteins give an individual cell, and therefore the plant, its form and function. Genes (words) are written using the four-letter alphabet A, C, G, T. The letters are abbreviations for four chemicals called bases, which together make up DNA. DNA is universal in nature, meaning that the four chemical bases of DNA are the same in all living organisms. Consequently, a gene from one organism can function in any other organism.

The ability to move genes into plants from other organisms, thereby producing new proteins in the plant, has resulted in significant achievements in plant biotechnology that were not possible using traditional breeding practices.

Clearly, the ultimate solution to each of these problems is reducing population growth, a difficult challenge that is further complicated by social, political, economic and religious considerations. In the hope that national and international efforts will help to stabilize world population in the next few decades, our challenge is to use the power of plant biotechnology towards the solution of the numerous problems caused by population growth by increasing productivity, by reducing crop losses, and by protecting and conserving the environment. Plant biotechnology is not a magic bullet that will solve all of these problems, yet it is becoming abundantly clear that it is the best tool that we have and it can, if used wisely and in a timely fashion, make significant contributions.

It is rather ironic that at a time when international agriculture is under increasing pressure to meet the food needs of the ever increasing population, and when plant biotechnology is beginning to make significant contributions to food productivity and environmental safety, it has become the target of well coordinated and sustained attacks by many environmental and self-appointed watchdog groups, particularly in Western Europe and in some of the developing countries. In the United States, we had faced similar attacks during the 1970's and 1980's.

However, after extensive public debate, protest demonstrations, court challenges and congressional hearings, a federal regulatory framework was developed that has served the public and the private interest well. It has allowed the plant biotechnology industry to grow and introduce its products into the market place. The American consumers and farmers have accepted and benefited from transgenic products. Transgenic crops are being grown this year on nearly 100 million acres of American farmland, accounting for 74 per cent of our soybean, 71 per cent of our cotton and 32 per cent of our corn acreages.

It is fortunate and encouraging that China and India, the two most populous countries in the world, with increasing demand for food and worsening environmental problems, have recognized the importance of plant biotechnology in agriculture and have established active and successful research and development programmes in plant biotechnology, targeting many regional vegetable and fruit crops, in addition to such staples as wheat, rice, maize, soybean, various pulses, canola, and cotton. It is not too far-fetched to expect that within the next few years these two countries will plant the largest acreages of transgenic crops in the world.

Argentina and South Africa are two other developing countries that are increasing their planting of transgenic crops. Indeed, at the present time, nearly 10 per cent of the global acreage of transgenic crops is planted in the developing countries. Thus the argument that plant biotechnology is a tool of the industrialized countries for the exploitation of the developing world is no longer sustainable. It is our hope that the success of plant biotechnology in these countries will encourage similar efforts in other parts of the developing world.

It is true that the first generation of transgenic crops, which contain genes for resistance to herbicides and insects, did not provide any direct benefits to the consumer. Nevertheless, there are numerous indirect benefits, such as the reduced use of pesticides and herbicides, reduced tillage leading to soil conservation, reduced use of natural resources such as petrochemical products and water for the manufacture, transport and application of agro-chemicals, and reduced labour costs. By producing more food on the same amount of land transgenic crops promote conservation and biodiversity by saving wildlife habitats and precious forests from being converted into farmland. Reduced use of pesticides has already shown a marked decrease in illness and death caused by pesticide poisonings in China and South Africa (nearly 500 cotton farmers in China die each year of acute pesticide poisoning).

The vital role of agriculture and food production in human health and nutrition, in poverty alleviation, and in social and political stability, is well known. This was recognized as far back as 1970, when Norman Borlaug was awarded the Nobel Peace Prize for his work that led to the Green Revolution and that helped to save hundreds of millions of lives in the developing countries. Plant biotechnology too can contribute to international peace and security by increasing food production, producing safer and healthier foods, protecting our rather finite natural resources and the environment, and improving human health. It is, therefore, morally and socially irresponsible and indefensible to prevent or delay the applications of plant biotechnology to problems of hunger, health and protection of the environment. The opponents of plant biotechnology would have us believe that it is an unnatural and unsafe process that produces harmful products, and that it is totally different from plant breeding and selection that account for almost all of our modern crops. The indisputable fact, however, is that humans have engineered crops for nearly 10,000 years. Almost all of our major crops-such as maize, wheat, potato, tomato and others-are man made. Indeed, none of our modern crops are capable of surviving in the wild without human care.

The molecular and genetic principles of plant biotechnology and plant breeding and selection are the same. Plant biotechnology is no different from breeding and selection or for that matter from radiation and chemically induced mutation breeding, except that it is extraordinarily precise and predictable, and is not restricted by taxonomic boundaries. No compelling evidence has ever been presented to show that transgenic crops are innately different the non-transgenic products of breeding and selection.

In retrospect, however, we must share some of the blame for the perception that plant biotechnology is different from plant breeding and selection. During the 1970's and 1980's, when there was a great deal of euphoria over the production of somatic hybrids, doubled-haploid breeding lines and transgenic plants, the plant biotechnology community made a serious error in strategy and judgement when it distanced itself from breeding and selection and established a separate identity for itself. It was us who placed the spotlight on the process and not the product. This has come to haunt us now as it has attracted undue and undeserved attention and opposition. It was also an error not to engage early in the debate on transgenic plants, and to permit the opponents of plant biotechnology to dictate the agenda.

As responsible members of the world community, and as scientists, we cannot, and should not, be silent observers of this debate. We must play an active role in the debate on plant biotechnology and make an informed and professional contribution to the public dialogue, emphasizing the many benefits of transgenic crops to human health and the environment. The opposition to transgenic foods in Europe and elsewhere is based exclusively on political and ideological differences rather than on any credible scientific evidence. On two rare occasions an attempt was made to present scientific arguments against the use of transgenic crops. These involved the allegedly harmful effect of pollen from Bt maize plants on the larvae of the Monarch butterfly, and the alleged transgene contamination of maize in Mexico.

More detailed investigations by several research groups have since refuted these claims and showed them to be of dubious scientific value. Indeed, in the Mexican maize story, the journal Nature, in an unprecedented action in its more than 100 year history, was forced to disown the paper published in its own pages. The consumer, the farmer and the biotechnology industry have been ill served by the sustained campaign of misinformation and unsubstantiated claims of dangers to public health and the environment by transgenic crops and their products. After more than ten years and thousands of field trials in many countries, after nearly a decade of commercial plantings on hundreds of millions of acres, and after transgenic food products having being used by hundreds of millions of humans and farm animals, there is not a single documented instance of illness reported in any human or animal, or of ecological or environmental damage. What then is the basis and rationale for the many restrictions still placed on the field planting and human use of transgenic foods?

The enviable and unblemished record of transgenic crops and their products is the strongest evidence for their safety and wholesomeness. The opponents of plant biotechnology should compare this record with that of the many drugs approved for human use in the United States. In an exhaustive study published recently in the Journal of the American Medical Association, it was reported that 20 per cent of the 548 drugs approved for human use during the past 25 years were later found to have serious or life-threatening side effects. Seven of the drugs possibly contributed to 1002 deaths, and 16 were forced to be withdrawn from the market. In comparison, not a single transgenic food product has ever been shown to have any harmful effects, and none has been withdrawn because of adverse reactions in humans or animals.

The plant biotechnology community has already done more for the environment and the developing countries than the self-proclaimed environmental groups and the so-called friends of the poor. Indeed, the opponents of plant biotechnology have done much harm to their professed cause by slowing down and/or preventing the planting and utilization of transgenic crops around the world. The contributions of the plant biotechnology community, on the other hand, are socially and morally responsible and of considerable humanitarian value. We have every reason to be proud of these contributions.

The rules and regulations adopted for transgenic crops in the 1980's were both prudent and necessary. At that time there were many unknowns about transgenic crops and about their possible effect on humans and the environment. There was a need to establish a database to satisfy the concerns of the general public as well as the scientific community. Three federal agencies, the United States Department of Agriculture, the Environmental Protection Agency and the Food and Drug Administration, were given oversight responsibilities for transgenic crops. The resulting open and transparent system established in the United States has worked well and has served its purpose. It has done much to gain the confidence and support of the American public for plant biotechnology and its products. In light of the demonstrated safety of transgenic crops to humans, animals and the environment, the question must be asked whether it is any more necessary, or even advisable, to continue the expensive, time consuming and burdensome requirements for the public release of transgenic crops and their products (field trials of transgenic crops are 10-20 times more expensive than of similar plants developed by conventional means).

We propose that based on our considerable experience and on the vast amount of information gathered about the safety of transgenic crops over the past decade, it is time for our regulatory agencies to consider whether some or all of the current regulatory requirements can be gradually relaxed and ultimately suspended, except in those rare instances where there is the clear likelihood of risk to human health and the environment. Genuine concerns about gene flow and development of resistance to antibiotics, and pests or pathogens, can be met adequately with currently available and emerging technologies.

The process of deregulation of transgenic crops, controversial and difficult as it may be, needs to begin now because the continuation of the present rules and regulations is entirely unnecessary, unjustified and counterproductive. In order to be effective and acceptable, the process should be open to all points of view. The decisions, however, must be based on science and facts and not on political or ideological considerations. Nearly two decades ago, the United States played a leading and useful role in establishing the rules and regulations for the field planting, evaluation and human use of transgenic crops. It should now play a similar role in having these restrictions relaxed and removed.

It is clear that the challenges we face in the 21st century are greater than those we faced in the last century. Of all the available technologies, plant biotechnology offers the best hope for producing more and better food, fibre and pharmaceuticals, and for protecting, preserving and improving the environment for the benefit of humankind. My own confidence in plant biotechnology comes from knowing that the science behind it is sound, that it is well tested and proven, that it benefits the consumer, the farmer and the industry, and that it protects and conserves the environment. It is for these reasons that I am convinced that plant biotechnology will within the next two decades become an integral part of the international agricultural system. With the United States, China, and lately India, three of the most populous countries in the world serving as examples, we have taken the first steps towards achieving that objective.

7

Phytochrome Regulation of Flowering

The observation that daylength regulates flowering in many angiosperms led to the discovery of phytochrome. However, regulation of flowering by phytochrome is part of a complex regulatory pathway, which has two components – light quality and the circadian clock. The perception of light and onset of dusk and dawn is mediated by phytochromes and by cryptochromes, blue light absorbing photoreceptors of plants. Phytochrome regulation of flowering has been extensively studied in Arabidopsis, and recently many of the components regulating flowering have been identified. A key link between the phytochrome and regulation of flowering genes is provided by the GIGANTEA gene. GIGANTEA (GI) is involved in control of flowering in response to daylength, the gi mutation causes late flowering under long days and makes the plant insensitive to daylength. The GI gene encodes a large protein, which based on homology to other proteins was predicted to be a membrane protein, since it contains five membrane- spanning domains in the N-terminus. The C– terminus of GI is hydrophilic and is not homologous to any known protein.

Ever since the discovery of phytochrome one has sought an answer to how phytochrome regulates floral initiation? The finding that both phytochrome and cryptochrome, which act as light sensors for photoperiodic light detection, can enter the nucleus invokes the possibility that these photoreceptors may directly regulate expression of flowering genes. The analysis of a mutant that shows reduced seedling de-etiolation under red light provided evidence that phytochrome regulates flowering by direct regulation of some of the key genes.

Cloning of the above mutant locus revealed that the mutated gene is the same as the GIGANTEA gene involved\ in control of flowering time. Using GUS–GI and GFP–GI fusion proteins, Huq et al.81 have showed that

GI is constitutively targeted to the nucleus in transient transfection assays. Moreover, the observation of green fluorescence in nucleus showed distribution of GFP–GI protein throughout the nucleoplasm. Interestingly, even though based on homology to other membrane proteins, it was predicted that GI would be a plasma membranelocalized protein, the above finding shows that it is a nucleoplasmically localized protein. This finding, though in conformity with its role in phytochrome B signalling, also adds a new dimension that photoregulation of flowering may be mediated by regulation of nuclear proteins.

Perspective

Finally, one may ask what use we can make of the information about phytochrome regulation in plant development. It may be still too early to answer this question as more has to be learned about phytochrome and its interaction with other regulators present in the plants. However, the analyses of transgenic plants overexpressing phytochrome have highlighted a few of the potential applications.

For example, transgenic tobacco plants, which over-express oat phytochrome A show improved harvest index82 by alleviating the shadeavoidance response in a densely planted plot. In potato, over-expression of Arabidopsis phytochrome B improved photosynthetic performance and increased life span, leading to higher yield of tubers83.

However, these approaches can have their own pitfalls; for example, though the number of tubers is increased in potato, the tubers were smaller than in the wild type83. It is hoped that information on phytochrome and its interacting partners and light-regulated genes would aid in our knowledge about plant development. This information would be helpful to regulate the crop yield in conjunction with other regulatory systems of plants.

INTRACELLULAR REDISTRIBUTION OF PHYTOCHROME

Plants are sessile organisms; therefore optimal adaptation to changes in the natural environment is an essential strategy. Light is a dominant but broadly variable environmental factor. It is not only an energy source but also an essential signal for adaptation and regulation of plant growth and development.

The most dramatic developmental switch occurs at the point of transition from skotomorphogenesis (development in darkness) to photomorphogenesis (development in light). Light-mediated signal transduction starts with the absorption of light by specialised photoreceptors. Several classes of photoreceptors that monitor the quality, quantity and the temporal and special pattern of light have evolved: the UVB photoreceptors (as yet unidentified), red/far-red photoreversible phytochromes, and the blue-light-absorbing proteins cryptochromes and phototropin.

Being an electromagnetic wave, light can easily penetrate plant tissues. Red and far-red light can even pass through compact organs, such as stems or leaves, and activate phytochromes deep within cells. Physiological experiments, in vivo spectroscopy, immunocytochemical studies and ubiquitous activities of phytochrome promoters in transgenic plants indicate that phytochromes are present in all tissue types examined. Therefore, each cell is likely to contain at least one type of phytochrome at any stage of development.

Until 1999, however, the intracellular localisation of most of these photoreceptors, including phytochromes, remained largely obscure. In the case of phytochromes, a controversial debate about their exact intracellular localisation erupted nearly 20 years ago and has lasted until very recently. Phytochromes and cryptochromes were generally thought to be cytosolic, and this view became even more prevalent after the first sequence information became available. Phototropin was thought to be an exception and localised in/at the plasmalemma.

Phototropin

Darwin's pioneering work revealed that phototropism of plants is mediated by a blue-light-absorbing pigment; however, the photoreceptor involved, phototropin, was identified only a few years ago. Physiological and biochemical data showed a strong correlation between phototropism and a blue-light-mediated phosphorylation of a 120 kDA protein associated with plasma membranes.

Using a genetic approach, it has been shown that the nonphototropic hypocotyl locus (NPH1) encodes a serine threonine kinase that contains two N-terminal domains characteristic of proteins mediating light-, oxygen- and voltage-dependent responses (LOV domains). Expression of NPH1 in baculovirus allowed the authors to demonstrate binding of flavin mononucleotide (FMN) to the LOV domains and blue-light-mediated autophosphorylation of the photoreceptor.

However, it remains to be seen whether phototropin is a membrane-associated protein, as was inferred by the early biochemical studies Jarillo et al. recently identified a new NPH1-like protein, NPL, that shows significant structural similarity to NPH1. We speculate that NPL1 and NPH1 together are probably the most important photoreceptors controlling phototropism in higher plants.

Signal transduction competes with proteolytic degradation of Pfr (PHYA) or dark reversion of Pfr back to Pr (PHYB).

Four different modes of action leading to regulation of gene expression can be distinguished: (1) the very-low-fluence response (VLFR), a response that is saturated at about 1% of Pfr and involves PHYA; (2) the high irradiation response (HIR), a response to continuous irradiation with maximal responsiveness in far-red (~720 nm) light that also involves PHYA; (3) the low fluence response (LFR), a response to red light pulses that can be reverted by far-red light and involves PHYB; and (4) the continuous red light response (cR), a response that can not be induced by red light pulses (not discussed in this review) and involves PHYB.

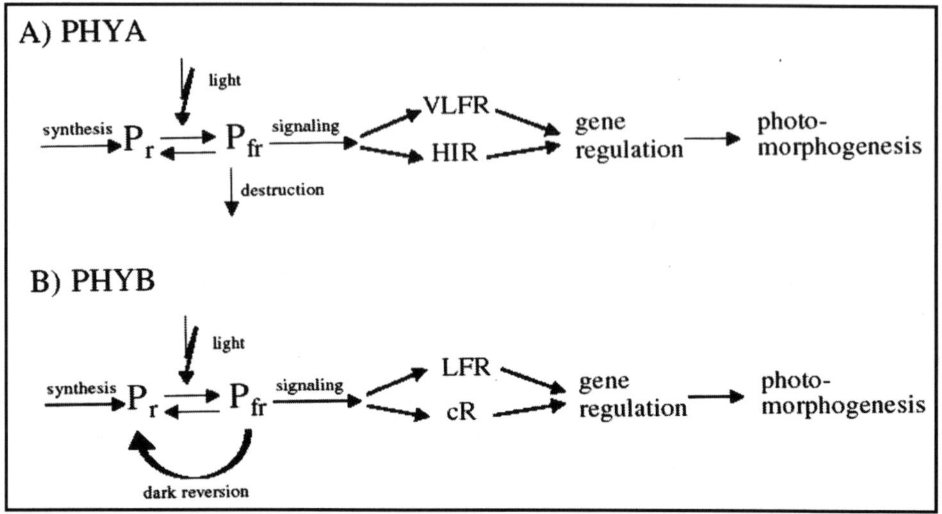

Fig. Modes of action of (A) PHYA and (B) PHYB. Red-light-absorbing (Pr) and far-red-light-absorbing forms (Pfr) of PHYA and PHYB are shown.

Cryptochromes

Ahmad and Cashmore cloned the HY4 gene, identifying the first member of the long-sought blue-light-absorbing photoreceptor family. HY4 encodes the cryptochrome 1 (CRY1) photoreceptor. CRY1 displays striking similarity to photolyases; however, it has a unique C-terminal extension and lacks photolyase activity.

FADH is its catalytically active chromophore; the second chromophore could be either a pterin or a deazoflavin. The second member of this subclass of blue-light photoreceptor, CRY2, has also been identified and contains a different C-terminal extension.

Results of biochemical studies employing CRY1- overexpressor lines were interpreted as indicating that this photoreceptor is cytosolic. In contrast, experiments on transiently transformed plant cells expressing the CRY1-GFP fusion protein indicated that CRY1 is localised in the nucleus in the dark. Moreover, analysis of the cellular distribution of the CRY2-GUS or CRY2-GFP fusion proteins in transgenic plants showed that the CRY2 photoreceptor is constitutively localized in the nucleus. A more complex picture is emerging in lower plants, especially in fern gametophytes.

Microbeam irradiation indicates that fern homologues of higher-plant CRY photoreceptors are present in the cytosol but also associated with the nucleus.

Recent studies showed that at least two members of the CRY family in Adiantum capillus-veneris are indeed localised in the nucleus and that the nucleo/cytoplasmic distribution of these receptors is not influenced by light.

Phytochromes

In Arabidopsis, phytochromes are encoded by five genes (PHYA-PHYE). PHYA (and probably all phytochrome types) binds an open chain tetrapyrol (phytochromobilin), which is covalently linked by a thioetherlinkage to the apoprotein, as a functional chromophore In general, the photosensory function of phytochromes is based on their capacity to perform reversible interconversion between the red-light-absorbing Pr form and the far-red-light-absorbing Pfr form following sequential absorption of red and far-red light.

However, the different phytochromes are not only encoded by different genes but also have different modes of action. PHYA is the most specialised phytochrome. It is responsible for the very low fluence rate response (VLFR) and the high irradiance response (HIR). The extraordinary responsiveness of PHYA allows it to control germination of seeds buried in the soil and to induce germination when seeds are exposed to star light. The other phytochromes control, to different extents, the classical red/far-red reversible induction, the so-called lowfluence- rate responses (LFRs) and responses to continuous redlight.

All phytochromes are synthesised in the biologically inactive, red-light-absorbing Pr forms. The half-life of the physiologically active form of PHYA is extremely short, and proteolytic degradation of the Pfr form is believed to be responsible for termination of signalling. In the case of the light-stable phytochromes, especially PHYB, termination of signalling is not yet understood. Regulation of the dark reversion of the Pfr to the Pr form is thought to be the most likely mechanism, since proteolytic degradation of the Pfr form of PHYB clearly cannot play a major role. Computational analysis of phytochrome sequences and structures from a variety of species, including ferns, mosses and algae, indicates that they are cytosolic: no motif, domain or even secondary structure that could facilitate insertion into or association with any type of membrane has been identified.

This contrasts with in vivo observations of lower plants, in which microbeam irradiation has revealed a very localized response and an action dichroism for phytochrome-mediated growth responses and chloroplast re-orientation. These observations suggest that phytochromes associate, at least in lower plants, with the plasma membrane and that this probably involves as-yet-unknown protein-protein interactions.

In higher plants, early immunocytochemical analysis of the subcellular distribution of PHYA demonstrated only cytosolic localisation of the photoreceptor. Therefore, until recently, in the absence of any other indication, phytochromes of higher plants were thought to be localized in the cytosol and/or associated with cell membranes.

Intracellular Localization of PHYB in Light and Darkness

PHYB is the most ancient type of phytochrome present in higher plants.

Given the observations outlined above, it was therefore fairly surprising when, in 1996, Sakamoto and Nagatani reported the enrichment of PHYB in nuclear fractions isolated from light-grown trabidopsis seedlings. They also showed that a fusion protein consisting of the C-terminal part of the A. thaliana PHYB fused to GUS is constitutively localised to nuclei in transgenic plants.

Despite the potential importance of these data, they were largely ignored. However, recent studies by Nagatani and co-workers and ourselves, which showed that chimeric proteins consisting of full-length PHYB fused to GFP can complement PHYB mutations, substantiated the earlier findings.

These later reports showed convincingly that the PHYB-GFP fusion protein is a biologically active photoreceptor, that the PHYB contains a functional NLS and that the NLS motif is localised in the C-terminal part of the photoreceptor.

More importantly, these reports documented unambiguously that the PHYB-GFP fusion protein is localised in the cytosol in the dark and that accumulation of the tagged photoreceptor in the nucleus is a light-driven process.

Nuclear translocation is induced by repeated pulses of or continuous red light and reversed by farred light. After light treatment, PHYB-GFP is detectable mainly in the form of speckles, and these speckles are present only in nuclei. Upon transfer to dark, the structures gradually dissolve, and nuclei become diffusely stained and finally free of any detectable fluorescence, which indicates that the PHYB-GFP fusion protein has disappeared.

Intracellular Localization of PHYA in light and Darkness

PHYA is a highly specialised photoreceptor that has physiology and biochemistry markedly different from those of other phytochromes. Therefore, it was expected that the lightdependent intracellular redistribution of PHYA, if it occurs and plays a role in PHYA-mediated signalling, would exhibit characteristics quite different from those of PHYB. It has been demonstrated that lightdependent intracellular redistribution of PHYA indeed takes place, and PHYA, similarly to PHYB, is imported into the nucleus. These studies clearly established that the PHYA-GFP fusion protein is a functional photoreceptor. In darkness it is localised exclusively in the cytosol, and light induces its translocation to the nucleus.

The kinetic, light-quality and -quantity requirements of the redistribution of PHYA are sharply different from those of PHYB. Import of the PHYA-GFP fusion protein is an order of magnitude faster, easily detectable within a minute and induced by pulses of or continuous far-red light. Although PHYA-GFP imported into the nucleus forms speckles, these speckles are smaller and more numerous when compared with those formed by PHYB-GFP.

In addition, the speckles are not restricted to nuclei: they are easily detectable in the cytosol, and formation of these structures in the cytosol precedes heir appearance in the nucleus. The PHYA–GFP-containing speckles formed in the cytosol are reminiscent of sequestered areas of phytochrome (SAPs) thought to represent intermediates or products of ubiquitin-mediated proteolytic degradation of PHYA.

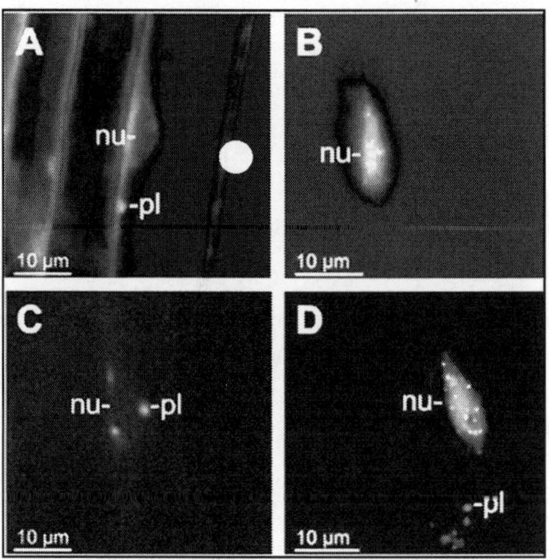

Fig. Localisation of PHYA-GFP (a and b) and PHYB-GFP fusion proteins (c and d) in Arabidopsis Seedlings grown for 7 days Darkness and kept in Darkness (a and c) Irradiated for 24 hours with far-red light (b) or Irradiated for 6 hours with red light (d), before Analysis by Epifluorescence Microscopy. Nuclei (nu) and selected Plastids (pl) are Indicated.

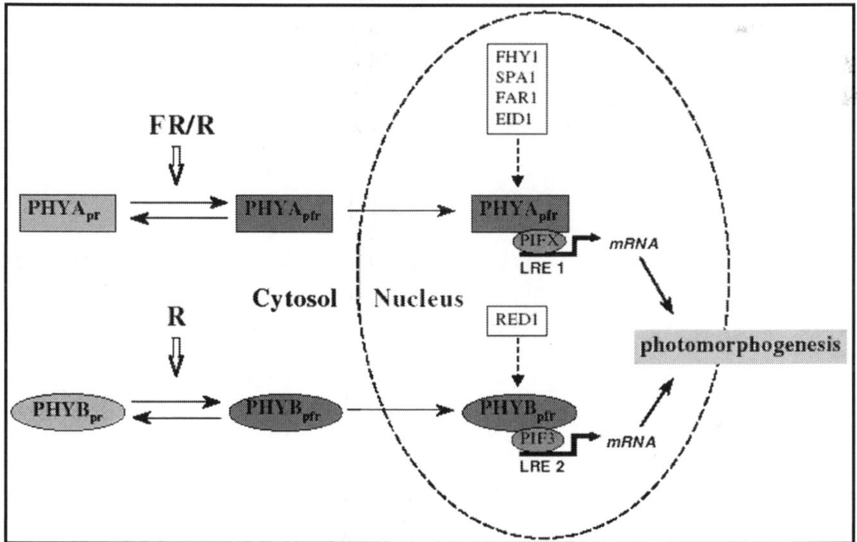

Fig. A model depicting molecular components and cellular processes involved in PHYA- and PHYB-mediated phototransduction in higher plants.

PHYApr and PHYBpr represent red-light-absorbing forms of phytochrome-A and phytochrome-B, respectively. PHYApfr and PHYBpfr represent far-red-light-absorbing forms of phytochrome A and phytochrome B, respectively. FHY1, SPA1, FAR1 and EID1 are shown to effect PHYA-mediated signalling; RED1 is known to modulate PHYB-mediated signalling; PIF3 binds to different LREs (cis-acting light regulatory elements) required for light-induced transcription of target genes and interacts selectively with the pfr form of PHYB. PIFX is a postulated nuclear protein that is involved in PHYA-mediated phototransduction.

Intracellular Localization of PHYC-PHYE

To analyse intracellular localisation of the other three members of the phytochrome photoreceptor family, we chose an approach similar to that used for PHYA and PHYB, analyzing the nucleo/cytoplasmic distribution of PHYC-GFP, PHYDGFP and PHYE-GFP fusion proteins in transgenic tobacco and Arabidopsis seedlings. Interpretation of the emerging data seems to be rather difficult for several reasons.

First, in sharp contrast to PHYA-GFP and PHYB-GFP, the PHYC-GFP, PHYD-GFP and PHYE-GFP fusion proteins are always detectable in the nuclei in dark-grown seedlings in all lines tested so far. Import of these proteins into nuclei might therefore be light independent and constitutive. Second, in etiolated seedlings, staining of the nuclei is always diffuse: speckles do not form.

Finally, red-light treatment induces formation of speckles in the nuclei (which can be reversed by far-red light) but does not significantly affect the diffuse fluorescence. The kinetics of the appearance/disappearance of these speckles is, however, comparable to that of PHYB–GFPcontaining speckles. These results might indicate that PHYCPHYE, in contrast to PHYA and PHYB, can be imported into the nucleus in their biologically inactive Pr (dark) form, and yet the speckle formation is dependent on the presence of the Pfr form.

STRUCTURE OF PHYTOCHROME

Sensing and responding to light is a central feature of plant biology that is critical for tuning plant metabolism to the light environment. To accomplish this vital task, plants use a number of different photosensory proteins that perceive different bands of the electromagnetic spectrum. For sensing red light– depleted (shade) and red light–enriched (full sun) conditions, plants use phytochromes, large proteins that possess the covalently linked linear tetrapyrrole (bilin) chromophore phytochromobilin.

First detected in 1959 by investigators at the USDA Plant Industry Station in Beltsville, MD, phytochromes share the characteristic that red light (R) irradiation converts the R-absorbing Pr state into the metastable, far-red light (FR)-absorbing Pfr state. This photoconversion is reversible, with Pfr returning to Pr either upon absorption of an FR photon or upon prolonged incubation in the dark via a thermal process known as dark reversion.

We now know that this reversible photochemistry reflects the unique environment of a bilin chromophore that is buried within a highly conserved N-terminal photosensory core region. Recently, Katrina Forest, together with Richard Vierstra and colleagues from the University of Wisconsin, captured a snapshot of the structure of the conserved photosensory core of a phytochrome, a result that has profound implications for the structural and functional basis of phytochrome signaling in plants.

Plant phytochromes (Phy family) are members of a more widespread family of photosensors that are found in cyanobacteria (cyanobacterial phytochromes Cph1 and Cph2) as well as in purple and nonphotosynthetic bacteria (bacteriophytochromes; BphP) and even fungi (fungal phytochromes; Fph family). Members of the extended family of phytochrome proteins share a common N-terminal photosensory core, with three blocks of homology sometimes referred to as P2, P3, and P4.

Phytochrome C-terminal domains mediate the transmission of photosensory signals perceived by the N-terminal region to signal transduction pathways within the cell. This region typically contains a histidine kinase–related domain that has been shown to confer ATP-dependent protein phosphotransferase activity in several cases. While bacterial and cyanobacterial phytochromes typically employ classical two-component phosphotransfer relays, the mechanism of plant phytochrome signaling appears considerably more complex, involving lightmediated nuclear translocation and regulation of transcription factor function.

The C-terminal regulatory domains of plant phytochromes have been shown to mediate homodimerization and light-modulated nuclear targeting, both of which are required for signal transmission. Signaling by plant phytochromes also involves protein–protein interactions with the N-terminal part of the protein, although it is not yet known whether these interactions are actually mediated via the conserved photosensory core (P2-P3-P4), via the N-terminal Ser/Thr-rich extension specific to plant phytochromes, or both.

In more primitive plants and algae, atypical phytochromes have been described in which the C-terminal region has been replaced by fortuitous gene fusions with phototropins and other eukaryotic Ser/Thr kinases. Bacteriophytochromes that lack recognizable kinase output domains have also been reported. It thus appears that the primary mechanism of light perception, arguably shared by the conserved photosensory core of all phytochromes, has been evolutionarily co-opted to regulate output domains with different m0olecular architectures.

The distinct bilin chromophores of the various phytochrome classes are all derived from the oxidative metabolism of heme. Heme oxygenases convert heme to biliverdin IXa (BV), which is the direct precursor of the chromophores of bacteriophytochromes and probably also those of fungal phytochromes. In cyanobacteria and algae, BV is converted to phycocyanobilin (PCB) via a four-electron reduction mediated by ferredoxin-dependent bilin reductases of the PcyA subfamily (Frankenberg et al., 2001). Higher plants instead convert BV to phytochromobilin (PFB) through the action of the homologous two-electron bilin reductase phytochromobilin synthase (HY2). In all phytochromes, the bilin chromophore precursor is covalently attached to the protein via a thioether linkage between a Cys residue and the bilin A-ring. Plant and cyanobacterial phytochromes utilize a conserved Cys residue in the P3 domain, whereas bacteriophytochromes (and probably fungal phytochromes) instead utilize a different

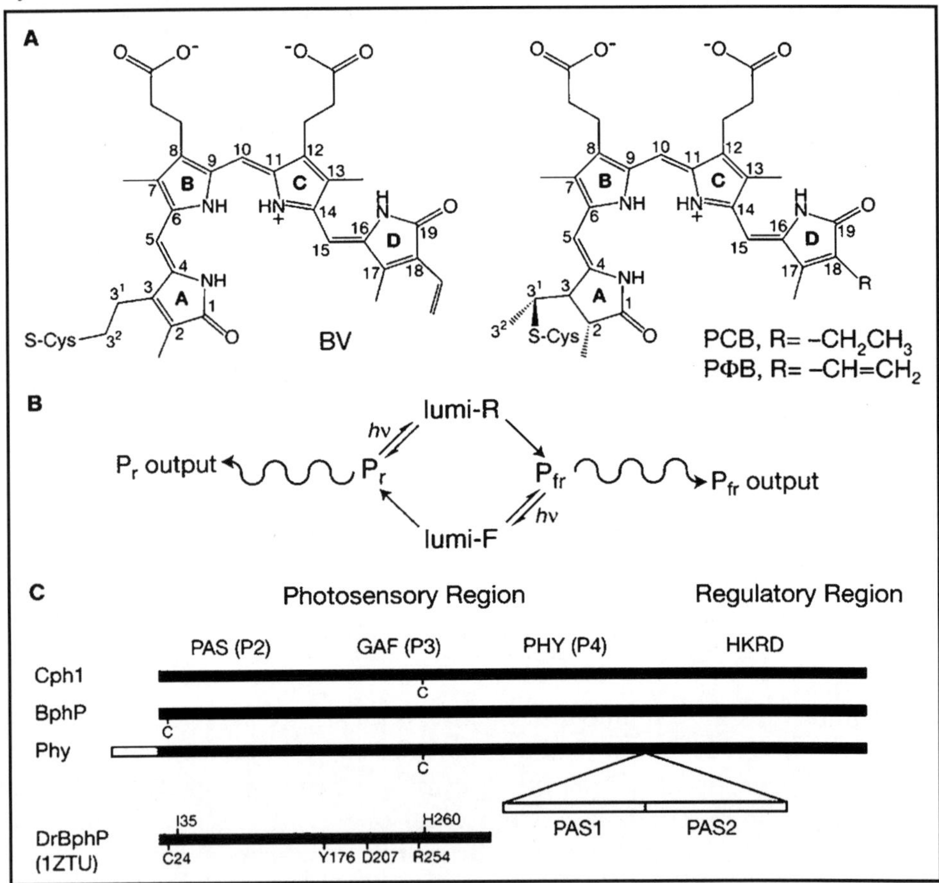

Fig. Bilin Chromophores, Photoconversion, and Domain Architecture of Phytochromes.

- The bilin chromophores derived from phycocyanobilin (PCB), used by cyanobacterial and algal phytochromes (right), phytochromobilin (PFB), used by plant phytochromes (right), and biliverdin (BV), used by the BphP subfamily (left), are shown along with their respective thioether linkages to the phytochrome apoproteins. Ring names (bold) and carbon numbering are indicated.

- The photoconversion between Pr and Pfr phytochromes is shown along with the primary photochemical intermediates. The conversions of lumi-R into Pr and lumi-F into Pfr are thermal processes that lead to changes in the protein moieties that initiate signal output.
- Conserved domains in the phytochrome primary sequence are shown. PAS and GAF domains are red and blue, respectively. The fragment of DrBphP whose structure has been reported is shown at bottom with the positions of key chromophore-interacting residues shown in the primary sequence. HKRD, histidine kinase–related domain.

Cys in the P2 region. Unlike the phycobiliproteins, which utilize C-S lyases for covalent attachment of their bilin chromophores, plant phytochromes do not require enzymes or cofactors to assist the covalent assembly reaction. The same is true for cyanobacterial, fungal, and bacteriophytochromes. While recent work implicates the importance of the P2 domain for proper holoprotein assembly of a cyanobacterial phytochrome, the more distantly related phytochromes of the Cph2 subfamily lack this domain altogether but are nevertheless able to support bilin attachment. This suggests that the P2 domain performs an accessory role in holoprotein assembly, possibly by stabilizing.

Current Perspective Essay

Proper folding of the chromophore binding pocket in the P3 domain. Interestingly, introduction of a P3 Cys into bacteriophytochromes supports assembly with PCB, arguing for structural conservation of the chromophore binding pocket. In contrast with the P2 and P3 domains, P4 domains of plant, cyanobacterial, and bacteriophytochromes are dispensable for covalent attachment of the bilin chromophore. The P4 domain is however necessary both for efficient photochemistry and for the normal Pr spectrum, which are measures of bilin chromophore conformation.

P4 therefore seems critical for optimizing the bilin chromophore environment to enable efficient photoconversion, perhaps by conferring rigidity to the bound bilin and thereby minimizing competing radiationless deexcitation pathways. It was previously reported that P2 and P3 are related to domains known to be involved in ligand binding and signal transduction: P2 domains of some phytochrome protein sequences have been identified by domain database searches to be PAS domains (named for period clock, ARNT, and singleminded proteins), while P3 is identified as a GAF domain (named for vertebrate cGMP-specific phosphodiesterases, cyanobacterial adenylate cyclases, and the transcription activator FhlA).

Interestingly, the three-dimensional structures of representative PAS and GAF domain proteins exhibit similarity between the two protein folds. The

P4 domain of phytochromes, classified as a PHY domain by domain family database searches, also exhibits some homology to GAF domains. However, the similarity between P4 and GAF domains is much weaker than that between P3 and GAF domains. Since P4 domains of representative members of the Cph2 subfamily of phytochromes bridge the sequence gap between GAF domains and PHY domains of other phytochromes, the P4 PHY domain is likely to adopt a GAF fold.

THE PHYTOCHROME MOLECULE

The research on phytochrome is currently centered around the mechanisms of the intramolecular light-signal transduction. The pigment is a homodimer. The monomers are each a large (ca. 124 kDa and ca.1130 amino acid residues) water-soluble chromoprotein with a covalently bound (via a thioether bond) open-chain tetrapyrrole. The monomer comprises a globular N-terminal chromophore-bearing light-sensing domain which carries determinants of photosensory specificities in phyA and phyB, and an extended C-terminal domain responsible for dimerization and for regulatory actions.

The structural elements in the latter recognise the signal perceived by the chromophore domain and transduce it via interaction with downstream signaling mediators. The C-domain contains also a motif similar to that in prokariotic histidine kinases. Within the C-terminal domain, there are two regions implicated in protein–protein interactions and also a nuclear localization motif.

The N- and Cterminal domains are linked by a highly conserved hinge region critical for productive interactions with the signal transduction circuitry. In spite of considerable efforts, the attribution of the roles of the different parts of the phy molecule is far from completed. In particular, Nagatani and co-workers suggest that, at least for phyB, the C-terminus only attenuates the activity of the pigment, whereas the N-terminal domain is responsible both for the initiation and for the completion of the photoresponse.

The Phytochrome Phototransformation

Phy function is based on the operation of the phy phototransformation which involves (i) a light stage, i.e., a cis–trans photoisomerization of the chromophore in Pr proceeding in picoseconds within the singlet excited state and leading to the formation of the primary photoproduct, lumi-R, and (ii) thermal processes including chromophore and protein conformational changes leading via low-absorbance intermediates (of meta-R type) eventually to the conformationaly restructured Pfr. The cycle is completed with photo- and/or dark reversion from Pfr to Pr.

This photochromism of the pigment is determined by specific chromophore–apoprotein interactions, which enforce the chromophore to adopt a close to linear configuration in the protein pocket favorable for photoconversion.Pfr was believed to be the physiologically active form (the central dogma). However, recent observations suggest that Pr from phyA cycled through Pfr (Pr*) may be also active.

Transduction of the Light Stimulus

According to recent fundamental findings, the signal transduction includes the following key processes: (1) Autophosphorylation of phy and transphosphorylation of its partners; (2) Events in the cytoplasm with the 'classical' second messenger pathways operating downstream; (3) Light-induced translocation of phy from the cytoplasm where it is synthesized to the nucleus where it operates; (4) Signaling in the nucleus via short chains of transduction or more complex pathways. The short chains involve a physical interaction of phy with transcriptional factors changing the rate of gene expression.The complex pathways include branches with (a) negative and positive regulators of phy activity and (b) early and late components specific for phyA and for phyB and common for both of them as well as for other photoreceptors (light signal integration).This interacting network accounts for very efficient sensing of the complex light environment.

PHYTOCHROME: POLYMORPHISM AND DIFFERENT MODES OF ACTION

Phytochrome Heterogeneity, Major Phytochromes A and B

Polymorphism of the pigment in the cell was directly shown by the detection of (1) its light-labile and light-stable pools, (2) immunologically distinguishable phys, (3) phy species in vivo with distinct fluorescence and photochemical properties, and (4) most importantly, different phy genes and their products (five phys in Arabidopsis). All phys show extensive amino acid sequence homology but different transcriptional and posttranslational regulation. The phyA synthesis is strongly downregulated in response to light. The Pfr form is rapidly degraded under light conditions. The other four phys are less strongly regulated by light. The complex behaviour of plants under different light conditions and the multiple functions of phy are believed to be connected, at least in part, with this heterogeneity and mainly with phyA and phyB.

Modes of Action of phyA and phyB

Three modes of phy action are generally considered with regard to the relationship between a response and the Pfr relative content, [Pfr]/[Ptot]. They are (i) very-low-fluence responses (VLFR), (ii) low-fluence responses (LFR), and (iii) high-irradiance responses (HIR). LFR are the classical red-induced/far-redreversible phy photoresponses, whereas VLFR do not show FR-reversibility and are active in the whole spectral range, including the FR region. The kinetics of VLFR and LFR are also different: VLFR are transitional, whereas LFR cause a more persistent effect. Pfr is the physiologically active form in these responses. The situation with HIR is, however, more complicated. HIR require prolonged activation by high fluences, the most active being FR and they do not occur at all under R. HIR were considered to be irreversible by R and to disobey the reciprocity law.

However, Shinomura et al.showed that high frequency brief far-red pulses (FRp) could produce the same response as continuous FR light (FRc) and in this case HIR follow the reciprocity law and are R-reversible. In the FR-induced/R-reversible HIR, the physiologically active form is suggested to be Pr*. This is an intriguing alternative to the classical explanation of HIR in terms of the physiologically active and light-labile Pfr form whose maximal content and effect are achieved under FRc. The use of phy null mutants has shown that VLFR and HIR are exclusively mediated by phyA, whereas LFR are mediated by phyB and by other less abundant phys.

Functions of phyA and phyB

Photoresponses with participation of phyA were clearly shown both in dark-grown and green plants in the VLFR and HIR modes.PhyA mediates (i) promotion and inhibition of seed germination, (ii) FR-induced de-etiolation and R-induced enhancement of de-etiolation, (iii) promotion of flowering and resetting of the circadian clock. PhyA modulates the activity of phyB and other minor phys and is also active in the modification of gravitropism and phototropism.

PhyA may have additional functions such as regulation of root development, control of transition from scoto-to-photomorphogenesis in specialized phloem cells, flowering and bud dormancy, and control of flower morphogenesis. The phyA functions are probably not strictly confined to VLFR and HIR. For instance, phyA makes a minor contribution to the promotion of cotyledon unfolding and seed germination in the low fluence region. PhyA plays also a minor role in the perception of Rc in Arabidopsis seedlings during deetiolation and induction of chlorophyll a/b binding protein (CAB) gene expression. It was shown that phyA initiates hypocotyl growth inhibition upon Rc onset and that the phyA effect was not saturated even at high Rc fluence rates (up to 1000 ìM m_2 s_1).

The R/FR reversible type of responses (LFR) with phyA participation was demonstrated for de-etiolation in transgenic Arabidopsis seedlings overexpressing oat phyA. It was also noted that the action spectrum for the night- break promotion of flowering in long day plants, which is mediated by phyA, show maximum effectiveness by R and reversibility by FR. PhyB, on the other hand, is active in light- and dark-induced germination of seeds. In de-etiolated and green seedlings, phyB mediates the shade avoidance syndrome by perceiving the reduction in R to FR ratio. PhyB is not involved in daylenght measurement, although it is responsible for delaying inhibition of flowering. PhyB modulates phototropism and gravitropism in a R/FR reversible way.

Interaction and Redundancy between phyA and phyB

Two types of interaction were established for phyA and phyB, i.e., antagonism and synergism. In general, the sign of the interaction between phyA and phyB depends on the mode of action of phyA: VLFR and HIR exert negative and positive effects, respectively, on the phyB action. PhyA is conditionally redundant to phyB since the response to R via phyA is only seen in the phyB background. Casal points out, however, that the redundancy between phyA and phyB is an example of generic redundancy rather than true redundancy because these photosensors have different photoperception functions.

PHYA STATE IN THE CELL AND LIGHT SIGNAL TRANSDUCTION

The molecular mechanism of photoperception by phyA can be thus essentially different from that of phyB. The basis for these differences is found at all levels of phy organization and operation by the analysis of (i) spectroscopic and photochemical properties and domain functions of the molecule, (ii) dynamics of the physiologically active phy population in the cell, and (iii) specificity of the transduction of the light stimulus.

Structural and Photochemical basis for the Functional Specificity of phyA and phyB

The photosensory specificities of phyA and phyB are generally believed to be determined by the N-terminal domain based on the domain-swapping data. In recent experiments it was shown that the phyA specificity is also connected with the C-terminal part of the protein.

All phys have the same chromophore and the mechanisms of its attachment are probably very similar for phyA and phyB. Spectroscopic and photochemical properties of phyA and phyB in vitro are also essentially indistinguishable.Only the kinetics of the Pfr Pr dark reversion, determined by the N-domain, appears to be significantly different although the capacity for the reversion is strongly influenced by the sample integrity and buffer conditions.

Transgenic phyB expressed in yeast and assembled with phycocyanin exhibited dark reversion in contrast to phyA, which showed no dark reversion. However Pfr Pr reversion in dicotyledons is independent of the phy type (phyA or phyB). In monocots, no Pfr Pr reversion is observed.On the other hand, a considerably lower extent of phyB phototransformation into Pfr than for phyA was observed in vitro by Furuya and co-workers and in our experiments in vivo.

In addition, not only Pfr but also Pr* were suggested to be the physiologically active forms for phyA, in contrast to phyB, implying differences between phyA and phyB at the level of the phy phototransformation.

Interestingly, recent observations indicate that the chromophore of phyA and phyB could also determine their functional specificity. Upon introducing chromophores into chromophore- deficient mutants of Arabidopsis, it was found that phytochromobilin, PÖB, is required for the phyA-specific activity under FRc, whereas phyB appears to be more flexible accepting both PÖB and phycocyanobilin, PCB, as a functional chromophore. The authors admit, however, that the existence of PCB in higher plants has not been demonstrated.

Phytochrome as serine/threonine-specific Protein Kinase Regulated by light and Phytochrome-Associated protein Phosphatases

Phytochrome is a light-driven Ser/Thr protein kinase capable of auto-(phosphotransfer between monomers within the phy molecule) and transphosphorylation of its partners. PhyA is autophosphorylated at residues Ser-7, Ser-17, and Ser- 598, the later being phosphorylated only in the Pfr form.20 Phosphorylation of serine-598 at the hinge region can turn on the inter-domain signal communication in the pigment molecule. 12 The key role of the phosphorylation/dephosphorylation events associated with phy was also demonstrated by the protein phosphatase-kinase coupling in a number of photoregulation processes.

A phy-associated protein phosphatase 2A (designated FyPP) was shown to dephosphorylate phys.It binds phyA and phyB with higher affinity for the latter and modulates in an opposite way both phys in flowering time control. M. Chen isolated phy-related phosphatase 1 (PRP1) exhibiting serine/threonine protein phosphatase activity.

PRP1 acts as a positive regulator of phyA signaling and a negative regulator of phyB signaling (cited by Campbell and Liscum. In general, biochemical and physiological activities of the phys are controlled by modulation of their phosphorylation status. Removal of the region of the molecule including Ser-7 or its modification brings about physiologically hyperactive phyA, whereas phosphorylation of this Ser inactivates it. It is also possible that the light-activated phyA is desensitized by phosphorylation at Ser-598 but sensitized by FyPP-mediated dephosphorylation.

Light-Induced Nuclear–cytoplasmic Partitioning of phyA and phyB

The functional specificity of phyA and phyB is clearly manifested at the level of light-induced import of the pigments to the nucleus. PhyA and phyB were found to localize in the cytosol of dark-adapted plants in agreement with earlier works. However, upon illumination a dramatic redistribution takes place with part of the pigment aggregating in the cytosol and part entering the nucleus. There is a close correlation between the light and time requirements of this process for phyA and phyB and their modes of action.

Initially, Sakamoto and Nagatani observed enrichment of phyB in the nuclear fractions of light-grown Arabidopsis and constitutive localization of the C-terminal part of Arabidopsis phyB in the nucleus. Later, it was shown that the phyB:GFP (phyB-green fluorescent protein) moved from the cytosol to the nucleus upon Rc or Rp, the process being reversed by FR, suggesting that it is regulated by LFR. In the dark, phyB:GFP disappears form the nucleus. Nuclear translocation of phyB takes hours and is too slow to explain rapid phy responses, suggesting that phyB might be functional also in the cytoplasm although recent experiments indicate that phyB is only active in the nucleus.

The nuclear import of phyA:GFP in Arabidopsis is 10-fold faster than that of phyB (within minutes) and induced by a single FR pulse or by FRc. Thus, nuclear import of phyA:GFP is controlled by VLFR and HIR. However, a large part of the pigment is retained within the cytoplasm and the appearance of phyA in the nucleus is only transient under R. PhyA is thus capable of translocation in the nucleus in its Pfr and Pr* forms.

The phyA:GFP nuclear translocation reveals strong species-dependence of its spectral sensitivity. Both VLFR and HIR mediate nuclear import of Arabidopsis phy- A:GFP as found in phyA-deficient Arabidopsis plants expressing Arabidopsis phyA:GFP, whereas only HIR (not VLFR) was employed in the nuclear transport of this protein in transgenic tobacco.

Both phyA and phyB form speckles in the nucleus. The process is preceded by the formation of the speckles in the cytosol. The speckles might be the sites of transcriptional activity where the photoreceptor and the nuclear factors interact with each other. For phyA, two patterns of nuclear speckle formation were observed (in Arabidopsis), i.e. nuclei with many tiny spots and nuclei with few small spots.

Regulation of the Phytochrome A Level in the Cell

The nuclear import of phyA depends on its abundance in the physiologically active form(s). The dynamics of the active phyA in turn is determined by the pigment degradation, dark reversion into the initial inactive form (in most dicots), and synthesis. Degradation proceeds via aggregation and formation of speckles or sequestered areas (SAPs) in the cytosol.

These SAPs are thought to be involved in phy ubiquitination and proteolitic degradation. The SAP formation by phyA is connected with its relative hydrophobicity distinguishing it from light-stable phyB. Determinants for phyA degradation reside at the 6 kDa N-terminal domain, between residues 742 and 790 (in oat) and also possibly near the chromophore (residues 323–360 in oat).Biochemical tagging is considered to be the first step in the phyA degradation process.

The tag could be a phosphate residue whose level is higher for Pfr and Pr* than for Pr.Interestingly, the process of phyA degradation alone may account for the HIR action spectrum without introducing the notion of Pr* as the physiologically active form.

The phyA level is also subject to transcriptional and posttranslational regulation. PHYA gene expression is negatively regulated by light, with monocots responding more strongly than dicots. PhyA and phyB are required for the negative regulation induced by FR and R, respectively. This negative regulation occurs predominantly in shoots, whereas in roots near constitutive expression is observed. PhyA is thus regulated by multiple signals, including environmental, organ-specific and developmental signals.

Divergent Signal Pathways Downstream from phyA and phyB

PhyA and phyB have specific signal transduction branches as revealed by perturbation in signaling from phyA and phyB8,10,31–34 and also by the different kinetics of phototransduction for a number of photoresponses. More to that, phyA signals in the VLFR and HIR modes are photobiologically and genetically distinct from one another.106–110 The differences between VLFR and HIR are associated with the phyA molecule itself. An important domain is the Δ6–12 stretch. Its deletion produces hyperactive or active phyA in VLFR but with a dominant negative effect for HIR as observed on oat phyA expressed in tobacco and Arabidopsis.

PHYTOCHROME LINEAGES IN FLOWERING PLANTS

Phytochrome genes have been detected in all major subclasses of flowering plants. Figure summarizes results of phylogenetic analyses of 172 full or partial sequences in GenBank. Data were either partitioned according to sequence length or combined in a matrix in which absent nucleotides were coded as missing data. Each of the sequences belongs unequivocally (91% or higher bootstrap support) to one of the four phytochrome lineages identified in analysis of full-length sequences: PHYA, PHYB/D, PHYC or PHYE. Three of these four gene lineages occur widely in flowering plants but, to date, homologues of PHYE have not been detected in monocots.

This distribution could result from the divergence of PHYE from PHYB/D having occurred after the divergence of monocots from dicots or from PHYE being lost from monocots. There is currently no DNA sequence data suggesting the presence of additional, widely distributed phytochrome lineages in flowering plants. In our analysis, tomato PHYF is a member of the well-supported (91% bootstrap value) cluster including eight PHYC homologues from dicots.

The interpretation of PHYF as a novel PHY subfamily was based on the degree of divergence between Arabidopsis PHYC and tomato PHYF and assumes that rates of nucleotide substitution among phytochrome subfamilies are equal. However, evidence is presented below that the PHYC subfamily is in fact diverging in sequence faster than the PHYA and PHYB subfamilies. Reports of additional genes in tomato (total of 9-12) inferred from Southern analyses have not been substantiated with sequence data. The possibility remains that such very high diversity may be the result of recent duplications within the PHY gene family in Solanum, or in some larger taxonomic group to which tomato belongs.

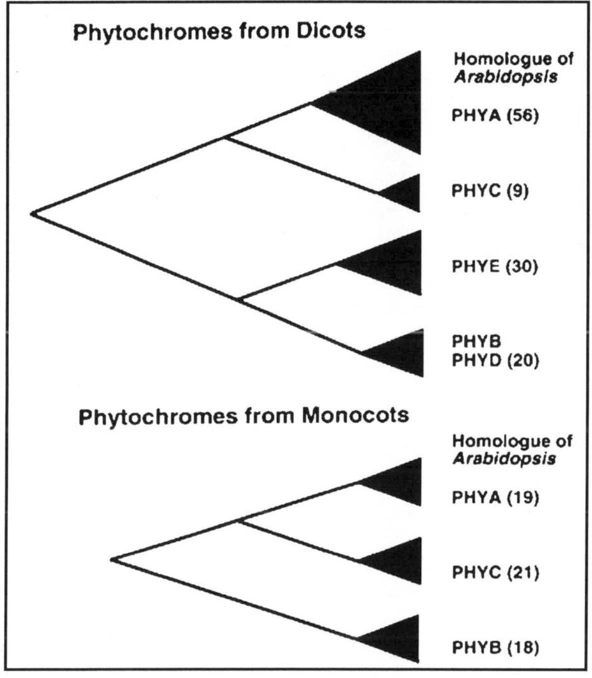

Fig. Distribution and Relationships of 172 Complete or Partial Phytochrome Nucleotide Sequences from Flowering Plants.

Numbers in parentheses represent the number of sequences sampled from GenBank; tomato PHYF occurs in the PHYC clade.

Recent diversification in the gene family is evident within dicot flowering plants, exemplified by gene duplications in the PHYA and PHYB/D subfamilies. Multiple homologues of PHYB/D are found in Arabidopsis, carrot and tomato, while multiple homologues of PHYA are found in carnation, legumes and Ceratophyllum, an aquatic angiosperm. In legumes, the duplicate PHYAs form a discrete lineage that is derived from within their PHYA subfamily.

In all cases, multiple sequences from a taxon are more divergent than one would expect if they represented merely allelic diversity. Conversely, evidence for recent diversification of PHY genes in monocots is limited. Hershey et al. suggested that four genes encode phytochrome in oat, and reported sequences of two PHYAs and a PHYA fragment that are nearly identical. This may not be strong evidence of diversification within a gene subfamily because Avena is a polyploid. Similarly, Christensen & Quail estimated the size of the gene family in maize to be from two to four members. They isolated clones representing three loci; one is a homologue of PHYA, another was interpreted as a pseudogene, and the sequence of the third was not reported.

Subsequently, partial sequences homologous with PHYB/D and PHYC have been isolated from maize, possibly accounting for the additional loci. Results from Southern blot analyses of genomic DNA of rice indicate that the three observed hybridization bands correspond to PHYA-, PHYB/D- and PHYC-like sequences. Finally, no evidence of multiple homologues was detected in any of the monocots sampled for phytochrome diversity, except from the grass Panicum which has two homologues of PHYA, one of which is very divergent from other PHYAs from grasses. At the nucleotide level, diversification in the PHY gene family in flowering plants is characterized by unequal rates of divergence. Relative rates of non-synonymous nucleotide substitution between all currently available full length PHY coding sequences and the reference PHY sequence from Selaginella were determined. Table shows that, in all possible pairwise comparisons of PHYA and PHYB sequences from single taxa, PHYAs are significantly more divergent from the common ancestor than are PHYBs. Furthermore, in Arabidopsis where full-length sequences are available, PHYC is significantly more divergent compared to PHYA, PHYB and PHYD, while PHYE is evolving rapidly relative to PHYB and PHYD. Thus, overall, PHYB/D appears to be the most evolutionarily constrained PHY sequence, followed by PHYA, then PHYC and PHYE.

Table. Results from pairwise comparisons of ralative rates of nucleotide substitutions (Wu & Li 1986) among full-lengh PHY sequences. The value d13 - d_{23} is the difference in non- synonymous nucleotide substitution between sequence 1 and sequence 2 ralative error: Single asterisks indicate differences significant at the 0.5 level and double asterisks differences at the 0.01 level. At = Arabidopsis, Nt = Nicotiana, St = Solanum,Os = Oryza, Gm = Glycine

Seq 1	Seq 2	d_{13}-d_{23}	SE	Direction
At A	At B	0·0454*	±0·0183	(A>B)
At A	At C	–0·0429*	±0·0202	(C>A)
At A	At E	–0·0242	±0·0198	
At B	At C	–0·0883**	±0·0254	(C>B)
At B	At E	–0·0696**	±0·0189	(E>B)
At C	At D	0·0770**	±0·0194	(C>D)
At C	At E	0·0187	±0·0207	
At D	At E	–0·0583**	±0·0191	(E>D)
Os A	Os B	0·0732**	±0·0188	(A>B)
Nt A	Nt B	0·0962**	±0·0176	(A>B)
St A	St B	0·0786**	±0·0181	(A>B)
Gm A	Gm B	0·0403*	±0·0192	(A>B)

Differences in the rates of divergence for two full-length phytochrome coding sequences, for instance PHYA and PHYC, are largely due to differences in the C-terminal half of the molecule. When the data are partitioned by exon, nucleotide substitution is more highly constrained in exons I, III and IV and is least constrained in exon II (data not shown). It has been proposed from mutational studies that sequences coded in exon II are critical for phytochrome's regulatory activity so it is possible that rapid evolution of PHYC and PHYE in this region reflects functional divergence.

MOLECULAR MAP OF THE PHYTOCHROMES

Spectral Activity

Chromophore Lyase

The seminal demonstration by Lagarias and coworkers that the PHYA apoprotein can autocatalytically attach the chromophore in vitro as well as in living yeast has been followed by similar demonstrations in vitro for phyB and phyC. Analysis of deletion derivatives of phyA and phyB expressed in yeast or transgenic plants has mapped the chromophore lyase activity sufficient for attachment towithin the NH2-terminal segment between positions 115 and 450.

In a more detailed site-directed mutational analysis of five conserved residues (positions 360, 369, 372, 375 and 377) surrounding the chromophore attachment residue (the cysteine at position 374), Song and colleagues have obtained evidence that none is essential for chromophore lyase activity, suggesting a structural rather than a catalytic role.

A schematic diagram of a 'consensus' phytochrome molecule derived from the alignment of multiple full-length phytochrome polypeptide sequences by Mathews et al. is shown at the top. The coordinates for this figure and throughout the paper are those for the consensus line of this alignment and are referred to as 'positions' within the consensus sequence. The extremities of the schematic (positions 1 and 1210) correspond to the NH2- and COOH-termini, respectively, of Arabidopsis phyB, the longest sequence in the alignment.

The indented positions at each end (positions 37 and 1195) correspond to the NH2- and COOH-termini, respectively, of Arabidopsis phyA. The location of the tetrapyrrole chromophore covalently linked to a cysteine is indicated at position 374. The residue substitutions indicated at various positions along the consensus molecule are a composite of phyA and phyB missense mutations causing loss of regulatory activity. The stippled area demarcates a region containing a high density of such mutations (the 'core' region) and the small black box, a short segment containing multiple substitutions of individual residues.

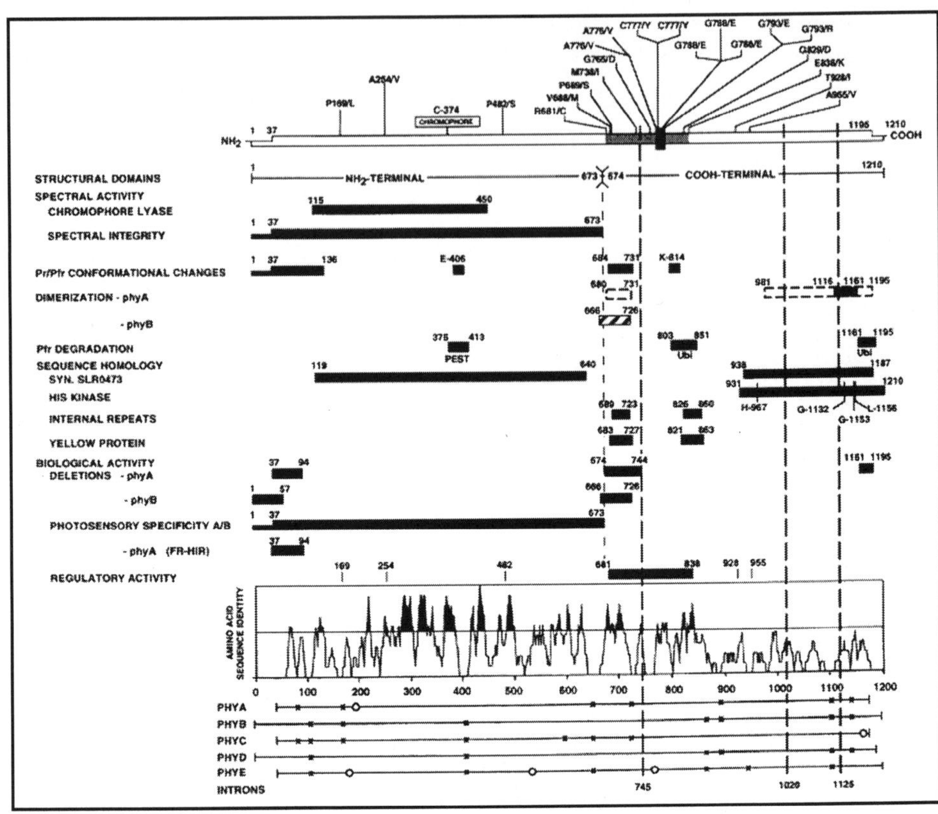

Fig. Molecular map of Phytochrome.

The horizontal bars in the body of the figure show the locations within the polypeptide of the activity, property, or feature of the molecule shown to the left. At the bottom is a plot of the distribution of amino acid sequence identity across the aligned polypeptide sequences of the five Arabidopsis phytochromes PHYA, B, C, D and E, shown below. The light vertical dashed line at position 673/674 represents the junction of the two major structural domains. The heavy vertical dashed lines represent the positions of the introns within the structural PHY genes of higher plants (except PHYC which lacks the intron at position 1020).

Spectral Integrity

Sequences necessary to sustain complete spectral integrity equivalent to that of the full-length molecule extend in both directions beyond that for chromophore lyase activity to encompass most, if not all, of the NH2-terminal domain (positions 1-673). Spectral integrity here refers to maintenance of photoreversibility and absorption spectra in both Pr and Pfr forms unaltered from that obtained with the native photoreceptor. Analysis of deletion derivatives of phyA in transgenic seedlings has shown that the full NH2-

terminal structural domain (positions 1-673) is sufficient for spectral integrity. A series of deletions covering a variety of segments within the first 115 positions at the NH2-terminus (residues 1-69 of the phyA sequence) has revealed subdomains necessary for stability and spectral integrity of the Pfr form alone (positions 51-94) or of both the Pr and Pfr forms (positions 51-115). These data are consistent with earlier proteolytic studies on native phyA suggesting interaction of this NH2-terminal segment with the chromophore, especially in the Pfr form. Similarly, determinants between positions 524 and 673 at the COOH-terminal end of the NH2-terminal structural domain have been shown to be necessary for spectral normalcy. Preliminary spectral analysis of some comparable deletion derivatives of phyB suggests a similar pattern to that of phyA.

Pr/Pfr Conformational Changes

Conformational changes in the protein moeity induced by photoconversion between the Pr and Pfr forms have been mapped in phyA to the NH2-terminal segment (within positions 37-136), to a segment around position 406 near the chromophore attachment site, to a segment in or near the linker region (positions 684-731), and to a region around position 814. Most of these data have come from older biochemical, immunochemical or physico chemical studies on extracted or purified, non-recombinant phytochrome.

Dimerization

Proteolytic fragmentation coupled with the use of monoclonal antibodies against mapped epitopes provided evidence some years ago that the COOH-terminal structural domain of phyA carries the determinants for dimerization, whereas the NH2-terminal domain is released as a monomer upon cleavage, indicating no stable intra- or intersubunit interactions. Transgene-encoded NH2-terminal domains of both phyA and phyB behave as globular monomers, confirming this latter conclusion, and the transgene-encoded COOH-terminal domain of phyB has been shown to associate into homodimers in Arabidopsis.

Using a conceptually elegant prokaryotic assay system, Edgerton & Jones presented data interpreted to indicate that two regions of the phyA COOH-terminal domain (positions 680-731 and 1116-1195, respectively) are capable of mediating dimerization. However, subsequent analysis of a COOH-terminal deletion series by Cherry et al. (1993) showed that simply removal of residues COOH-terminal of position 981 was sufficient to cause loss of dimerization capacity.

This result indicates that neither the initial (positions 680-731) nor revised (positions 652-741) central segments are sufficient to mediate dimerization in the context of the native molecule. Moreover, since internal deletion of a similar region of phyB (positions 666-726) has been shown not to affect dimerization when expressed in transgenic Arabidopsis, it would appear that this region is neither necessary nor sufficient for dimerization in the context of the native molecule. By contrast, the COOH-terminal deletion analysis by Cherry et al. indicates that residues between positions 981 and 1161 are necessary for dimerization of native phyA in transgenic plants. Combined with the proposal from the prokaryotic assay that positions 1116-1195 are necessary for dimerization, these transgenic plant data might suggest that critical residues for dimerization lie between positions 1116 and 1161.

Pfra Degradation

The rapid intracellular degradation of phyA induced upon its photoconversion to the Pfr form (representing a 100-fold greater turnover rate for PfrA than PrA) is one of the earliest molecular properties of the photoreceptor recorded in the literature. The other phytochromes (phyB, C, D and E) do not display this dramatic isoform specific difference in degradation rate and are therefore considered to be more 'light stable. Recent domain swapping experiments using transgenic Arabidopsis have provided evidence that the determinants specifying recognition of phyA for degradation in the Pfr form reside in the NH2-terminal structural domain.

However, these determinants are alone insufficient for degradation because the NH2-terminal domain alone of phyA is light stable in vivo, requiring a contiguous COOH-terminal domain for degradation to proceed. Indeed, deletion of the extreme COOH-terminus of the phyA polypeptide has been reported to stabilize the molecule as Pfr, localizing one determinant necessary for degradation to a small stretch of amino acids at the extremity of this domain. Because the full COOH-terminal domain from either phyA or phyB supports degradation when fused to the phyA NH2-terminal domain, the necessary determinants are apparently common to both family members.

Proposed mechanisms of selective PfrA degradation have focused on the possible involvement of the ubiquitin- mediated proteolytic pathway and/or the hypothesized PEST-sequence mechanism. There is evidence that phyA is rapidly ubiquinated in vivo upon conversion to Pfr and kinetic data are at least partially consistent with subsequent degradation ensuing via the ubiquitin pathway. The proposed location of the ubiquitin target site (one or more of the three lysine residues between positions 803 and 851) in the COOH-terminal domain, as well as the conservation of these three lysine residues between phyA and phyB sequences would be consistent both with the need for a contiguous COOH-terminal domain for degradation and with the interchangeability of the phyA and phyB COOH-terminal domains in this process.

Unfortunately, site-directed substitution of arginines for the three lysines apparently did not affect the rate of PfrA degradation, indicating that if ubiquitination does occur at these sites it is not necessary for proteolysis. Thus, although still clearly a viable possibility a direct link between ubiquitination of PfrA and degradation remains to be demonstrated. The presence of a PEST sequence and its associated cluster of acidic residues in phyA, but not phyB, C, D or E sequences, as well as its location adjacent to the chromophore (positions 375-413) in the amino terminal domain, and the photoconversion-induced exposure of this region in the Pfr form, are all consistent both with the selective degradation of phyA following Pfr formation and the location of the determinants for this selectivity in the NH2-terminal domain.

One possibility would be that the PEST sequence provides specific recognition of PfrA by the degradative machinery which then initiates degradation via the ubiquitin pathway involving determinants in the COOH-terminal domain.Unfortunately, no direct experimental test of the relevance of the PEST sequence to PfrA degradation has been reported to date and the importance of this sequence to the degradation of several other PEST-containing proteins has been questioned.

Sequence Homology

Schneider-Poetsch and colleagues provided the first provocative evidence from computerized database searches that the distal half of the COOH-terminal domain of the phytochromes (positions 931-1210) exhibits a degree of sequence similarity to the sensor histidine kinase module of bacterial two-component signalling systems, thereby raising the possibility that the photoreceptors function as a family of photoregulated histidine kinases. Subsequently, three considerations tended to detract from the likelihood that this simple extrapolation from the prokaryotic systems is accurate.

First, the histidine residue that is autophosphorylated in the majority of the bacterial kinases is conserved at that site (position 967) only in the monocot phyAs and Arabidopsis phyC, and not any of the other reported phy sequences. Secondly, attempts to detect autophosphorylation of histidine residues in preparations of purified phyA were unsuccessful. Thirdly, site-directed mutagenesis of four single residues [positions 967 (histidine), 1132 (glycine), 1153 (glycine), 1156 (leucine)] in oat phyA (the latter three residues conserved among all phytochromes and the majority of histidine kinases) failed to abrogate the activity of the molecule when tested in transgenic Arabidopsis.

However, the recent sequencing of the entire genome of the cyanobacterium Synechocystis sp. strain PCC6803 has strongly revitalized this proposal. One of the open reading frames (ORFs) in the Synechocystis genome (Cyanobase database), designated ORF SLR0473, has regions of striking sequence similarity both to the phytochromes and to the sensor histidine kinases.

ORF SLR0473, encoding 748 amino acids, consists of an NH2-terminal domain with 36% identity (60% similarity) to the NH2-terminal domain of phyE between positions 119 and 640, and an immediately adjacent COOH-terminal domain with 20% identity (52% similarity) to the COOH terminal domain of phyE between positions 938 and 1187. This latter region of phyE is within the proposed histidine kinase-like domain of the phytochromes.

In turn, the COOH-terminal domain of ORF SLR0473 shows convincing sequence similarity to the histidine kinase module of established bacterial two-component systems (e.g. 31% identity and 55% similarity to B. subtilis KinA), including the presence of the conserved signature motifs, designated H, N, G1, F and G2. These data suggest that the Synechocystis SLR0473 protein may be a photoregulated histidine kinase and that the plant phytochromes may be evolutionary remnants of these prokaryotic molecules.

Strong support for at least part of this notion has been provided recently by Lagarias and coworkers who have shown that the SLR0473 protein can attach tetrapyrrole chromophores, including phytochromobilin, with the formation of a chromoprotein that is fully photoreversible by red and far red light analogous to plant phytochromes. In addition, a genetically defined locus for chromatic adaptation in the cyanobacterium Fremyella has also recently been shown to encode a protein with some sequence similarity to the phytochromes in its NH2-terminal domain and to the histidine kinases in its COOH-terminal domain.

Two other related observations arising from sequence comparisons are that the COOH-terminal domain of the phytochromes contains two repeats of a sequence (positions 689-723 and 826-860, respectively), and that these repeats have weak sequence similarity to a segment of photoactive yellow protein from the purple bacterium Ectothiorhodospira. The significance of these findings is yet to be determined.

Biological Activity

Deletion Analysis

A number of investigations have used overexpression of various deletion derivatives of phyA or phyB in transgenic plants to identify regions of the photoreceptors necessary for biological activity. Although subtleties exist between the effects of some individual constructs and plant systems, the basic conclusion from these combined studies is that deletion of relatively short segments at either terminus of the polypeptide either eliminates or modifies detectable biological activity. The results indicate that determinants within the first 60 residues or less at the NH2-terminus of phyA (positions 37-94) or phyB (positions 1-57), and within as few as 35 residues at the COOH-terminus of phyA (positions 1161-1195) are necessary for normal biological activity.

In addition, internal deletions near the linker region at the proximal end of the COOH-terminal domain of phyA (positions 674-744) or phyB (positions 666-726) also eliminate normal activity, indicating the necessity of this region. The limitation of such data is that is unknown whether these regions are directly involved in photoreceptor activity or the deletions indirectly perturb the function of another critical part of the molecule. One interesting 'gain-of-function' effect is that either deletion of the serine-rich region between residues 6 and 12 of phyA (positions 42-51) or site directed substitution of alanines for all the serines in this region result in enhanced biological activity of the transgene-encoded phyA. This result suggests that the serines in this region may mediate attenuation of phyA activity.

Photosensory Specificity

Like other receptors, the phytochromes can be considered to exhibit dual molecular functions: a sensory function involving perception and interpretation of the incoming light signals, and a regulatory function involving biochemical transfer of the perceived information to downstream transduction chain components. The phytochrome system is well known to monitor multiple parameters of the light environment (wavelength, fluence rate, etc.), to do so in multiple sensory 'modes' (LFR, VLFR, HIR, etc.), and to regulate multiple facets of growth and development throughout the plant life cycle via modulated gene expression.

Studies with photoreceptor mutants of Arabidopsis have shown that some of this sensory complexity, at least in early seedling development, is attributable to the fact that phyA is predominantly, if not exclusively, responsible for FRc perception in the FR-HIR, whereas phyB is predominantly responsible for Rc perception in the R-HIR, and that the two phytochromes transduce mutually antagonistic signals in response to Rc or FRc enrichment.

To begin to map the determinants involved in this photosensory specificity, reciprocal domain-swap experiments have been performed in which the NH2-terminal domain of phyA fused to the COOH terminal domain of phyB at position 673 (phyA/B) and the converse fusion protein (phyB/A) have been overexpressed in transgenic Arabidopsis. The data show that the wavelength-dependent seedling responses to Rc or FRc mediated by the phyA/B fusion resemble those of phyA, whereas the responses mediated by the phyB/A fusion mimic those of phyB.

The evidence therefore indicates that the determinants for the photosensory specificity of phyA and phyB to FRc and Rc, respectively, reside in the NH2-terminal domains. Because deletion of the NH2-terminal 52 residues of phyA (positions 37-94) produces a trans gene-encoded molecule defective in FRc perception, but retaining Rc responsiveness, it is possible that one or more of these determinants of phyA photosensory specificity reside in this short-terminal segment.

On the other hand, the NH2-terminal domain of neither phyA nor phyB when expressed alone exhibits normal biological activity despite being fully photoactive and having apparently normal spectral properties. This result indicates that a contiguous COOH-terminal domain provides determinants necessary for transmission of perceived photosignals. Because the COOH-terminal domains of phyA and phyB appear to be fully interchangeable in this function, the critical COOH-terminal determinants appear to be common to both phytochromes. From these data alone, it cannot be determined whether this domain plays a purely structural role or is directly involved in signal transfer. However, at least for phyB, the COOH-terminal domain is not capable of autonomously performing phyB function. This result indicates that the phytochromes appear to be unlike some biological molecules, such as receptor kinases, in which removal of the ligand-binding (receptor) domain results in constitutive activation of the signalling (catalytic) domain.

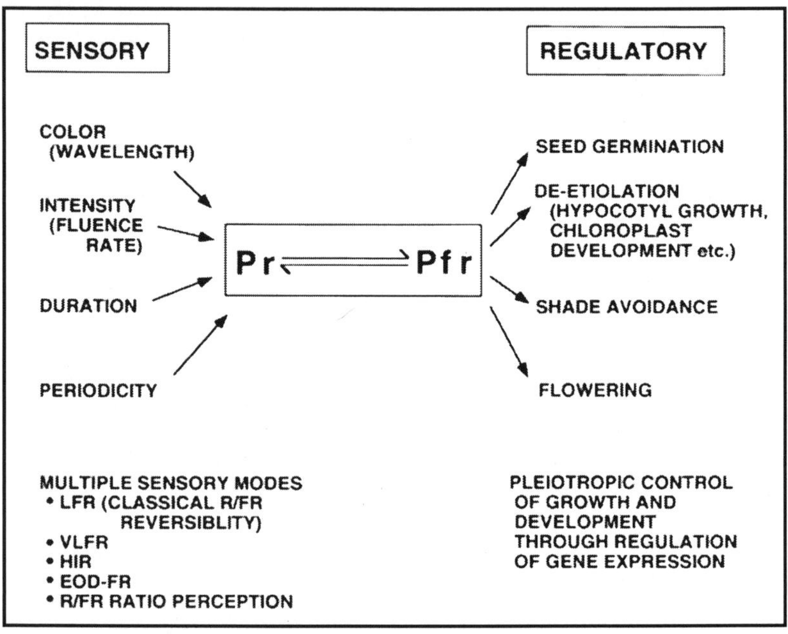

Fig. Dual Molecular Functions of the Phytochromos.

The photoreceptor molecular can be considered to have two sequential functions: sensory, defined as perception and interpretation of signals from the light environment, and regulatory, defined molecular as biochemical transfer of the perceived signals to transduction pathway components that ultimately regulate morphogenesis through altered gene expression. The principal parameters in the light environment knows to be monitored by the phytochrome system are shown to the left, as are the multiple sensory 'modes' displayed by the photoreceptor: LFR = low fluence response: VLFR = very low fluence response; HIR = high irradiance response; EOD-FR end-of-day far red; R/FR ratio = the ratio of the fluence rates of red to far red light (the parameter used for vegetative shade or proximity detection). Major facets of plant growth and development controlled by the phytochromes are indicated to the light.

Regulatory Activity

One strategy for identifying determinants in the phytochrome protein potentially involved in downstream signal transfer, as distinct from signal perception, is to screen for missense mutations within the polypeptide that do not affect expression level, spectral properties, or gross structural properties (dimer formation, apparent molecular cross-section), but do disrupt biological activity. Such mutant molecules can be considered to be competent in photoperception but defective in regulatory activity.

The mutated resides detected in a series of such screens involving phyA and phyB are compiled at the top of Fig.. A majority of these mutations (76%) cluster at the proximal end of the COOH-terminal domain (positions 681-838), designated the 'core'region, with multiple substitutions occurring at four positions within an 18 residue sub-segment (positions 776-793).

The data indicate therefore that this restricted 'core' region of the polypeptide contains determinants necessary for effective communication of perceived light signals to the cellular transduction circuitry. Because the COOH-terminal domains of phyA and phyB are functionally interchangeable and because the regulatory missense mutations cluster in the same region for both phyA and phyB, it would appear that the samefunction is disrupted in both molecules by these mutations. If the region between positions 681 and 838 is directly involved in the transfer of signalling information from photoreceptor to recipient transduction component, the biochemical mechanism of this transfer would appear to be the same for phyA and phyB.

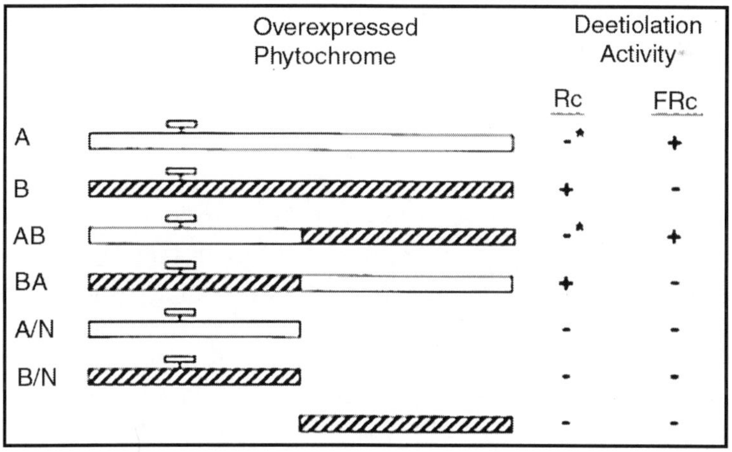

	Overexpressed Phytochrome	Deetiolation Activity	
		Rc	FRc
A		-*	+
B		+	-
AB		-*	+
BA		+	-
A/N		-	-
B/N		-	-
		-	-

Fig. Activity of Transgene-encoded Phytochrome Derivatives in Enhancing the Deetiolation process when Overexpressed in Arabidopsis.

How could the specificity of signal perception determined by the NH2-terminal domains of phyA and phyB be reconciled with a putative common biochemical mechanism of downstream signal transfer specified by the 'core' region? One possible formal model is depicted in Figure. In this model, the NH2-terminal domains of phyA and phyB carry different determinants that allow each to recognize (bind to) its own cognate reaction partner(s) (target selection) upon appropriate signal perception. The biochemical modification of the bound reaction partner involving the 'core' COOH-terminal region would then be identical in all cases (e.g. phosphorylation of a residue in the reaction partner).

Subsequent steps in the signal cascade could then either converge immediately or remain independent to the point where the common cellular functions controlling the growth and developmental responses to phyA and phyB are affected. This model is analogous to that of many other families of receptor molecules, especially the receptor kinases.

Constructs: A, oat phyA; B, rice phyB; AB, chimeric protein with NH2-terminal domain of oat phyA (positions 37-673) fused to COOH-terminal domain of rice phyB (positions 673-1210); BA, chimeric protein with NH2-terminal domain of rice phyB (positions 1-673) fused to COOH-terminal domain of oat phyA (positions 673-1195); A/N, NH2-terminal domain of oat phyA only (positions 37-673); B/N, NH2-terminal domain of rice phyB only (positions 1-666); B/C, COOH-terminal domain of rice phyB only (positions 647-1210). Deetiolation activity = capacity of transgene encoded phytochrome to enhance deetiolation (suppress hypocotyl elongation) of transgenic Arabidopsis seedlings in Rc or FRc. + = enhanced deetiolation relative to non-transgenic Arabidopsis; - = no enhancement of deetiolation relative to non-transgenic Arabidopsis; -* = quantitatively less effective than rice phyB in enhancement of deetiolation in Rc on a per mole basis of overexpressed phytochrome.

An Integrated Picture

Several interesting insights emerge from considering the various features or activities that have been defined for a given region of the phytochrome molecule.

- The locations of the introns in the higher plant PHY genes do not appear to correlate in any obvious way with the structural domains or various functional activities described. In addition, the algal phytochromes have larger numbers of introns than their higher plant counterparts, and Arabidopsis PHYC lacks the intron at position 1020. The functional significance, if any, of these observations is unknown.
- The region of the higher plant phytochromes that has the highest sequence similarity to the Synechocystis SLR0473 ORF (positions 119-640) corresponds extremely well to the region most highly conserved among the higher plant phytochromes themselves. Indeed 67% of the invariant residues in this region of the plant phytochromes are also invariant in the Synechocystis sequence (data not shown). This

observation is consistent with the demonstrated functional homology between the cyanobacterial and plant sequences in chromophore attachment and photoreversibility.

- Each of the regions that undergo photoinduced conformational changes overlaps with other activities. The

NH2-terminus (positions 1-136) and the region surrounding position 406 are each within the larger domain defined as determining the differences in the photosensory specificity and Pfr-specific degradation between phyA and phyB. As each of these functions requires phototransformation, each of these conformational changes is potentially involved in either or both functions. Similarly, the regions from positions 684-731 and surrounding position 814 both lie within the relatively conserved 'core' region, defined functionally as being necessary for downstream signal transmission. This correlation would be consistent with the notion that photoconversion induced conformational changes expose surfaces in this region necessary for biochemical transfer of perceived informational signals to phytochrome reaction partners.

NH2-terminal domains of phyA and phyB are postulated to contain distinct determinants which recognize specific cognate determinants (protruding triangle or diamond) on separate reaction partners (X1 and X2, respectively) in response to photosignal perception (induced target selection).

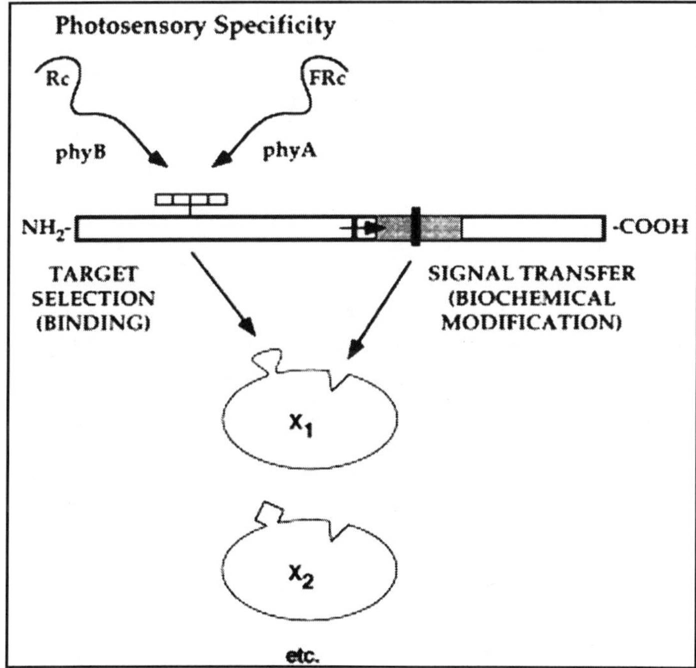

Fig. Two-point Contact model of Phytochrome Action..

Biochemical transfer of the perceived signal to the reaction partner is postulated to be functionally identical for the two photoreceptors and to involve interaction of the 'core' region (stippled area) of the COOH-terminal domain of each with a common determinant (indented triangle) present on all reaction partners (X1, X2, etc.). Thus, the specificity in signal perception exhibited by phyA and phyB involving the capacity to discriminate between Rc and FRc signals is postulated to be transduced at the first step via selection of different cognate molecular targets that undergo the same local biochemical modification in the process of signal transfer from the photoreceptor.

It should be noted that, despite the attractiveness of the possibility that the phytochromes may function as photoregulated histidine kinases, none of the mutations affecting regulatory activity thus far reported falls within the postulated kinase domain, and site-directed mutations targeted at conserved residues within this domain failed to disrupt phytochrome activity in vivo. Although there are many possible explanations for these observations, caution is still needed in attempting to transpose the bacterial two-component model directly to the phytochrome system.

PHYTOCHROME: A SERINE KINASE ILLUMINATES THE NUCLEUS

PLANTS being sessile organisms need stringent mechanisms to overcome changes in their ambient environment.To optimize their survival, the higher plants have evolved several mechanisms to sense environmental changes.

Among these, the mechanism to sense light is the most elegant. Plants can detect almost all facets of light such as direction, duration, spectral quality, quantity, solar angle, etc. This feat is achieved by plants by evolving specialized photoreceptor molecules that allow them to sense light throughout the visible spectrum. The examination of spectral sensitivity of different plant photoresponses has revealed that these responses are basically elicited using three specific wavebands of the solar spectrum: UV-B (280–320 nm), UV-A/blue (320–500 nm), and red (R)/far-red light (FR) (600–750 nm)1. The extensive studies conducted on the physiology of these photoresponses have divulged that each of these wavebands is sensed by a specific group of photoreceptor(s).

Many of the photoresponses elicited by the red/farred waveband show a unique feature, i.e. the reversibility of light action.

Several photoresponses such as redlight- induced seed germination can be reversed, if followed immediately by a far-red irradiation. This characteristic photoreversibility of response aided researchers to purify and characterize the responsible photoreceptor molecule – phytochrome.

In fact, for several years, phytochrome was the only plant photoreceptor that had been purified and characterized. In the plant kingdom, phytochrome is ubiquitously present in all species and in cyanobacteria3.

The distribution of phytochrome has now expanded beyond photosynthetic organisms with the discovery of phytochrome-like proteins in the eubacteria Deinococcus radiodurans and Pseudomonas aeruginosa4.

Phytochrome acts as a light-regulatory switch throughout the plant life cycle, right from seed germination to vegetative and reproductive development till senescence. It also enables the perception of shade and detection of neighbouring plants. The discovery of phytochrome in the early fifties was soon followed by several hypotheses about its likely mode of action in regulating photomorphogenesis. In many cases, the investigators sought a strict correlation with the photoreceptor level and the final photoresponses, and responses deviating from this were termed as 'phytochrome paradoxes'. However, the results obtained in the past decade using a combination of physiology, genetics and molecular biology have led to a totally new perspective in our understanding of the mode of phytochrome action in higher plants. The present review summarizes the current status of the mode of action of phytochrome.

Molecular and Physiological Properties of Phytochrome

The realization that phytochrome accumulates in bulk in dark-grown seedlings allowed its purification with relative ease due to lack of photosynthetic pigments in the seedlings. Based on various physiological evidences, it was predicted that phytochrome exists in vivo in two photoreversible forms5. The purification of phytochrome confirmed this view and showed that in dark-grown plants, phytochrome is present in the Pr form.

On exposure to red light, the Pr form is converted to the Pfr form, which is considered as the biologically active form. The Pfr form on absorption of far-red light is converted back to the Pr form. This photoconversion of phytochrome is correlated with the change in absorption maxima of these two forms: the purified phytochrome in the Pr form is blue in colour and absorbs maximally at 666 nm, whereas the Pfr form is olive-green in colour and absorbs maximally at 730 nm6. Phytochrome is a chromoprotein similar to algal biliproteins, with a molecular mass of ~120 kDa.

The open-chain tetrapyrrole chromophore called as phytochromobilin (PB) is linked to the protein moiety through a cysteine residue6. The synthesis of phytochromobilin uses heme as the first step, which is converted by the action of heme oxygenase7,8 to biliverdin IX. The biliverdin IX is converted to 3Ephytochromobilin by PB synthase, and finally an isomerization forms 3Z-phytochromobilin.

Wavelength (nm)	Photoreceptor	Molecular properties	Functions
600-750	Phytochrome(s)	Serine kinase with linear tetrapyrrole as chromophore	Photoperiodism, seed germination, shade avoidance
	Cryptochrome(s)	Flavoprotein, resembles microbial DNA photolyase	Photoperiodism, hypocotyl elongation

320-500	Phototropin	Serine/threonine kinase with flavin as chromophore	Phototropism
	Rpt	Novel protein with putative NLS and phosphorylation site	Phototropism
280-320	UV-B photoreceptor	Not known	Flavonoid biosynthesis

The phytochrome molecule exists as a dimer in vivo with each monomer having a single chromophore. The photoreversibility of the molecule can be attributed to cis–trans isomerization of the chromophore, which also causes changes in the protein conformation generating the Pr and Pfr forms6. Physiological and biochemical analyses of light- and dark-grown plants suggested the existence of two different pools of phytochrome species:

A light-labile species predominantly present in dark-grown plants and a light-stable species characteristic of light-grown plants. Physiological studies led to the recognition that many of the photoresponses, though elicited by red or far-red light, lack photoreversibility.

Additionally, several photoresponses in plants can be mediated by extremely low fluence of light, whereas many other photoresponses require continuous irradiation with high fluence red or far-red light. The diversity of the photoresponses elicited by phytochrome led to the speculation that the variation is due to existence of multiple phytochrome species9.

The approaches using newer tools and moleculargenetic analyses of photoresponses confirmed the presence of multiple species of phytochrome. These studies first conducted in Arabidopsis revealed that phytochrome is encoded by a small multigene family. In this plant at least five phytochrome genes (PHY) have been identified, namely PHYA, PHYB, PHYC, PHYD and PHYE encoding distinct apoproteins.

Among these, the PHYA gene encodes a phytochrome species, which accumulates predominantly in etiolated seedlings, and is rapidly downregulated on exposure to light. The remaining phytochrome species are relatively more stable in light and therefore are predominantly present in lightgrown mature plants. Based on biochemical and physiological properties, these phytochrome species can be grouped in two types. The type 1 (light-labile) is found in etiolated seedlings and is phytochrome A that accumulates in bulk in the seedlings.

The red-light exposure triggers decline of phytochrome A by inhibiting its transcription, and degradation of its mRNA. In addition, the Pfr form of phytochrome A protein is also quickly degraded by ubiquitination. The remaining phytochrome species, i.e. phytochrome B to phytochrome E constitute the type 2 (light-stable) phytochrome, which is the predominant pool of phytochrome in green plants. Unlike type 1, the type 2 phytochromes are more stable and their levels do not decline on light exposure.

Since type 2 phytochrome species are present in very low amounts in plants, these have not been purified to homogeneity using conventional biochemical techniques, except phytochrome B. The phytochrome B, purified from transgenic Arabidopsis, showed spectral properties similar to phytochrome A13. Although plants synthesize and use phytochromobilin (PB) as chromophore in vivo, phytochrome apoprotein can also covalently bind and use phycocyanobilin (PCB) as chromophore.

This finding has allowed the constitution of photoreversible phytochromes by expressing recombinant phytochrome proteins in yeast and assembling them in vitro. A recent study has shown that the yeast-assembled phytochrome C has 661/725 nm and phytochrome E has 670/724 nm as the red/far-red absorption maxima. The spectral characteristics of phytochromes C and E were found to be very different from those of phytochromes A and B.

Role of Different Phytochrome Species in Plant Development

The realization that different photoresponses have different requirements for light, followed by discovery of multiple phytochrome species indicated that these phytochrome species may have distinct and overlapping function in plant development. The mutant studies showed that phytochrome A is essential for deetiolation of seedlings in far-red light and mediates photoresponses to very low fluences of blue, red and far-red light. Nevertheless, phytochrome B is the major phytochrome species regulating many of red/far-red light reversible responses, whereas phytochrome D and E probably aid phytochrome B action. This is evident because the function of phytochrome D and E can be seen only in phytochrome B mutant. It also illustrates redundancy between phytochromes B, D and E, with phytochrome B being the main regulatory molecule18–20. The genetic analysis of Arabidopsis phytochrome mutants has shown that the major phytochrome is phytochrome

The mutations in PHYB gene have substantial effects on Arabidopsis development throughout the life cycle21. So far, no phytochrome C mutant has been isolated; nevertheless, the overexpression of phytochrome C in transgenic plants suggests a role in primary leaf expansion. The analysis of photomorphogenic mutants has highlighted a complex network of interactions not only among phytochromes, but also with other photoreceptors such as cryptochromes and phototropin.

Is Pr the Active form of Phytochrome

The proposal that the Pr form of phytochrome may also be the biologically active form had been made several times with equally vehement rejections by others. Since in dark-grown plants phytochrome accumulates in the Pr form, it is presumed to be biologically inactive. At the same time, the realization that several of phytochrome A mediated responses can be induced by far-red light is inconsistent with the idea that Pfr is the active form, at least for phytochrome A. Now a possible explanation has emerged from the finding that the Pr form of phytochrome A after undergoing photo-cycling between the Pr and Pfr forms, shows greater autophosphorylation activity. In contrast, the non-cycled Pr form, which is present in dark-grown tissue, shows less amount of autophosphorylation.

It is likely that this actively cycled Pr species may be responsible for far-red mediated effects of phytochrome A23. This notion is supported by recent experiments of Shinomura et al.24 demonstrating that brief cyclic pulses of far-red light with three-minute intervals inhibit hypocotyl elongation to the same extent as continuous far-red light. Since the response is reversible by red light and inducible by far-red light, it is most likely mediated by a form other than Pfr. In all likelihood, this active form is the Pr form of phytochrome that was generated by far-red pulses. It is expected that more experiments on this line would resolve this issue.

Phytochrome is a Light-Regulated kinase

Soon after the discovery of phytochrome, the Beltsville group proposed that it might be an enzyme. However, the above group quickly abandoned this view and favoured the idea that phytochrome may act by regulating membrane properties. The former view was rekindled by Lagarias who proposed that purified phytochrome has a protein kinase activity. But controversy erupted again when this finding was quickly contested by other researchers who could not find a kinase activity in phytochrome preparations or could separate both the activities biochemically. The analysis of phytochrome A amino acid sequence also did not show any motif for a typical protein kinase.

However, comparison of gene sequences of phytochrome species with other known proteins led to the proposal that phytochrome C has motifs similar to bacterial sensor proteins of the 'two-component' regulatory system. Bacteria constantly regulate their physiology and behaviour to respond and adapt to external environment and a typical 'two-component' regulatory system consists of a sensor protein and a response regulator protein.

The sensor protein detects a change in the external environment and communicates this information to the response regulator protein, which in turn either controls the expression of specific genes or initiates other appropriate cellular functions to respond to environmental stimuli. The communication between the sensor protein and the response regulator protein occurs via phosphorylation–dephosphorylation steps.

It is well known that these sensor proteins act as histidine kinases, autophosphorylating themselves. The phosphate group is then transferred to a regulator molecule, leading to a cascade of events that modulate gene expression. This finding led to an intensive re-examination of phytochrome sequence to deduce the putative motifs for kinase activity. It is believed that phytochrome has evolved from the biliproteins present in cyanobacteria. The sequencing of the complete genome of Synechocystis, a unicellular colony-forming cyanobacterium31, facilitated examination of the presence of a phytochrome-like open reading frame in this cyanobacterium.

The phytochrome-like Cph1 gene from Synechocystis PCC6803 was discovered by homology search of the above database. The deduced amino acid sequence of Cph1 gene product showed several similarities with higher plant phytochrome, particularly towards the C-terminal domain, which has a histidine kinase motif. Later in vitro studies showed that the Cph1 holoprotein can indeed act as a R/FR-dependent histidine autokinase and transphosphorylate an aspartate residue of Rcp1 regulator molecule, which is encoded 15 bases downstream of Cph1 gene on the Synechocystis chromosome. Both autophosphorylation and transphosphorylation activities were higher in the Pr form than in the Pfr form3,32. This finding lent strong support to the view that phytochrome protein may have autokinase and phosphotransferase activities, characteristics of transmitter molecules.

The finding mentioned above stimulated a fresh search for kinase activity in higher plant phytochromes.

The new results suggest that these phytochromes also function as kinases. Their C-terminal signalling ends consist of two repeats of a domain called the PAS and another histidine kinase-related domain, which bears weak similarity to the histidine kinase domain of cph1. Studies on site-directed mutagenesis of higher plant phytochrome genes have revealed that the phytochrome molecule consists of two major parts: a highly conserved photosensory domain at the N-terminus and a Cterminal domain, which plays a key role in onward transmission of the light signal.

However, the higher plant phytochromes do not show a histidine kinase motif at their C-terminus, similar to the cyanobacterium Cph1 protein. Nevertheless, purified or recombinant oat phytochrome A can autophosphorylate as well as phosphorylate histone H1. It can also phosphorylate the cph1 substrate.

However, it is autophosphorylated on serines rather than on a histidine, which is the case for cyanobacterial phytochrome. Likewise, it phosphorylates serines or threonines on the substrates rather than any aspartate residue. Based on biochemical analysis of recombinant phytochromes, it was suggested that during the course of evolution the cyanobacterial phytochrome has been modified, yielding a higher plant phytochrome species that functions as a serine/threonine kinase. In case of higher plants, phytochrome autophosphorylation is somewhat higher in the Pfr form. The cycling of phytochrome between Pr and Pfr increases the autophosphorylation activity of Pr. It has also been found that phytochrome A has at least two in vivo phosphorylation sites and in light grown plants one of these sites is preferentially phosphorylated.

Phytochrome A and B Migrate to Nucleus on Activation by light

The experiments on phytochrome localization using the technique of subcellular fractionation always favoured the view that it is predominantly located in the cytosol of the plant. In contrast physiological studies on

Mougeotia supported the view that phytochrome is either bound or is in close proximity of the membrane33, whereas Arthur Galston34 detected phytochrome in pea nuclei by microspectrophotometry. These subcellular locations were subsequently not favoured as potential sites for phytochrome localization and phytochrome was believed to be a cytosolic protein. However, the notion that phytochromes reside in the cytosol turned out to be only half-true because, when activated by light, phytochromes do accumulate inside the nucleus.

The recent advances in gene technology have permitted a re-examination of the question of phytochrome localization by using fusion proteins. The study of intracellular localization of phytochrome B by fusing it to GUS reporter protein showed the presence of phytochrome B in the nucleus. These results indicated that the Cterminal region of phytochrome B possesses a functional nuclear localization signal. That phytochrome B indeed moves to the nucleus was confirmed by immunoblot analysis of isolated nuclei. Since the level of phytochrome B is higher in the nucleus of light-grown plants than in the nucleus of dark-grown plants, it indicates that the migration of phytochrome B to the nucleus is light dependent. The dark-adaptation of plants reduces the level of phytochrome B in the nucleus, confirming that nuclear localization of phytochrome requires light. In fact, the phytochrome B level in the nucleus isolated from dark-grown plants is below the detection limit; however, a brief pulse of red light triggers accumulation of phytochrome B in the nucleus.

Since far-red light causes a decrease in the level of phytochrome B from the nucleus, this process appears to be autoregulated by phytochrome.

This is further evident from the observation that a mutated phytochrome B lacking a chromophore-binding site, which is needed for photoreversibility, failed to enter the nucleus. These fusion proteins were biologically active as transgenic plants showed enhanced light responses. However, the accumulation of phytochrome B in the nucleus was very slow and required several hours after transfer from dark to light.

Similarly, phytochrome A too is localized to the nucleus on light exposure. In accordance with the molecular properties of phytochrome A, its nuclear localization was promoted by very low fluences of either red light or by far-red light, a characteristic of physiological responses regulated by phytochrome A. Interestingly, phytochrome A fusion protein was detected in the nucleus after about 15 min compared to phytochrome B that takes several hours to accumulate in the nucleus. Thus, both phytochrome species move to the nucleus in their active Pfr form, and this nuclear migration may be linked to their capacity to regulate light-mediated gene expression in the plants.

Phytochrome Activation Involves associations with Partner Proteins

Protein–protein interactions are necessary for many signal transduction cascades. It is reasonable to expect that phytochrome also interacts with some partner protein(s) to pass on information about light environment in the cells. The presence of kinase activity in phytochrome and its migration to the nucleus suggests that, on activation, phytochrome may phosphorylate proteins and/ or it may also take a partner to the nucleus.

Such interacting partners for phytochrome were searched using yeast two-hybrid screens with sequences from signalling regions of phytochrome molecules as baits. The phyto chrome interacting factor 3 (PIF3), a basic helix–loop– helix protein, was isolated using the entire C-terminal domain of phytochrome B as bait.

PIF3 interacts in vitro with C-terminal domains of both phytochromes A and B38. The binding of phytochrome B shows red/farred reversibility, with phytochrome B binding to PIF3 only in its active Pfr form. This reversible binding is also consistent with the fact that the phytochrome B is involved in regulating red/far-red light reversible responses in plants. Therefore, PIF3 qualifies as a potential candidate for being a direct messenger of phytochrome B signals.

The proof that PIF3 protein indeed plays some role in phytochrome signalling came from the finding that Arabidopsis poc1 mutant showing altered red-light responses lacks this protein. The PIF3–GUS fusion protein in onion epidermal cells is present in the nucleus, both in light and dark, indicating that it is a constitutive nuclear protein. It is likely that the PIF3 protein might be a transcription factor.

The transgenic plants over-expressing PIF3 gene in the sense direction showed only a little increase in sensitivity to red light, whereas antisense plants showed a long hypocotyl phenotype under both red and far-red light. Similar to PIF3, using a portion of the second repeat of the cryptic sensor kinase domain of phytochrome A as bait in yeast two-hybrid screen, another protein PKS1 (phytochrome kinase substrate1) protein was discovered. This protein too interacts with both phytochromes A and B, similar to PIF3.

Although PKS1 is present in vivo in a phosphorylated form, red light exposure to plants further stimulates the magnitude of PKS1 phosphorylation. It is also phosphorylated in vitro by phytochrome A, which is slightly higher with the Pfr form of phytochrome A. In contrast to PIF3, transgenic plants over-expressing PKS1 showed elongated hypocotyls in red light, but antisense plants had a phenotype like the wild type. It has been proposed that PKS1 may negatively regulate phytochrome B responses. While it is suggested that phytochrome B, acting as a kinase may phosphorylate PKS1, its role in phytochrome signaling is unclear.

The PKS1 protein is localized in the cytoplasm and not in the nucleus. It is possible that PKS1 interacts with phytochrome B in the cytoplasm, but gets dissociated when it migrates to the nucleus.

A third partner for phytochrome was discovered by Song's group, a nucleoside diphosphate kinase 2 (NDPK2), which performs many functions in eukaryotes such as suppression of tumours in animal and human cells. This protein too was screened from Arabidopsis using yeast two-hybrid screening.

The study of localization of NDPK2 in tobacco stomatal cells shows NDPK2–GFP fusion protein in both cytosol and nucleus. The activity of NDPK2 kinase is increased in vitro in the presence of purified phytochrome A by red light, but not by far-red light. However, this evidence indicates that NDPK2 is most likely a positive regulator of phytochrome B signalling.

A role for NDPK2 in the phytochrome signal transduction is evident also from the observation that T-DNA insertion in an intron of the NDPK2 gene results in a defect in red-light-mediated processes such as cotyledon opening and greening. In addition to interacting with proteins involved in the signal transduction cascade, phytochrome also interacts with another photoreceptor cryptochrome1, which detects blue light. The recombinant cryptochrome1 could be in vitro phosphorylated by recombinant oat

phytochrome A. The magnitude of phosphorylation of cryptochrome1 by phytochrome A was higher in red light or blue light than in the dark. Regarding in vivo phosphorylation of cryptochrome1, it was higher in red light and was reversible by far-red light. The physiological significance of phytochrome-dependent phosphorylation of cryptochrome1 remains to be established.

Phytochrome Regulates gene Expression by Binding to a Transcription Factor Complex

One of the attractive proposals for phytochrome action envisages regulation of gene expression by direct interaction of phytochrome with the transcription system. A detailed investigation into the function of

PIF3 has indeed demonstrated the existence of such a mechanism in Arabidopsis. The transgenic plants expressing the PIF3 gene in the antisense direction showed reduction in light-inducible expression of a subset of light-inducible genes. Since PIF3 has the motif of a transcription factor, it is likely that this reduction results due to retardation of transcription arising from depletion of phytochrome B–PIF3 complex. In a recent paper, Martinez-Garcia et al., have made an important breakthrough in understanding the mechanism by which the phytochrome-perceived signal is transduced to the nucleus to modulate gene expression. They showed that

PIF3, which is a basic helix–loop–helix transcription factor, specifically binds to various light-regulated promoters containing a G-box DNA-sequence motif44. Phytochrome B binds to this G-box-bound PIF3, particularly after redlight mediated formation of biologically active Pfr form. Importantly, this interaction is photoreversible and phytochrome dissociates upon FR-induced conversion of phytochrome to the inactive Pr form.

These results support the view that phytochrome may function as an integral part of a transcription regulating complex, which can be turned on or off depending upon light availability. This provides a very rapid signaling mechanism where phytochrome itself binds to protein– DNA complexes, to regulate target gene expression. Such a mechanism would allow plants to continuously monitor their light environment and react to changes in light availability by concomitant changes in lightregulated gene expression.

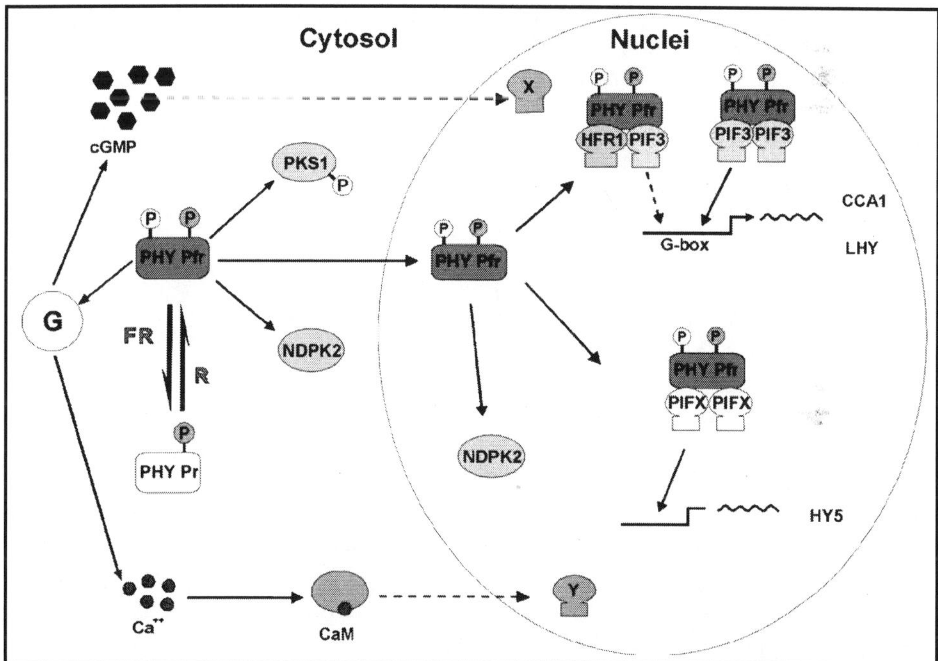

Sketch illustrating the possible pathways of light-regulated gene expression modulated by phytochromes. The red light (R) alters the conformation of the Pr form to the Pfr form and triggers migration of the Pfr form of phytochrome A (PfrA) or phytochrome B(PfrB) from cytosol to the nucleus. The light-induced formation of Pfr is also accompanied by autophosphorylation of phytochrome and transphosphorylation of PKS1 in the cytosol.

The formation of Pfr form also triggers, through yet unknown steps, activation of heterotrimeric G-proteins (G-P) and leads to changes in the levels of cGMP and Ca2+. These second messengers may then set a signal chain activating hypothetical nuclear proteins (X and Y), which may regulate transcription. In the nucleus, phytochrome A and B may regulate expression of light-regulated genes by directly or indirectly interacting with nuclear proteins such as SPA1, FAR1, GI and PIF3.

The light may also control transcription regulated by transcription factors such as HY5, by modulating the level of COP1. In darkness, the COP1 with help from COP/DET/FUS complex may regulate the HY5 level by proteolysis. The exposure of light may inactivate the COP1 protein and also induce its translocation from the nucleus to the cytosol and dissociation of the COP/DET/FUS proteosome complex. The expanded inset inside the nucleus shows that phytochrome B (PfrB) imported in the nucleus, binds to PIF3, a nuclear transcription factor, bound to G-box motifs in light-sensitive promoters. The PfrB–PIF3 complex then activates/represses transcription by modulating the transcription machinery. The far-red light reverts phytochrome B to the PrB form, triggering its dissociation from PIF3 complex, leading to inhibition/enhancement of transcription of light-regulated genes. The dashed lines indicate that additional steps in the signal transmission are yet to be discovered.

Phytochrome Activation Causes Redistribution of Proteins between Nucleus and Cytoplasm

One of the ways by which phytochrome regulates photomorphogenesis appears to be the modulation of the subcellular distribution of key regulatory factors. In dark-grown suspension culture cells of parsley,

CPRF2 – a member of the common promoter-binding transcription factor family (CPRF) – is almost exclusively in the cytosol. The N-terminus of the CPRF2 protein has two domains responsible for its retention in the cytosol45. However, the cytosolic retention of the protein is abolished by red light treatment, which promotes its nuclear translocation in red–far red reversible manner. The CPRF2 protein is a phosphoprotein in vivo, which is phosphorylated on the C-terminal half by a cytosolic protein serine kinase. The phosphorylation of CPRF2 does not alter its DNA-binding activity; so it may not be related to transcription, but may have a role in nucleocytoplasmic partitioning.

One of the best characterized examples is lightregulated distribution of COP1 protein in Arabidopsis during photomorphogenesis between nucleus and cytosol. Although phytochrome itself migrates to the nucleus after activation, it also regulates the level of certain proteins in the nucleus, which in turn may allow further modulation of light-mediated gene transcription.

The analysis of Arabidopsis mutants which had a lightgrown phenotype in darkness, has led to the identification of a novel regulator protein COP1. Since the mutant seedlings displayed constitutive photomorphogenesis and the mutation was recessive in nature, it indicated that COP1 might act as a negative regulator of photomorphogenesis. COP1 protein shows a typical bipartite NLS and three other structural motifs, viz. an N-terminal RING-finger, a coiled-coil domain and a C-terminal WD-40 repeat region48. These three domains are involved characteristically in protein– protein interactions required for COP1 function. Interestingly, COP1 also possesses a cytoplasmic localization sequence (CLS), which is perhaps needed for moving COP1 out of a nucleus on exposure to light.

Transgenic seedlings over-expressing full-length COP1 protein show an etiolated phenotype in different light conditions, substantiating the notion that COP1 is involved in repressing photomorphogenesis and acts downstream of photoreceptors. Similar to phytochrome, COP1 also shows accumulation to a high level in nuclei of hypocotyl cells, but this high level is seen only in dark-grown seedlings47.

In light-grown plants, the COP1 level in nuclei is very much reduced. Light appears to cause a massive redistribution of COP1 between the nucleus and cytosol. The search for the partners of COP1 protein has led to identification of additional components of light-signalling pathway. In the cytosol of plants, COP1 interacts with a cytoskeleton- localized protein CIP1, which perhaps acts as a cytoplasmic anchoring mechanism for regulating partitioning of COP1 between nucleus and cytosol.

Studies on factors regulating nuclear localization of COP1 show presence of a complex network involving several proteins. Apparently, interaction among them is responsible for nuclear accumulation of COP1. The constitutively photomorphogenic phenotype similar to cop1 mutant is seen for at least ten other mutants which are mutated at different loci, such as det1, fus4–fus6, fus11, fus12, cop8–cop10, and cop16. The essentially similar phenotype of these mutants suggests that these are constituents of a network that functions co-operatively in regulating photomorphogenesis.

In fact, an examination of GUS–COP1 localization in this group of cop/det/fus mutants has shown that the darkgrown seedlings do not accumulate GUS–COP1 fusion protein in the hypocotyl nuclei52. This reinforces the view that this network of genes in the wild type has a role in nuclear localization of COP1. Based on these observations, it has been proposed that light might be regulating the nuclear level of COP1 by controlling the activity of several COP/DET/FUS proteins, which may exist as part of a large protein complex.

In fact, a multisubunit COP9 complex, renamed as signalosome and localized within the nucleus contains many of the COP/DET/FUS proteins as its constituents53. The COP9 complex has eight core subunits that exhibit sequence similarity to non-ATPase subunits of the 19S regulatory particle of the 26S proteasome. It is now believed tha COP/DET/FUS genes operate by repressing the default pathway of photomorphogenesis in darkness. The COP/DET/FUS gene family seems to be evolutionarily conserved and its homologues are present even in mammals53. Based on this homology some of the COP/FUS proteins have been recently renamed as CSN (COP9 signalosome), with new names such as CSN1 (COP11, FUS6), CSN4 (COP8, FUS4), and CSN7 (FUS5) and CSN8 (COP9).

In the nucleus of plant cell, the COP1 protein is distinctly associated with other nuclear proteins. Interestingly, the first nuclear target identified for COP1 represents a protein, the loss of which gives a phenotype opposite to that of cop1 in Arabidopsis. The hy5 mutation in Arabidopsis is epistatic to cop1 mutation, and displays a phenotype showing etiolated seedlings in light with elongated hypocotyls. The cloning of HY5 gene showed that it encodes a bZIP-type transcription factor54, which is involved in light-mediated transcription regulation by the promoters containing the G-box.

The HY5 protein acts as a positive regulator of photomorphogenesis and suppresses cop1 mutations. The recent studies show that the COP1 negatively regulates HY5 activity. The direct interaction between

COP1 and HY5 was observed both in yeast two-hybrid assays and in vitro binding assays. The co-localization studies have provided evidence for an in vivo interaction between COP1 and HY5.

In onion epidermal cells, COP1 and HY5 colocalize to specific nuclear spots, apparently in close contact with each other, as evident by the transfer of fluorescence resonance energy between them55. The abundance of HY5 in the nucleus is directly correlated with the extent of photomorphogenic development, and that the COP1–HY5 interaction specifically targets HY5 for proteasome-mediated degradation in the nucleus56. These results show that photomorphogenesis of plants is facilitated by light-dependent stabilization of a transcription factor that is rapidly degraded in darkness. It also shows that in photomorphogenesis too, similar to cell division and circadian rhythms, gene regulation may be governed by proteolysis.

Most likely, the suppression of photomorphogenesis by COP1 in the dark involves its interaction with multiple transcription factors to repress gene expression. This view is supported by the identification of CIP7, a transcriptional regulator, as another protein that interacts with COP1. The transgenic plants having CIP7 gene in antisense direction show reduced expression of several light-inducible genes. Since the expression of CIP7 is upregulated by light, it may be a downstream target of COP1. In essence, the accumulation of COP1 in the nucleus is supported by the action of COP/DET/FUS gene products. In the nuclei of darkgrown plants, COP1 interacts with transcription factors and inhibits photomorphogenesis. Exposure to light reduces the abundance of COP1 in the nucleus and relieves the inhibitory effect on transcription factors regulating photoresponses. genetic screens for light insensitive/hypersensitive phenotypes.

Several mutants such as fhy1, fhy3, spa1, fin2, pat1 and far1 have been identified as being disrupted in phytochrome A signal transduction pathway, of which spa1, far1 and pat1 have been characterized at the molecular level58–60. The PAT1 gene encodes a cytoplasmic protein of 490 amino acids with sequence homologies to the plant-specific GRAS regulatory protein family.

The PAT1 gene is localized in the cytoplasm and acts as a positive component downstream of phytochrome A signal transduction pathway60. The SPA1 is a suppressor of a weak phytochrome A mutant, encodes a nuclear protein, possibly a transcription factor that functions as a negatively acting factor at an early step in phytochrome A-specific signal transduction. The cloning of the SPA1 gene showed that it encodes a WD (tryptophan–aspartic acid)-repeat protein that also shares sequence similarity with protein kinases58. Likewise, FAR1 protein also contains a nuclear localization signal and is targeted to the nucleus in transient transfection assays59. A class of mutants specifically affected in red-light signalling, such as red1, pef2 and pef3, has also been isolated in Arabidopsis61,62, which shows phenotype similar to phytochrome B mutants. The analysis of the molecular nature of these genes in future would aid in our understanding of phytochrome B signalling.

Second Messengers may Mediate Phytochrome Responses

The multiplicity in the phytochrome-regulated responses signifies that this diversity results from operation of multiple signal transduction pathways triggered by the phytochrome. One attractive proposal is that phytochrome activation modulates the levels of certain secondary messengers in the plants.

The second messenger concept was developed after the discovery that cyclic adenosine 3¢,5¢- monophosphate (cAMP) is an intracellular mediator of glycogenolyic hormones in the liver. In recent years, enough evidence has accumulated to support regulation of photoresponses via a set of second messengers. Phytochrome-deficient mutants, such as tomato aurea (au) mutant, display a phenotype showing elongated pale-green seedlings63. These seedlings also lack anthocyanin and have poorly developed chloroplasts.

Gunther Neuhaus and colleagues could successfully restore the photomorphogenic responses in seedlings of aurea mutant by microinjection of oat phytochrome A into single subepidermal hypocotyl cells. The restoration of photomorphogenesis by microinjection of phytochrome provided a convenient assay for finding components of the phytochrome signalling pathway. One of the early components in the signal transduction chain appears to be a G-protein, as evident from the observation that inhibitors of heterotrimeric G-protein completely blocked the action of injected phytochrome. This view is strengthened by emulation of phytochromemediated responses in au cells by injecting GDP-g-S which activates G-proteins. Beyond the G-protein, the signalling splits into two parallel and/or interacting pathways:

- The calcium/calmodulin pathway controlling chloroplast development;

- The cGMP pathway controlling anthocyanin induction. These pathways show a reciprocal control and have a crosstalk among them such as shut-off of the photosynthetic complex synthesis by high levels of cGMP and switch-off of the anthocyanin pathway by high levels of calcium.

The most ubiquitous second messenger is the Ca2+ ion, which on activation is translocated from sequestration sites in the cell or from outside into the cytoplasm where free Ca2+ concentration is very low. In maize leaf protoplast, red light stimulated calcium uptake within two minutes of light exposure and was reversed by far-red light. Additionally, 5-hydroxytryptamine also enhanced calcium uptake in dark, indicating a role for phosphoinositides in light-regulated responses.

Besides regulating calcium uptake and distribution, phytochrome may also control phosphoinositide metabolism in plant cells. There are additional steps in signal transmission such as phosphorylation of proteins by protein kinase C, which in turn regulates gene expression of nitrate reductase. Additionally, the metabolite sucrose also appears to play an important role in regulating phytochrome-controlled responses.

Interaction between Plant Hormones and Phytochrome

It is logical to assume that phytochrome which regulates several processes in plant development, and plant hormones which control and coordinate several plant development responses, must interact closely at least at a few points. For example, many phytochrome mutants show elongated seedlings under specific light regimes.

It is known that plant hormones also regulate stem elongation, suggesting a cross-talk between the two pathways. The evidence in favour of such interaction has come again from analysis of hormonal levels in

photomorphogenic mutants. The dark-grown seedlings of brassinosteroid mutants of Arabidopsis show the phenotype of light-regulated plants, with short hypocotyls and expanded cotyledons as if seedlings are growing in light. The adult phenotype of these mutant dwarf plants with short stems and petioles, dark-green leaves and delayed senescence is essentially the opposite of phytochrome B mutants. Apparently, these two processes appear to be unlinked, as light does not seem to regulate the brassinosteroid biosynthetic genes. Recently, a link has been found between phytochrome B and brassinosteroids by isolation of an extragenic, dominant, gain-of-function suppressor of a phyB missense mutation called BAS1-D, which codes for a cytochrome P450 catalysing the inactivation/degradation of brassinosteroids.

The bas1-D phyB-4 double mutant has no detectable brassinolide (BL), the most active BR. The bas1-D suppresses a phyB-null allele, but not a phyA-null mutation. Though it is still not known how light regulates BAS1 activity, a link between photoreceptors and brassinosteroid metabolism is apparent.

A few of the photomorphogenic mutants show a phenotype similar to phytohormone mutants, for example, the spindly mutant defective in gibberellin (GA) signalling resembles phytochrome B mutants with long stems, pale leaves and early flowering.

Phytochromes can regulate the transcription of GA biosynthesis genes. Similarly, a role for auxins in photomorphogenesis is indicated by the study of hy5 mutation, which results in seedlings with long hypocotyls under light.

The HY5 gene encodes a bZIP transcription factor54 that is involved in auxin signalling. A link between auxin signalling and photomorphogenesis is provided by cloning of SHY2 gene. The shy2 mutant suppresses the longhypocotyl phenotype of hy2 mutant73 and this dominant mutation resides in the auxin-induced gene IAA3, indicating a link between light and auxin in regulation of hypocotyl elongation74. The search for a suppressor of COP1 led to identification and characterization of a new light regulated gene, FIN219, that represents a novel locus whose mutation results in a long hypocotyl only in far-red light. FIN219 plays a role in phytochrome A mediated FR inactivation of COP1 and in promoting photomorphogenesis.

The sequence of FIN219 gene is highly similar to a family of proteins defined by the soybean early auxin-inducible gene GH3. Similar to GH3, the expression of FIN219 is rapidly inducible by auxin. This finding indicates a molecular cross-talk between auxin response and light regulation, suggesting that by regulating FIN219 expression, auxin may in turn play a role in light regulation of development. Such an interaction between phytochrome and auxin signaling is also seen for the negative regulator ATHB-2, a transcription factor upregulated by far-red light. The transgenic plants over-expressing ATHB-2 confer a shade avoidance phenotype, which is partly due to interference with auxin transport.

LIGHT SENSING AND SIGNAL TRANSDUCTION: NEW INSIGHTS

The crystal structure of the chromophore binding domains of DrBphP provides a wealth of new information about the nature of the chromophore binding pocket and, by analogy, those of plant phytochromes. The paramount question of chromophore configuration in the Pr state is now conclusively answered: the chromophore adopts a C5-Z,syn/C10-Z,syn/C15-Z,anti conformation. In addition, the DrBphP structure exhibits many interactions between the chromophore and conserved residues.

The entire structure of the chromophore is buried, surprisingly including the hydrophilic carboxylate side chains. The burial of the water-soluble carboxylates appears counterintuitive on first principles but for the observation that one carboxylate interacts with the highly conserved Arg254 that lies within the knot region. Such a buried salt bridge is likely to be energetically favorable for the assembly process. It is particularly interesting that three ordered water molecules are found within 4A° of the BV chromophore, one of which is in close contact with its conjugated ring system. Since the exclusion of solvent water from the chromophore binding pocket is likely a key factor in ensuring efficient photochemistry, it will be interesting to assess how the position of these waters is affected by the presence of the PHY domain that is missing from the DrBphP structure.

The chromophore ring system of DrBphP makes a number of contacts with conserved residues, including the universally conserved Asp207-Ile208-Pro209 motif. Strikingly, the D-ring is less tightly packed by protein residues than other parts of the chromophore, presumably making it easier for this ring to rotate during the photoconversion process. The hydrophobic contacts that the D-ring does make are primarily with conserved aromatic residues; these may provide rigidity to modulate the photochemical reaction.

One of the residues lining the D-ring cavity, Tyr176, is known to be critical for proper photochemistry, perhaps by gating the rotation of the chromophore D-ring. The photochemical gating role for this conserved Tyr residue in the bacteriophytochromes was more recently questioned.

Primary photochemistry is generally accepted to involve a Z-to-E isomerization about the C15,16 double bond. Consistent with this hypothesis, a recent theoretical study of PCB provided evidence for considerable double-bond character about the C15,16 bond in the ground state but not in the excited state. The DrBphP structure provides new insight into the process of photoconversion.

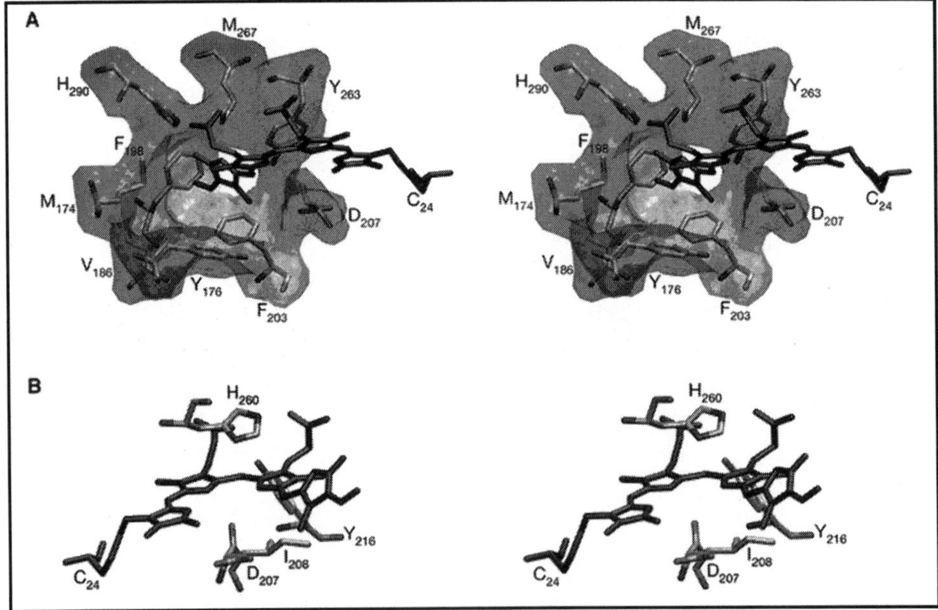

Fig. Stereo Views of Interactions between the BV Chromophore and Nearby Residues in the DrBphP Apoprotein.

(A) Residues within 4.5A ° of the chromophore D-ring are shown colored by atom type and as a single solvent-accessible surface prepared with MSMS colored by similarity as in Figure. Cys24 and the BV chromophore are shown in purple. The colour highlighting of the protein residues ranges from dark blue (absolutely conserved) to light blue (strongly conserved). No red-shaded (variable) residues are located near the chromophore. (B) Tight packing of His260, Ile208, and Tyr216 about C10 is shown with Cys24 and chromophore in purple. Amino acid residues are shown colored by atom type as in Figure. Since the bilin A-ring is covalently attached to the protein matrix and deeply buried between two a-helices, rotation about the C4,5 double bond is unlikely.

The B- and C-rings are tightly packed by His260, Asp207, Ile208, and Tyr216, making it difficult to envision rotation about C10. These structural constraints explain the observation that holophytochromes assembled with phycoerythrobilin, a bilin that is missing the C15,16 double bond, are intensely fluorescent, nonphotoconvertible biliproteins. In the DrBphP structure, the BV D-ring is twisted out of plane relative to the A-, B-, and C-rings. This torsional strain would reduce the energetic barrier for rotation of the D-ring by destabilizing the ground state, thereby facilitating photochemistry.

It is also conceivable that photoconversion might proceed via a hula twist mechanism involving simultaneous rotation about two bonds rather than via a single-bond rotation mechanism, as has been described for other photoprotein systems. However, a hula twist about C5 would introduce a steric clash between the A- and B-rings, since both are tightly packed by protein residues and the A-ring is covalently anchored to the protein matrix. A hula twist about C10 also would be energetically disfavored due to close interactions between the BV chromophore and the side chains of the conserved His260, Ile208, and Tyr216 residues. Moreover, assembly of the closely related bacteriophytochrome Agp1 from Agrobacterium tumefaciens with synthetic, sterically locked biliverdin derivatives strongly supports the conclusion that the Pr chromophore adopts a C15-Z,anti configuration, while the Pfr chromophore assumes a C15-E,anti configuration.

For this reason, as well as the observed bathochromic shift of the primary Lumi-R intermediate, a hula twist about C15 also can be ruled out in the Pr-to-Pfr photoconversion pathway.

8

Seed Propagation Techniques

Seed or sexual propagation is the most frequently used means of producing new plants. Seed propagation offers several advantages over other methods: It is economical, fewer diseases are transmitted, the seeds can be easily and inexpensively stored for long periods and usually very little personal effort is needed. Seeds are not always used for propagation because of some disadvantages. Many seeds require too long to germinate or require special pretreatment to overcome internal dormancy. Some seed require physical abrasion before germination can take place. Genetic variation is another problem which may arise from seed propagation. Commercial seed companies go to great effort to develop plant cultivars uniform in flower colour. When hybridizing, they go to even greater costs to be sure the crosses will breed true each time. When you purchase hybrid seed, you are planting the first generation of the parental cross. Should you save the seed from the hybrid plants and sow them next year, the results will be a genetic scrambling, producing an array of flower colours and sizes.

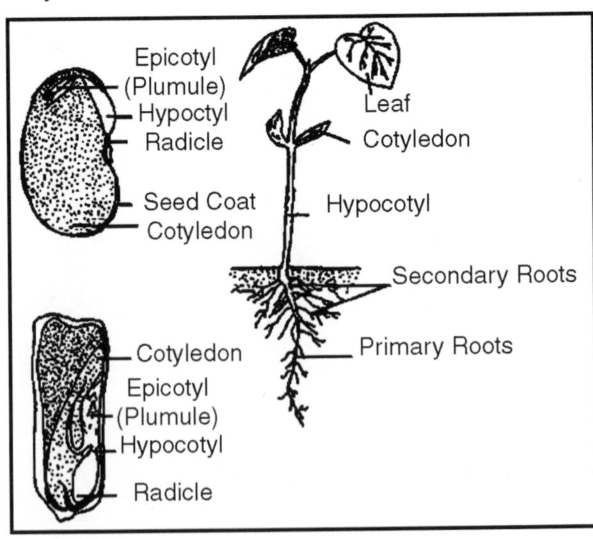

Fig.: Seed and Seeding Anatomy.

While seeds range in size from the large coconut to the very small petunia seed, they all have the same basic structure: seed coat, embryo and endosperm, which contains stored food. The embryo is the result of fertilization, and is made up of a radicle (embryonic root), hypocotyl (stem region below cotyledons), cotyledons (embryonic leaves) and epicotyl (stem region above cotyledons)

Seed germination begins with water being absorbed. Enzymes convert starch to sugar and combine with oxygen to carry on the building of new cells and expansion of others. Eventually, the radicle emerges from the seed coat and penetrates into the soil, forming the primary root. Next, the hypocotyl emerges and pushes the cotyledons above the ground. This point is witnessed as germination.

SEED PROPAGATION

Start with good, clean, fresh seed. Next, be particular about the medium in which the seeds will be germinated: This should be something which will allow sufficient air and water to reach the seeds. Typical garden soil is a poor choice. Milled sphagnum moss, which may be difficult for homeowners in some areas to obtain, is an excellent material for seed germination because it is highly acidic. Weed seeds, insects, or disease microorganisms are essentially eliminated. Other materials to consider for starting seeds are perlite, vermiculite, peat moss and commercial mixes such as Jiffy-Mix. Containers for starting seeds should be clean, inexpensive and readily available. Homeowners use a wide array of containers — old milk cartons cut in half, wooden flats, old butter tubs, or recycled salad trays from some fast food establishments. The choice is really yours. Just make sure it is clean and can be freely drained.

Fill the container to the top with the growing medium of your choice. Firm to about one-half inch from the top and score off some shallow rows for planting seeds. At this point, a mistake is commonly made; the seeds are planted too deeply. When in doubt, plant shallower than you think necessary. As a rule of thumb, try to space the seeds about one-fourth to one-half inch apart in the rows and make the rows about 2 inches apart in the absence of any packet instructions. All of this fussiness is to cut down on the incidence of disease, especially the fungus causing damping-off. This disease will girdle and kill seedlings. If everything is clean the disease will not be present, so spacing is less important. If the disease is present, spacing as suggested may only slow the spread.

To avoid the spread of this disease, be careful watering. Subirrigation is the best approach for the homeowner, allowing the water to enter in through the drainage holes. When the media surface glistens (appearance may vary with mix used) with moisture, remove the container and allow it to drain freely. Then cover with either a clear plastic bag or a piece of glass.

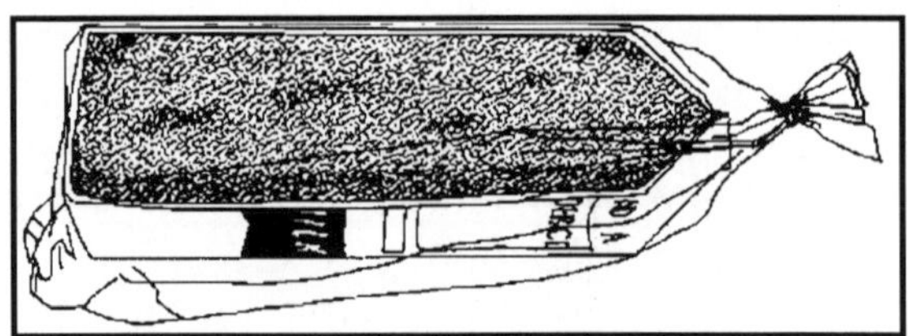

Fig: Seed Propagation: Cover Container with Plastic Bag.

After most of the seeds have germinated, promptly remove the cover to allow air circulation. The first leaves to appear are the seed leaves or the cotyledons. When the next leaves or first true leaves have opened, most seedlings are large enough for transplanting.

LIGHT QUALITY AND INTENSITY IN SEEDLING DEVELOPMENT

Light quality and intensity are important for proper seedling development. The indirect light from a window may often be sufficient for seedling development. If the indirect light is too low, the seedlings may become leggy. In that case, the seedlings would benefit from exposure to direct sunlight. Drafts from windows in the cold winter months and the unilateral direction of the window light often create problems, first with germination and then with the straightness of seedling growth. In this situation, the homeowner is not in complete control: It may not be sunny, or it may be very cold. The result is lowered success. Fluorescent lighting and heating cables put more control of seedling development into the hands of the grower. Generally, bottom heat from thermostatically controlled cables set to about 72 F and standard fluorescent light positioned directly overhead about 6 to 9 inches from the surface will result in swift germination and strong, healthy development of most garden plants.

TRANSPLANTING AND POST TRANSPLANTING CARE

To transplant, gently lift beneath the seedling with your finger or a wooden label and lightly pull by holding onto one of the leaves. Grasping the stem and pulling up provide a greater chance of injuring the seedling. If the delicate seedling stem is crushed, the plant dies. If one of the leaves is damaged from rough handling, then usually only that leaf is lost and the plant has a chance to recover. Make a hole in the new pot deep enough to contain the roots. Gently but firmly pack soil around them and water in to eliminate air spaces (drying out) of roots.

Once transplanting is complete (either into individual pots or flats) light is very important. Direct sunlight is best, but if this is not possible, then use the fluorescent lamps suggested for germination. Houseplants can be maintained at relatively low light intensities once they become established. Garden plants which are intended for flower or vegetable production will need an excess of 1,000 footcandles to grow satisfactorily without stretching or discolouring. This intensity can be maintained by keeping the fluorescent light source 6 to 9 inches from the surface of the growing plants.

Most flowering plants will benefit from a pinching, which is the removal of the tip of the stem above a leaf axis after at least three sets of leaves have developed. This will cause the plants to branch and become compact in their growth habit. If you desire to have the plants tall and single-stemmed, then do not pinch. Plants intended for the outdoors should generally be started four to six weeks before moving outside. The transition should not be sudden. Going from an interior where the temperature and light conditions are consistent to the shifting conditions in the garden could destroy the plants.

Begin the move by gradually acclimatizing the plants: An initial 30 minutes in the sun, building up to a full day in a protected area; a night in an unheated porch, watering less frequently. All or any of these steps should begin 10 to 14 days before the actual move-out date. Be aware that some plants are considered "tender;" a mere brush with near-freezing temperatures could set them back or kill them. Others, being more hardy, can tolerate occasional brief exposures to cold with no visible harm.

Keep a wary eye on the weather. If it appears that we are in a cycle of frosts when the set-out date is near, delay for a few more days when the weather appears to be more amenable to plant establishment.

ASEXUAL PROPAGATION TECHNIQUES

Many members of the plant world are capable of reproducing themselves from small pieces of root, stem or leaf. This characteristic allows the home propagator another resource for increasing plant numbers when reproduction from seed is not practical. Asexual propagation gives the home propagator another advantage that seeds do not; exact genetic replication of the parent.

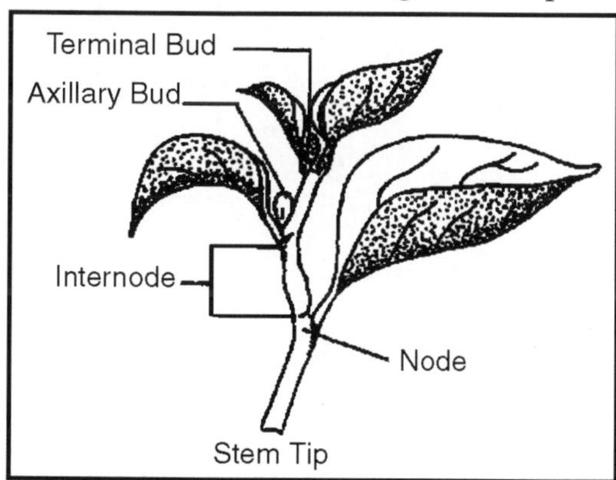

Many methods of asexual propagation exist for the homeowner: cuttings, bulbs, divisions, offsets, runners, layering, grafting and micropropagation. Perhaps the easiest and most common method of asexual propagation is the cutting. Plants like coleus, African violets and Swedish ivy are examples of plants commonly reproduced in this manner.

Coleus propagation is usually accomplished via stem cutting 4 to 6 inches long just below a node and bearing at least two nodes. The basal end of the cutting is inserted in some rooting medium, and with care, roots will develop suffficiently in three weeks to support a new plant. Sufficient root system development is the key to success in propagating with cuttings. Buds or areas where leaves arise on stem cuttings are the likeliest areas for good root development. Make the cuts about one-quarter inch below the last bud. In some cases, a proliferation of parenchyma cells (living cells arising from the cambium, forming callus tissue) may form prior to root emergence; in other cases, no callus formation preceeds root emergence. In contrast to roots arising in response to wounding are a classification of root cells known as preformed or latent root initials. They are found in such species as the easily rooted poplars *(Populus)*, willow *(Salix)* and currant *(Ribes)*. These root initials elongate into an established root system when a plant part (stem) is removed from the parent plant.

With some plant species, further wounding is beneficial to rooting. This involves making an additional light cut on either side of the cut stem at the base to expose more cambial area (the single layer of meristematic tissue which gives rise to the functional tissues — bark and xylem) and, in some cases, to break a physical barrier which may be impeding the formation of root initials and thus the emergence of such roots. This is evident in carnation cuttings where no further wounding results in the emergence of roots from only the base of the cutting.

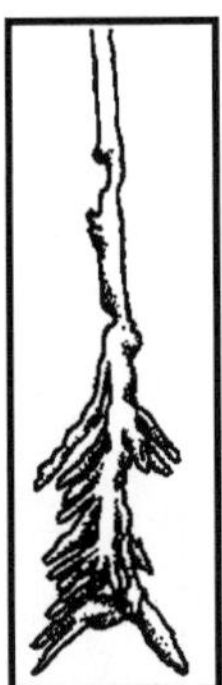

Fig: Good Root Emergence from Wounding.

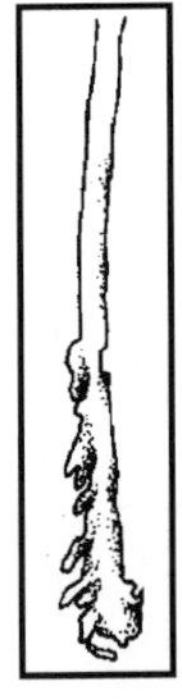

Fig: Poor Root Emergence from Wounding.

Another type of cutting infrequently used by home plant propagators is the root cutting. Where adventitious roots develop on stem cuttings, adventitious shoots develop on root cuttings. Plants which are propagated in this manner are: apple *(Malus* spp.), poplar *(Populus* spp.), and raspberries *(Rubus* spp.). It is important that propagation be made from disease-free plants in all situations, but especially with raspberries. Red raspberries sometimes carry a virus which they can tolerate, but will be fatal to the black raspberries. Propagating from virus-free stock is the best way to avoid this trouble. Taking sections from actively growing roots of these plants usually results in success. It is important when working with stem and root cuttings that the sections be properly oriented. Make sure that the tops of all pieces are cut differently than the base, to keep from inverting them in the rooting media. Failure to keep track of this detail could result in failure of the cutting to become properly established.

MEDIA FOR CUTTING PROPAGATION

Any medium which will lend physical support, provide moisture and oxygen in the right balance, and is free of pathogens can be used for propagating stem or root cuttings. Sand, perlite, sphagnum moss (unmilled), vermiculite and various combinations of these materials have been used to successfully establish cuttings. Obtain whatever is available and relatively inexpensive. As with seed propagation, the container the medium is placed in should be well drained to allow good aeration for rapid root development. Push the cuttings into the medium just far enough to be self-supporting. Cuttings pushed too deeply into the medium may develop roots similar to those shown in sketch 6 because there is insuffficient oxygen at the lower level for rooting to take place.

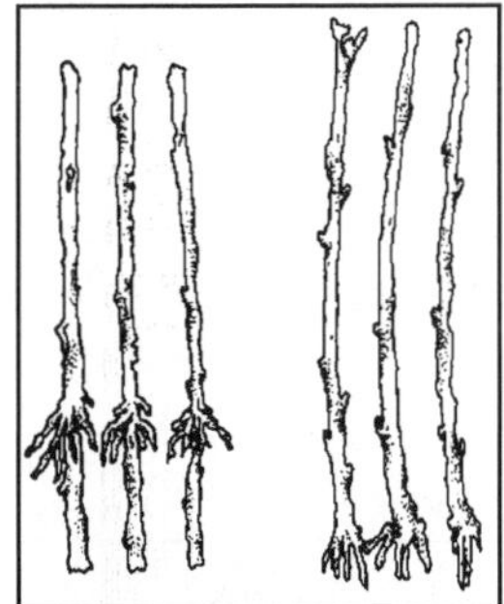

Fig: Influence of Oxygen Concentration Affecting Root Formation.

There are at least as many variations in the rooting procedure as the imagination can provide. Simply use whatever is convenient and effective.

Fig: Cuttings: A Stem Cutting Rooting in Water.

Fig: Cuttings: Plastic Bag Closed.

Fig: Cuttings: Stem Cuttings Placed in Perlite.

VARIATIONS IN CUTTINGS

Anyone who has ever grown African violets, jade plants, peperomias or wax begonias knows that these beautiful plants are easily propagated by using leafpetiole cuttings. For these and other plants, always use the best looking leaf — not too old or too immature, but one which is healthy, vigourous, and of a dark-green colour. Using a sharp, clean knife, carefully cut a choice leaf from an inner circle of the rosette of leaves. Then before placing it in the medium for rooting, cut the end again to make it into a "V." This will provide more plantlets from the base of the cutting.

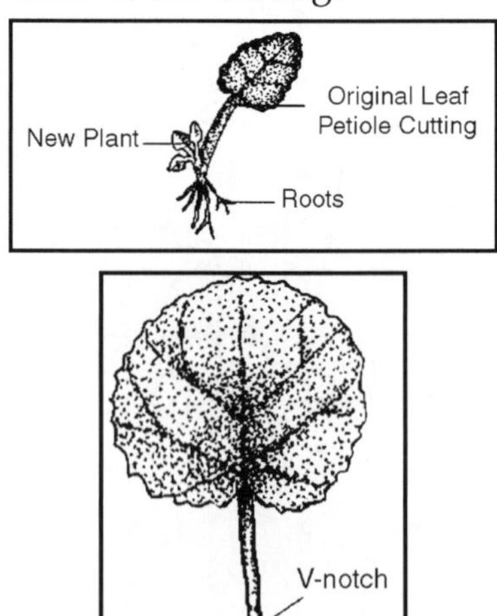

If you follow the accepted practice of rooting the African violet in water, opt to use rainwater, water from a dehumidifier or distilled water rather than tap water. Most tap water contains some chlorine and dissolved salts which inhibit rooting. Look for roots to form in four to six weeks. As the water level goes down, add fresh water and change completely once per week to prevent stagnation. Some plants such as the Rex begonia and Sansevieria can be propagated by using just leaf sections. As long as the leaf cutting sections each have a major vein, a new plant can be generated at that point. This method cannot be used to propagate certain forms of variegated snake plant because the variegation is not reproduced in the new plant.

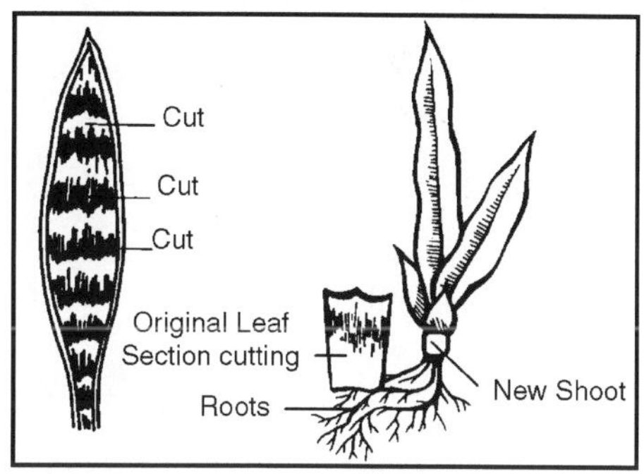

Runners and stolons are similar stem-like structures which emerge from the crown of the plant and spread horizontal to the ground surface. They both root at the nodes and can be easily separated from the parent plant to propagate a new individual. Examples of plants with these structures are the strawberry, spider plant(*Chlorophytum* spp.) and Ajuga.

Offshoots are lateral shoots that develop from the base of the main stem in certain plants. These occur in monocots such as pineapple, banana or date palm, which are not found naturally in the Midwest. Generally, these can be removed from the parent plant after they have been notched and new roots have formed. Suckers are well known to most gardeners. They are actually vigourous shoots arising from underground roots. Raspberries, some dogwoods and Canada red cherry(*Prunus virginiana* 'Schubert') are examples of plant species which sucker easily.

BULBS, CORMS, RHIZOMES, TUBERS AND TUBEROUS ROOTS

Landscape plantings commonly include tulips, hyacinths, and daffodils, which are all propagated by bulbs. Bulbs are modified storage organs consisting of fleshy, compressed stem axis and flower primordium enclosed by thick, fleshy scales. A corm, while bulb-like in appearance, is a compressed stem structure with nodes and internodes.

The scale-like covering over this stem is known as a tunic and functions to protect it from injury or dehydration. The most common example of a corm is the gladiolus. A rhizome is an underground stem structure which grows just below the soil surface. Examples of rhizomes are iris, bamboo and lily-of-the-valley. Being a true stem, a rhizome possesses nodes and internodes.

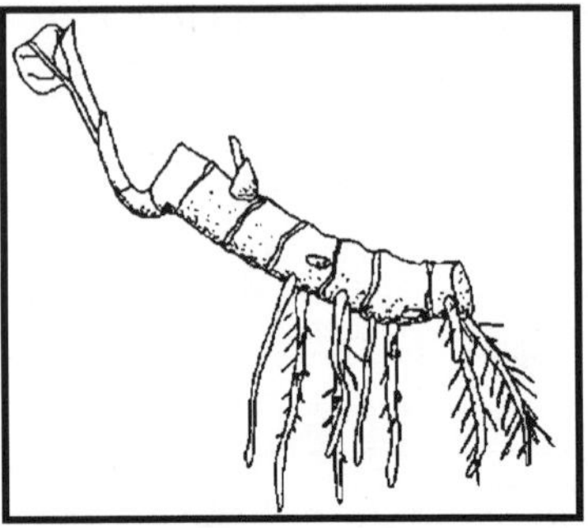

Fig: Rhizome Section.

Tubers are swollen, underground stems, possessing nodes in a greatly enlarged structure. The Irish potato is a conspicuous example.

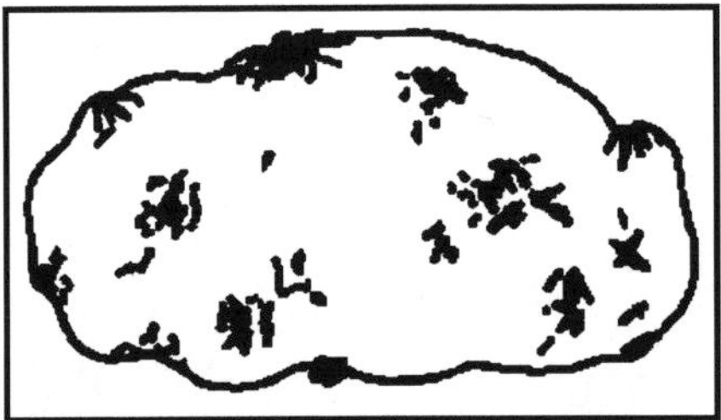

Fig: Underground Stem: Potato Tuber.

Tuberous roots are true roots which develop swollen, underground structures. Sweet potato and dahlia are structures of this nature, producing adventitious shoots on the proximal end (nearest crown) and adventitious roots on the distal end (farthest from crown). These structures contain no nodes or internodes.

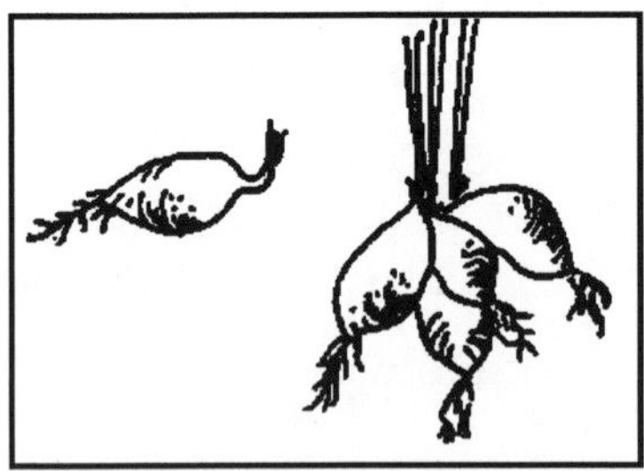

Fig: Dividing Tuberous Roots.

The tuberous begonia, often grown as a tender annual in shady locations, is a tuberous stem. These structures are vertically oriented with one or more vegetative buds produced on the upper end or crown. Propagation of these structures is shown and briefly described.

Table. Propagation of some Underground Structures.

Structure Name	Description of Propagation Method
Bulbs	*Tulips, Daffodils, Hyacinth, Lilies*: Allow foliage to remain intact after flowering. Dig in September, separate bulblets or scales (lilies) and replant. *Hyacinths*: Scoop, scale and plant. Remove bulblets second year and plant.
Corms	*Gladiolus*: Plant in spring, dig in September, separate new corms (larger) from cormels (smaller). Replant following spring. Corms will flower, cormels will require two seasons to flower.
Rhizomes	*Iris*: Remove from parent plant. Cut into pieces, each one containing a node. Replant.
Tubers	*Irish Potato*: Plant whole tuber or cut up with at least one eye remaining on each piece.
Tuberous Roots	*Dahlia, Sweet Potato*: Plant whole. *Dahlia*: Dig in fall and separate the tuberous roots with a shoot bud on each.Sweet Potato — after shoots emerge, separate from mother root and plant.

LAYERING

For the homeowner, two types of layering are worth considering: simple layering and air layering. For simple layering, select a woody shrub you wish to propagate (grape, raspberry, forsythia, etc.) and early in the summer after new growth appears, pull a branch of this shrub down to the ground (works best in sandy soil), burying part of the branch with several nodes three or four inches deep. Anchor the tip in place with a stake so wind movement will not disturb it. To accelerate root development and improve chances of success, try wounding the branch and dusting with a rooting powder before burying. In September (or the following spring), carefully dig up to see if any roots have formed at some of the nodes. If so, separate it from the parent plant and a new plant is produced.

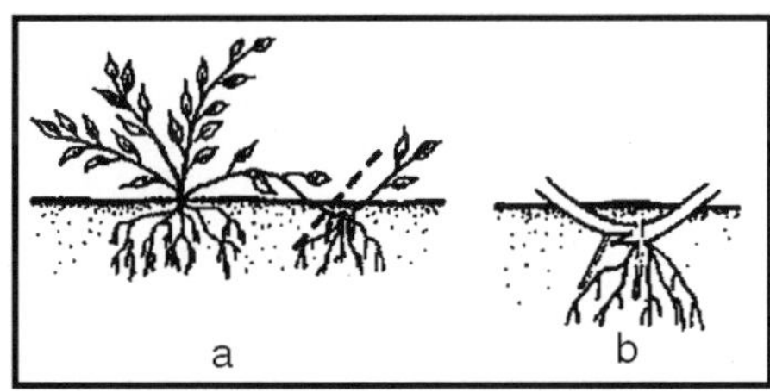

Fig: Simple layering.

Air layering is perhaps the most useful technique of plant propagation for the homeowner to know. With this method, roots can be produced on stems above the soil. If, for example, a rubber plant has grown to the point that it reaches the ceiling, air layering is a way to save the top section of the stem and start a new plant. Or, if the plant has lost most of its lower leaves, looks unsightly, but continues to grow, air layering can be used to start a new plant. Air layering involves one of two processes, depending on the plant species. The first process involves making an upward, slanting cut about onethird of the way through the stem, about one foot from the tip. (This is commonly carried out with Dieffenbachia spp.) The cut is held open with a bent toothpick, dusted with a rooting powder, and packed with damp sphagnum moss (unmilled). Wrap the sphagnum moss with shrink film, plastic or a split sandwich bag. Hold in place with twist ties or electrician's tape.

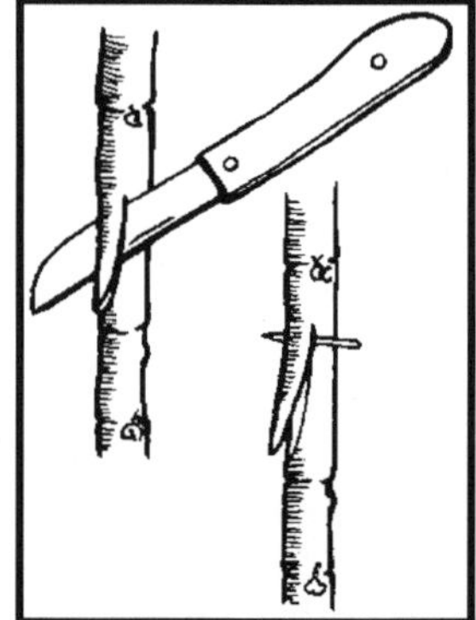

Fig: Air Layering: Hold the Wound open by Inserting a Toothpick in it.

Fig: Air Layering: Cover the Moss with Clear Plastic and Fasten it in Place.

The other process, more frequently used with dicots, is to cut a ring of bark about one inch wide and remove completely from the stem. The xylem cylinder is then carefully scraped to remove any cambial tissue so that a bridge of callus tissue will not form. Continue with same procedure as described earlier with the slanting cut.

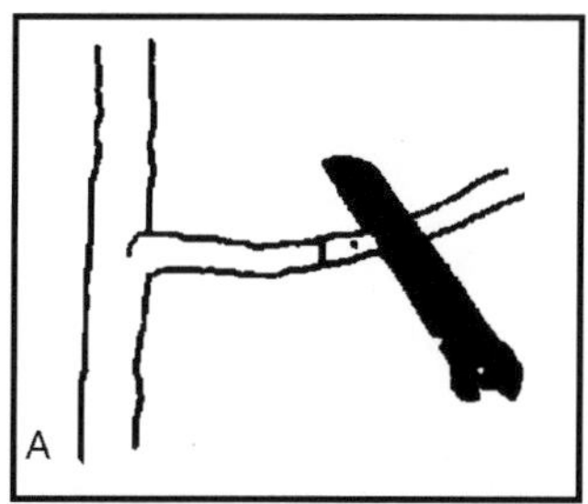

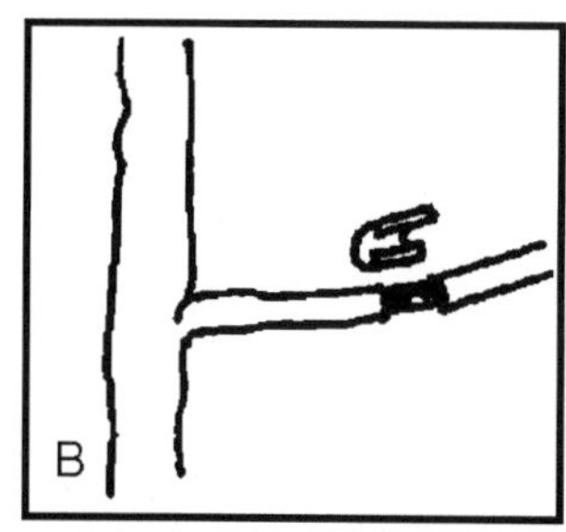

In about four to six weeks, roots should be apparent. Once the roots have filled the ball of sphagnum moss, remove the plastic covering and sever the stem beneath the moss. Pot the new plant using an appropriate soil mix, water well, and cover loosely with a clear plastic bag, keeping it out of direct sunlight. This bag covering will help retain moisture around the plant leaves and keep it from wilting.

It is generally needed for only a couple of weeks. Unless more air layers are desired, the parent plant from which the air layer was made can probably be discarded. Lateral buds often develop after the air layer is removed, but their growth is usually ungainly. Another alternative is to remove the top and root as a tip cutting in a plastic bag. The remaining basal portion can often grow into a new, attractive plant.

DIVISION OF PROPAGATION TECHNIQUE

The simplest propagation technique is division. This involves dividing up a plant into a series of pieces. Plants from African violets to peonies can be divided. The only basic requirement is that each division must have some roots and eyes. Then simply plant, give away or discard the pieces divided from the original crown or plant.

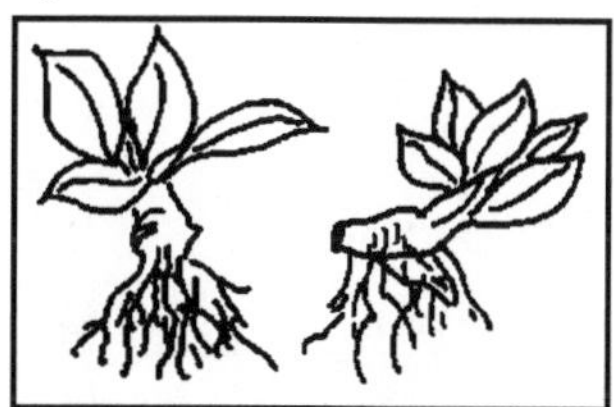

Houseplants such as African violets can be divided anytime the crown becomes crowded; outdoor plants such as phlox or peonies can be dug and divided in the fall.

GRAFTING

This process of propagation involves two separate plants; one provides the root system and the other provides the desired foliage, flower or fruit. The root system is referred to as the understalk, while the branch system is known as either budwood or scion wood. Stories abound about the success of grafting in the nursery profession, from saving grapes from root eating insects, to imparting hardiness to roses and tree fruits, to producing dwarfing or multiple varieties on one rootstock.

With grafting or budding (a form of grafting), success or failure depends primarily on three conditions:

- The compatibility of the two plant parts,
- The closeness of fit, and
- Cambial contact.

This process of asexual propagation has definite limitations; it is limited to dicots and must be carried out between varieties of the same species and genus. In all grafting techniques, a sharp knife is essential. Be sure to have it carefully sharpened on a stone as illustrated in *Sketch 23* to get as smooth a cut as possible for good tissue contact. Dull knives result in wavy cut surfaces and more cut fingers.

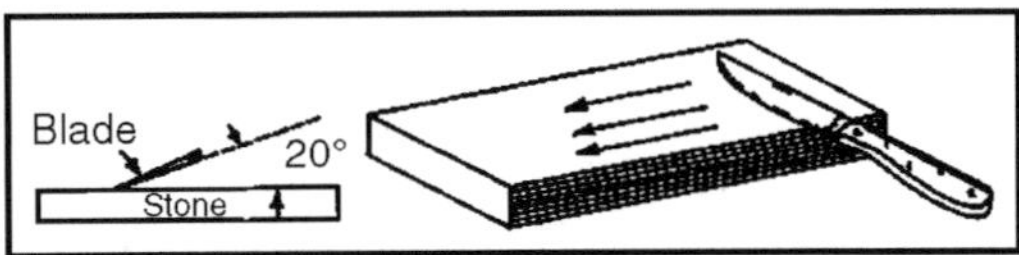

Bark grafting, using the steps shown in sketch 24 the easiest technique for the homeowner to carry out when topworking a tree is desired. For smaller material a side graft is used. The illustrations in sketch 25 show the steps in this process. A small cut about 1½ to 2 inches long is made on the side of the understock, lifting a flap from the stem but leaving it attached at the base. The scion is then cut in a similar manner on both sides, but the pieces in this instance are removed. The scion base is then trimmed obliquely and placed against the side of the understock. The outer cut on the scion is covered by the flap which was left for this purpose on the understock. Once inserted, the graft is bound together with electrician's tape.

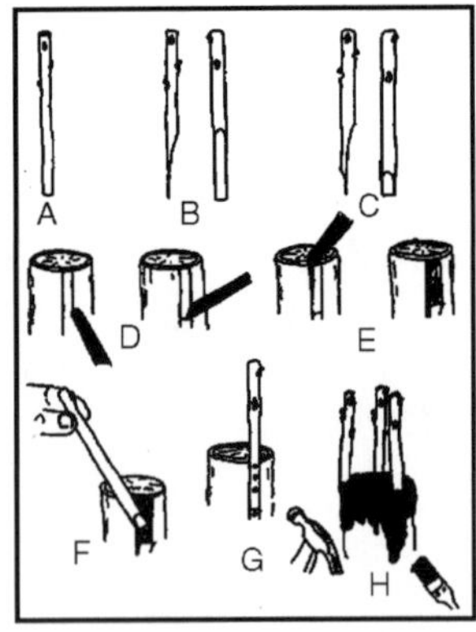

Fig: Bark Grafting.

ASEPTIC-MICROPROPAGATION TECHNIQUES

While keeping plant parts clean for disease free culture requires diverse procedures in a commercial operation, a simplified general procedure and culture medium can be used in the home. The few basic supplies are available from local grocery, drug, or health food stores. The home tissue culture medium described here was tested on African violet leaf/petiole sections, Boston fern rhizome tipes, and wandering Jew shoot tips. The medium was compared to a commercial complex medium and had the same results.

For one pint of medium, mix the ingredients in *Table* together. While constantly stirring, bring the medium to a boil. Continue to boil it until the agar flakes dissolve completely, approximately two to five minutes.

Table. Home Tissue Culture Medium.

Medium Constituent	Amount
Table sugar	1/8 cup
Tap water	1 cup
All-purpose soluble fertilizer	1 cup of stock solution
Inositol tablet (250 mg)	½ tablet
Vitamin tablet with thiamine	¼ tablet
Agar flakes	2 tablespoons

1 This mixture makes 1 pint of medium.

2 Prepare the fertilizer stock solution by mixing

¼ Tablespoon of a balanced, water-soluble fertilizer, such as 10-10-10, in a gallon of water.

3 lnositol is sometimes sold as myo-inositol.

Once the agar flakes dissolve, pour the medium into heat-resistant glass receptacles, such as baby food or canning jars. Pour medium at least 1 inch deep into each jar. Loosely screw on the lids when all of the jars are filled. They are now ready to be sterilized, along with water, tweezers, and razor blades that will be used later. Place the water in heat-resistant jars with lids loosely screwed on or with the foil covers. Wrap the tweezers and razor blades in paper towels.

Sterilize by placing the items in a pressure cooker at 15 psi for 15 minutes. Slowly release the pressure after 15 minutes and place all the items in a clean area. Tighten the lids on the jars with the medium and allow it to cool and solidi*. If the jars won't be used for several days, place a clean layer of foil over them to prevent dust from gathering.

Plant Culture

Micropropagation will only succeed in a sterile environment. Do all transferring of plant material into the jars as quickly as possible in a clean area. Scrub your hands and counter tops with soap and water just before you begin to disinfect plant material. Rubbing alcohol or a 10 per cent bleach solution (one part commercial bleach to nine parts water) can be used to wipe down the working surface. After the medium is cool, prepare the plant material for culture. Plants must be disinfected before placement on the medium because plants usually harbor bacterial and fungal spores. Bacteria and fungi can grow on the medium faster than the plants and dominate the culture. Various plant parts can be cultured, but small, actively growing portions usually result in the most vigourous plantlets. Ferns, for example, are readily propagated by using only one-half inch of a rhizome tip. For other species, one-half to one inch of the shoot tip is sufficient.

If using the shoot tip, remove the attached leaves and discard. Place the tip into a 10 per cent bleach solution for 10 minutes. Be sure that all plant tissue is submerged in the bleach solution. Rinse off excess bleach by placing the shoot tip in sterile water.

Once the plant material has been disinfected it should be touched only with sterile utensils. Sterilize utensils with the medium in the pressure cooker or soak them in isopropyl alcohol (rubbing alcohol). Make sure all alcohol has evaporated from the utensils before they touch the plant material. Hasten evaporation, if necessary, using a burning alcohol lamp. Remove any bleach-damaged tissue with a sterile razor blade after rinsing the plant material. Remove the lid of a culture jar and place a plant part in the jar, making sure that it is not completely submerged under the medium. Recap the jar quickly.

Growing On

Place the cultured plant material in a warm, well-lighted location (not in direct sunlight) to encourage growth. Any contamination of the medium will be obvious in three to four days. Remove and wash contaminated culture jars as quickly as possible to prevent the spread of disease to other uncontaminated cultures. Transplant the plantlets into a standard growing medium when they reach a sufficient size. Handle the plantlets carefully because they are moving from a warm, humid environment to a harsher, drier one. After transplanting, water the plants thoroughly and place them in a clear plastic bag for several days. Gradually remove the bag to acclimate the plants to their new environment.

Although this home tissue culture medium and procedure has worked for some plants, success of the process may vary. The precision of the aseptic procedures and types of fertilizer, vitamins, water, and plants may affect the results. If you enjoy intriguing projects, give home cloning a try.

PROPAGATION TO REPAIR DAMAGED WOODY PLANTS

BRIDGE GRAFTING

Our landscape plants, especially the fruit trees, get damaged during winter months from rodent activity. Or, because the turfgrass is growing right up to the trunk, damage may be initiated from mower impact. Wrapping the trunk with galvanized wire — "hardware cloth" — will prevent rodent damage, and removing the sod from the base of the tree for at least two feet will eliminate mower damage. Bridge grafting, while not a propagation technique in the strict sense, takes advantage of some of the principles of propagation to save valuable trees. Damage usually becomes evident as the snow melts in the spring. As soon as the "sap flows" or the "bark slips," usually in late April or early May, bridge grafting may be carried out.

Water sprouts or suckers which are about pencil size in diametre can be collected from the tree. Before an area is to be bridge grafted, the ragged edges must first be trimmed away. This is accomplished by cutting into living tissues above and below the point of injury. Scions are then cut so that when in position, both ends will extend into healthy tissue. The scions (at least six will be needed to repair a complete girdling) are held in the place they will occupy over the injury, and a channel for each end is marked on the trunk to facilitate a perfect fit once nailed into place. The scion wood is beveled on both ends and the marked positions on the trunk are cut slightly shorter than the length of the scion. The scions are then inserted into the cuts, keeping the tops oriented to the top of the tree. If a power stapler is handy, use it to secure the bark flaps over the scion ends. If not, wire brads can be hammered in. After inserting, the scions should have a slight bow to them to allow for normal tree movement. Once all scions are in place, grafting wax or grafting paint is applied over the surface to prevent drying. If buds should sprout on the inserted scions, simply rub them off. The scion inserts will act as conductors of the products of photosynthesis, moving them into the root system for utilization the following spring. Details of the Bridge Graft. A and B, the Injured Trunk before and after the Scion, Showing Method of Shaping Cuts; D and E, the Complete Graft Prior to Waxing.

PROPAGATION FROM PREEXISTING MERISTEMS

Four methods for in vitro propagation of plants:

1. Shoot culture (enhanced auxiliary shoot proliferation) – The most frequently used
2. Node culture
3. Shoot organogenesis
4. Nonzygotic embryogenesis

Propagation for:

- Production of pathogen eradicated (PE) plants, and
- Preservation of PE germpalsm

SHOOT APICAL MERISTEMS

Define apical meristem: specialized region of little differentiation; retains embryonic capacity for unlimited division. Highly organized and genetically stable. Not connected to vascular system of the stem. *How does it obtain nutrients, then?* At the tip, get developing leaf primordium and lateral buds within the axils of subtending leaves.

IN VITRO CULTURE OF SHOOT MERISTEMS

Ball first detailed procedures for isolation plants from shoot meristem tips = Father of Micropropagation. Procedure used for fungal and bacterial pathogen eradication. Typically done by putting donor plant through thermo or chemotherapy and excising a meristem tip for explantation. (= meristem tip culture). Cytokinins repress apical dominance and are used to enhance axillary shoot production.

SHOOT AND NODE CULTURE

Shoot culture = propagation via repeated enhanced formation of axillary shoots from tips or buds, which are then subdivided into shoot tip and nodal segments for further secondary proliferation or for rooting as microcuttings. Node Culture = promote shoot growth using either intact shoots placed horizontally on the medium or using node segments. *What are node segments?* Shoots are then rooted or acclimatized to ex vitro conditions. *Ex vitro – very intuitive. I like it.* Associated with the least genetic variation.

MICROPROPAGATION STAGES

Murashige in 1974 positied three stages for successful micropropagation, though we now recognize five stages.

Stage 1:

- *Donor plant selection*: In vitro responsiveness influenced by phytosanitary and physiological conditions of donor. Best if maintained in lower humidity, drip irrigation and antiobiotic sprays. *In animal culture, they repeatedly tell us to avoid use of antibiotics. I wonder why this is different.* Trimming, cytokinin or gibberellic acid sprays and forcing solutions (2% sucrose and 200mg/L 8-hydroxyquinoline citrate) will all increase explant responsiveness.

Stage 2:

- *Establishment of aseptic cultures*: Use surface disinfested apical meristems, meristem tips or shoot tips from terminal or lateral buds. Screening for internal microbial contaminants is essential. Stabalization in this phase can take 3-24 months. *Wouldn't that be a great way to spend your year. No wonder this is expensive.* The time can be reduced by initial culture in liquid media.

  Factors affecting stage 2 establishment include:
  - *Explantation time*: In reference to period of growth
  - *Explant position on stem*: Terminal bud explants better than lateral *How are they so different that this matters? What else is going on there?*

- *Explant size*: Juveniles do better, smaller more dependent on medium supplements (NEED auxin or cytokinin, where usually just enhance – BA most commonly used).
- *Polyphenol oxidation*: Accumulation of these will kill explants. *Clearly that would not be good.*

PGRs for this stage must promote establishment and shoot growth but limit callus formation and adventitious shoot formation.

Stage 3:

- *Proliferation of axillary shoots*: Characterized by repeated enhanced formation of axillary shoots. Cultures are routinely subdivided for further proliferation. Often the most expensive stage. Inverting shoot explants can double or triple axillary shoot production. Selection of cytokinins and the amount for use is done via trial and error it seems, watching multiplication rate, shoot length, and frequency of genetic variation. Auxin may be used, but can be risky because it increases the chance of callus formation. Additionally, when selecting a cytokinin, possibility of adverse carryover to Stage IV should be considered. *After all this work, you wouldn't want to set yourself up for failure!*

Stage 4:

- *Pre-transplant (rooting)*: Preparing for successful transfer to soil. All about elongating the shoots. Expensive. *That seems to be a theme here...*Often try to eliminate this stage due to its high costs. Pre-treating in aqueous auxin prior to transplanting can be beneficial. For woody plants auxin is a must.

Stage 5:

- *Transfer to natural environment*: Ultimate success of the whole procedure rests here. Two reasons underlie most plant failure when transferring from in vitro to greenhouse conditions:
  - Heterotrophic mode of nutrition must rapidly switch to photoautotrophic for survival.
  - Poor water loss control due to epicuticular wax reduction, poorly differentiated mesophyll, abnormal stomata function or poor vascular connection between roots and shoots.

These reasons make good sense. I hadn't thought about the heterotropic to photoautotrophic switch yet, but that would be hard on a poor little transplant. Acclimitizing the transplants through gradual humidity reduction and higher light intensities over 1-4 weeks will promote later survival.

MICROPROPAGATION OF SYNGONIUM BY SHOOT CULTURE

Micropropagation has been successfully used for the commercial propagation of tropical foliage plants (*read* house plants – *never know that is what they were classified as*) such as *Syngonium podophyllum*, or arrowhead vine. *Behold, one is sitting on the shelf above my head here at home as I type. Neat.*

Four laboratory exercises follow that illustrate the principles:

1. Use isolated shoot tip explants to establish Stage I cultures
2. Index cultures for contaminants *This word index keeps cropping up...what does it mean in reference to plants and pathogens?*
3. Propagate plants via axillary shoot proliferation
4. Root and acclimatize microcutting transplants.

Important concepts to grasp in the process:

- Initial shoot tip explant size correlates to response and culture contamination (Bigger is more successful, *that is why it is called meristem tip explants – meristem plus some)*
- Importance/Essentiality of Stage I adaption prior to Stage II axillary shoot proliferation
- Direct rooting and acclimatization of microcuttings.

GENERAL CONSIDERATIONS

Use lateral bud scales or vining stems, or terminal shoot tips and lateral buds if nonvining plants are used. S. podophyllum can be purchased from a garden shop and maintained in soilless potting mix; Vining can be promoted by fertilized 2x per week. Media for Stage I and II (as per Miller and Murashige, 1976 with phytagel as solidifying agent) only differ by the removal of IAA from the media for stage II. *I'll bet I could make this media! pH of 5.7 seems to be common.*

Protocol for establishing Stage I cultures from meristem on pg 91. Don't use broken buds. *The incisions they direct you to make in the protocol sound daunting. I'm still rather awkward with my tweezers and scalpel.* Make explants 1-5 mm in length. Anticipate that smaller than 2 mm won't make it. We go through step I cultures so that contaminated cultures aren't mass propagated in stage II. Protocol for Stage I indexing, Stage II shoot multiplication and Stage IV acclimatization.

Objectives of this protocol include:

- Familiarizing students with indexing procedure
- To examine explant response and successfully root following stage succession.

MICROPROPAGATION OF DIEFFENBACHIA

Tissue culture propagation of *Dieffenbachia,* an aroid, is a commercial success story. Aroids were first produced free of dasheen mosaic virus in 1974. Four years later, researchers developed a method for liberating *Dieffenbachia* from systemic viral and bacterial pathogens. In 1982, it was hailed as the first cultivar free of systemic pathogens.

TC methods have also cut production time from 5-10 years down to 2-3. *Occasionally I wonder what the purpose of all this is. Excellent. Companies can make house plants for less. Why do we need that? Why don't we just go outside more?*

Two experiments are outlined (pg 100 and 101), one of which initiates cultures from a healthy vigourously growing plant (up to 6 months), while the second obtains shoots from established commercial cultures (3-4 months).

- *Initiation experiment*: Explant = axillary buds from potted plants; induce for adventitious bud production. Careful maintenance and regular transfers are a must. *I'll bet...over 6 months, you wouldn't want to have to start over at month 5...*
- Multiplication of cultures
  - Experiment demonstrates effects of cytokinin levels on growth and development of adventitious buds and root and shoot formation.
  - Raising 2iP levels will decrease root production.
  - Transfer callus base regularly so adventitious buds continue forming

TISSUE CULTURE OF LILACS

Traditionally lilacs have been grown from seed, suckers or by grafting superior clones on to compatible rootstocks. Lilac seedlings of several species, privet stem cuttings as well as ash seedlings/root cuttings have all been utilized as rootstocks. In recent times there has been a move away from using rootstocks for various reasons, and nurserymen have been trying to produce cultivars on their own roots. While cuttings taken at the right time are often successful for many species, this is not always a sure recipe for thevulgaris clones.

For more than two decades some of the leading nurseries in the US and Europe have been using tissue culture for propagating lilacs on their own roots. While the main reason for resorting to micropropagation is to produce large numbers of selected clones within a short period, we have discovered it to be no less effective in producing small numbers ofvulgaris cultivars which are recalcitrant to initiating roots by conventional means. The following is a short photo-essay on the process.

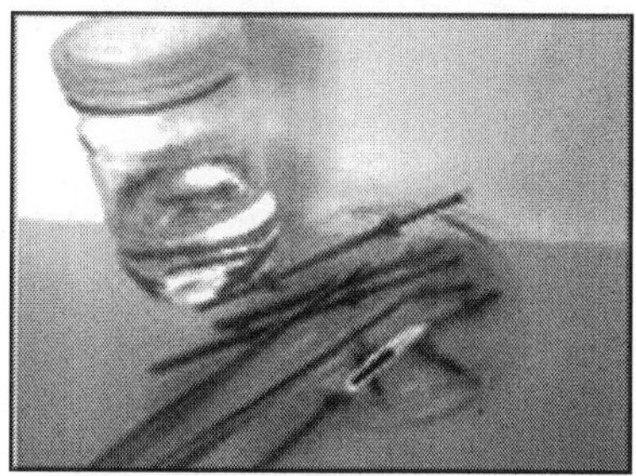

The most difficult part of the operation is obtaining material free of bacterial and fungal contamination. Succulent terminal cuttings are divided into single node 'explants' and agitated in dilute chlorine for varying lengths of time to try eliminate pathogens.

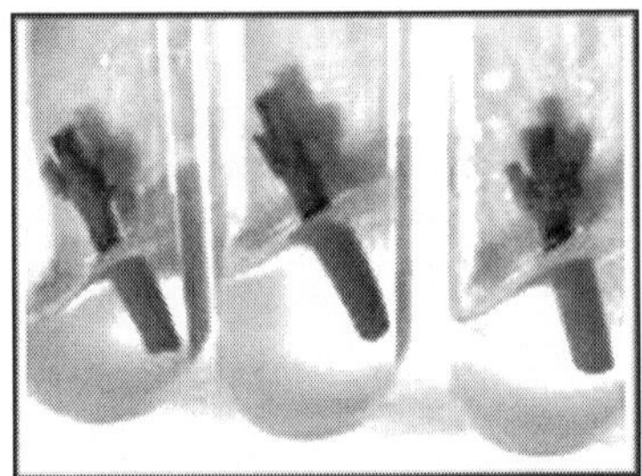

Cleaned explants are placed in nutrient agar under aseptic conditions and stored under lights at a constant fairly high temperature for several weeks.

Explants free of contamination make 20-50 mm of growth over several weeks and are ready for weaning and rooting, or for further division to repeat the process until the required quantity is achieved. Leaves at this stage are tiny, resembling those of young seedlings than that of mature plants.

A tray of cuttings after root initiation and several months' growth in the greenhouse, around 150 mm height.

Our first micropropagated lilac, standing all of 125 mm after a season's growth in the greenhouse. Note the sucker emerging from the left of main stem. A second season's growth should see this on the market.

Well grown tissue cultured *S. v.* 'Sensation' towards the end of the second season at leaf-fall, standing nearly 1 m high. Some cultivars such as*S. x hyacinthiflora* 'Maiden's Blush' branch more easily and make less height.

MICROPROPAGATION OF DIEFFENBACHIA

Dieffenbachias are very popular and hardy shade loving plants, commonly grown in pots for indoor decoration and also in shade and semi-shade in gardens of warm humid regions. It has the common name Dumb Cane because of the poisonous nature of their sap, which causes intense pain and swelling of the tongue, if it is taken internally. The leaves of this fleshy plant are oval-oblong, gradually tapering towards the apex, arising from the central stem and spotted, striped or speckled with white or yellow, fifteen to forty cm or so in length. They grow in any type of well drained soil but organic matter certainly helps to make the leaves more attractive. The plant can be propagated by suckers from the base, by top cuttings or by planting of stem node pieces with a good bud on it. A large number of plantlets are also produced through micro propagation. This plant has got a natural tendency to be lanky and top heavy. In that case cut the stem from where you have new buds or eyes. New shoots will start from that point. Do not discard the top part, put it in a potting mix and keep it moist. new root will start after a few days, and will get one more plant. This plant is not disease prone. Sometimes they can be infested by mites. Treat them as you do for other plants.

DIEFFENBACHIA AMOENA CALCUTTA

It is a sport of D. Amoena, large and thick stemmed. Leaves large, arching, lamina predominantly yellow-white with irregular green spot, margin branded green.

TROPIC SNOW

The scientific name is Dieffenbachia Amoena. It has compact growth, leaves leathery, glossy deep green, highly variegated form, centre with cream and green, border deep green.

CAMILLA

It's scientific name is Dieffenbachia Exotica. It has compact growth, freely branched. Its leaves are small, ovate, pointed apex, predominantly creamy yellow, border green.

DIEFFENBACHIA MACULATA

It is a fairly large hardy plant with glossy green leaves, oval with chordate base, blotched or marbled with white pale green.

DIEFFENBACHIA JUPITER

An outstanding and attractive variety, came from Dieffenbachia Maculata It has many ovate leaves and leaf size is smaller compared to others. Young leaves are yellowish cream and border light green.

PROPAGATION FROM NONMERISTEMATIC TISSUES AND ORGANOGENESIS

Meristematic tissues are responsible for the division of new cells... they are zones of actively dividing cells.

Before we precede, we should first recognize that growth in plants includes two stages:

1. First the production of new cells and,
2. Secondly the expansion of these cells via uptake of water by the vacuole.

Cell division occurs solely in meristematic regions, while expansion may occur anywhere. Thus in a single plant there are zones of young dividing cells, maturing cells, and mature cells. We will go into more detail on meristematic regions when we go over the stem and root individually, but for now an introduction is sufficient

Let us recognize that there are 3 meristematic regions in the plant:

- *Apical meristems are located at the apices or tips*: At root and shoot tips and are directly involved in their elongation They create derivatives which form primary growth.
  1. The protoderm which forms the outer dermal layer of tissues,
  2. The ground meristem which forms the cortical cells and
  3. The procambium which forms the vascular tissue.

Ferns, Gymnosperms and Angiosperms are all terrestrial plants. Tissues therefore arise as a result of a cell or a group of cells dividing, thus giving rise to a large number of cells. In flowering plants these cells undergo changes that change the structure and function of the cell; differentiation has take place. As a result of differentiation, a cell or group of cells can have a specialized function; i.e. specialization takes place. *Thus, a tissue is a group of cells that have differentiated to perform a specific function.* If a group of cells are more or the less the same, then it is known as a Simple tissue e.g. epidermis, parenchyma, collenchyma and sclerenchyma. If, however, the tissue is composed of a number of different cells, then it is known as a Complex tissue such as xylem and phloem.

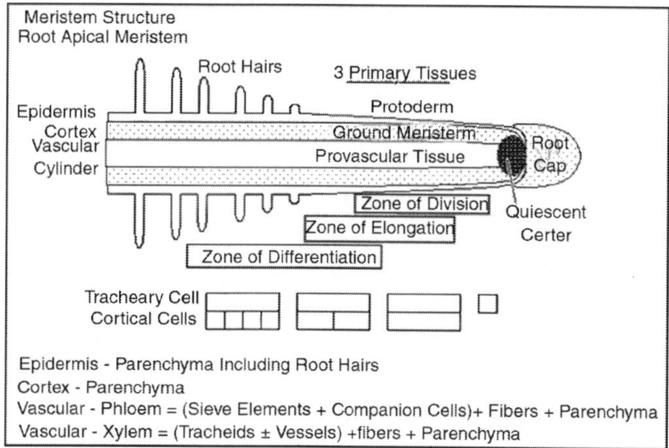

Fig. in shoots, plants protect their meristems with young leaves and by forming dormant reserve meristems (i.e., buds), that protect their apical meristems with a root cap

In Angiosperms the most important tissues can be classified as shown below:

	Tissues of Flowering Plants		Meristematic
Tissues	**Permanent**	**Tissues**	
Primary Meristems	Secondary Meristems	Simple Tissues	Complex Tissues

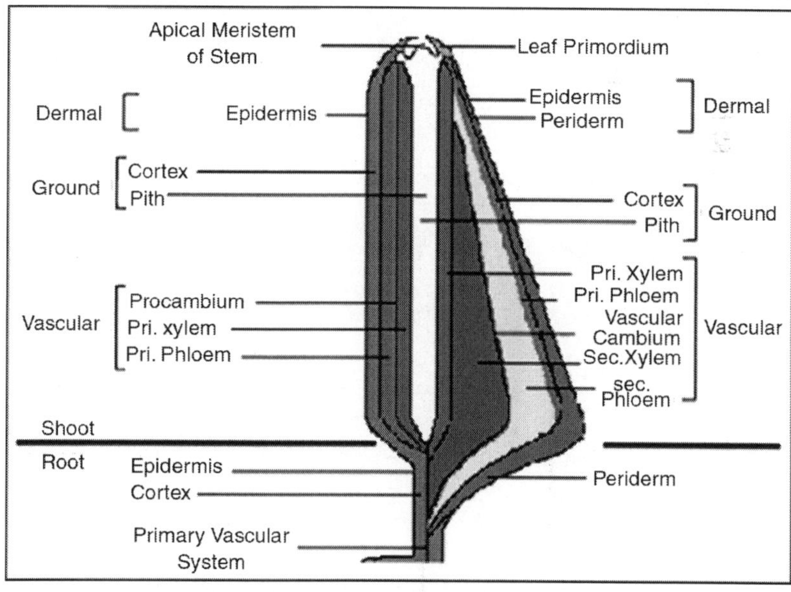

DIAGRAMMATIC REPRESENTATION OF A PLANT BODY

Ground Tissue is simple non-meristematic tissue (non-dividing tissue)

made up 3 cell types:paranenchyma, collenchyma and sclerenchyma.This tissue generally forms either the pith, cortex or bulk of leaf (mesophyll).

Parenchyma Cells

- Mostabundant cells in plants;
- Spherical cells which flatten at point of contact;
- Alive at maturity;pliable, primary cell walls;
- Large vacuoles for storage of starch, fats, and tannins (denature proteins);
- Primary sites of the metabolic functions such as photosynthesis, respiration, and protein synthesis; they are "ready reserves" from which a plant makes specialized cells to meet its changing needs.

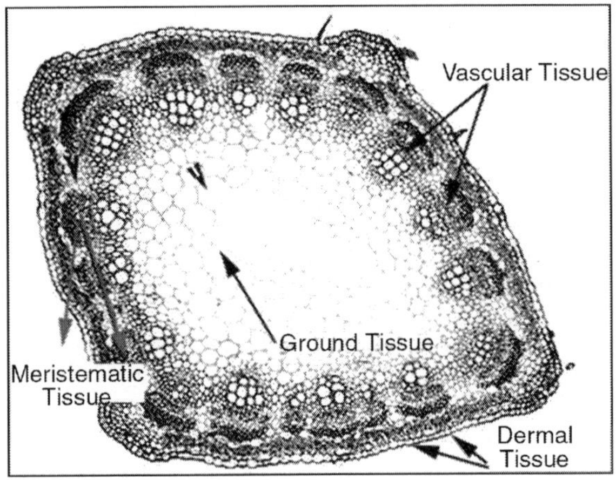

SPECIALIZED PARENCHYMA

- *Chlorenchyma*: Photosynthetic cells; have high density of chloroplasts

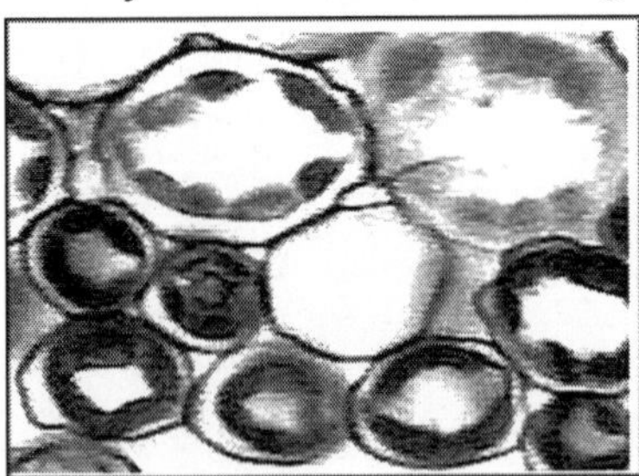

- *Aerenchyma*: Prominent intercellular spaces that improve gas exchange capacity of the tissue; provide maximum support with a minimum metabolic requirement
- *Transfer cells*: Specialized for short distance transfer of solutes between cells; have secondary cell walls; they are inner extensions of wall that increase surface area.
- Transfer cells occur in areas of high solute transport, such as

secretory tissues, which release substances that are produced within the protoplasm and are moved outside, i.e., nectar cells, mucilage in sundews, and resins.

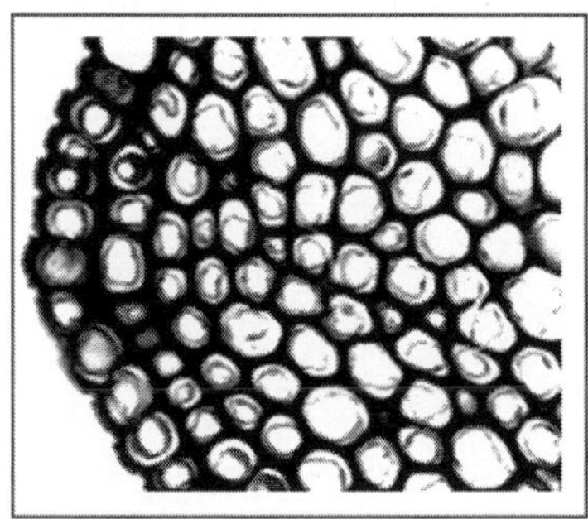

Collenchyma Cells- Colla = Glue, Glisten

- Living protoplasm;
- Unevenly thickened primary cell walls; elongate cells;
- Longer than wide; just beneath epidermis; function to support growing organs, grass, floral parts, and border veins; their nonlignified cell walls can stretch

Sclerenchyma Cells- (Skleos = Hard)

- Most are dead at maturity;
- Rigid, thick, lignified, nonstretchable secondary cell walls. there are 2 types of sclerenchyma cells:

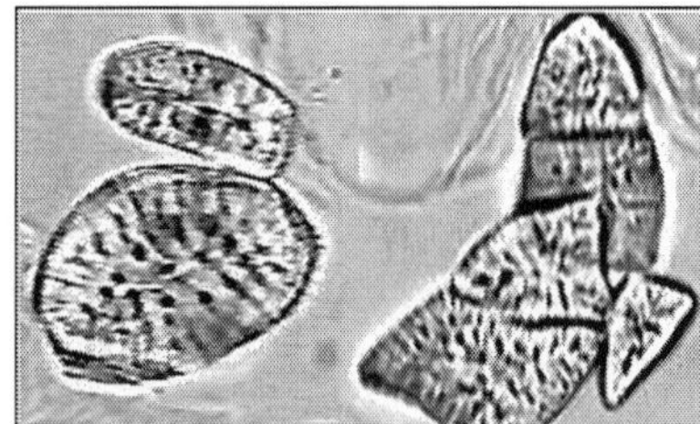

Sclereids or stone cells we saw in lab- short; variable shape; form hard layers such as the shells of nuts and seed coats; produce the gritty texture of pears. Fibres- long, slender; occur in strands or bundles; tiny cavity or lumen; the different hardnesses of fibres are used to make coarse rope, linen cloth, etc.

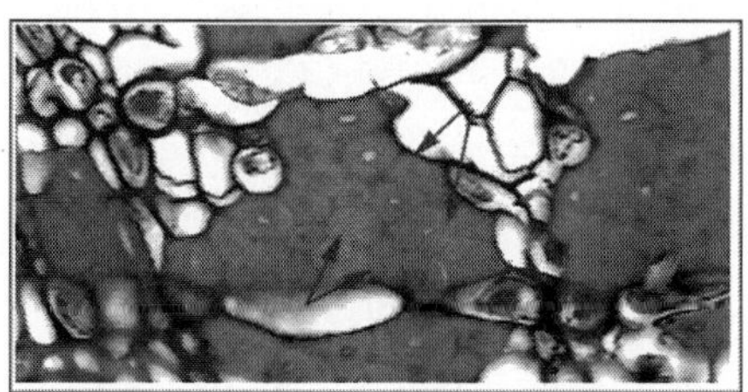

MICROPROPAGATION AND IN VITRO FLOWERING OF ROSE

The present invention is directed to compositions and methods of micropropagation of a rose plant and in vitro flowering of the rose plant. Young shoots are induced to produce buds in an enclosed vessel containing a first culturing medium comprising benzyladenine, an auxin and 2% sucrose as carbon source.

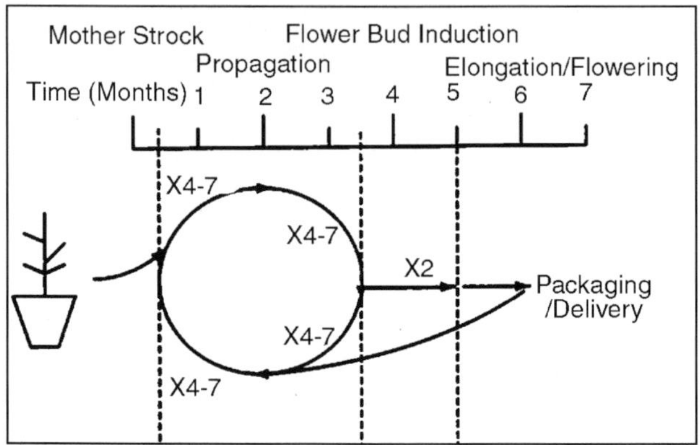

Buds are excised and cultured on a medium for propagation and multiplication of plantlets. Plantlets are then transferred to a medium comprising thidiazuron, an auxin and myo-inositol for induction of flower buds, followed by culture on a medium comprising benzyladenine and an auxin for plantlet elongation and finally culture in the fifth medium without phytohormone for flowering. Alternately, after propagation and multiplication, plantlets are transferred to a medium comprising zeatin, auxin and myo-inositol for induction of flower buds followed by culture on a medium without phytohormone for elongation and flowering.

SHOOT ORGANOGENESIS

Shoot regeneration from leaf and stem explants of 19 Chrysanthemum cultivars has been studied. Adventitious shoots could be regenerated from leaf and stem explants of 16 cultivars. Most cultivars showed the best shoot regeneration on Murashige and Skoog (1962) medium supplemented with 8.9 μM BAP and 2.7 μM NAA (0.5 mg/l). The number of regenerated shoots per explant was more than 10 in some of the cultivars studied. It has not been posssible to regenerate shoots from leaf explants of cultivars 170 and 260. The cultivars 140 and 8010 produced the highest number of shoots. Regenerated shoots can easily be rooted on half strength Murashige and Skoog medium supplemented with 0.98 μM IBA (0.2mg/l).

After potting in soil and transfer to the glasshouse normal plants develop. Apparently no somaclonal variation has been observed among the regenerants.

Biolistic particle bombardment is used to transfer desirable genes into difficult to transform plants, but research is lacking on factors that can affect transformation success. This study was conducted to determine if manipulating culture conditions forRhododendron 'PJM Hybrid' and 'Iridon' chrysanthemum leaf explants could alter transient gene expression of β-glucuronidase (GUS) after bombardment. Both rhododendron and chrysanthemum leaf explants were bombarded with 1.0 ìm gold particles coated with 2 μg of linear plasmid DNA, containing the GUS gene.

Different explant treatments before bombardment were tested, including time in culture on regeneration medium, growing in the presence or absence of 0, 0.09, 0.2, 0.4, or 0.6 M sucrose, and growing in the dark for 0, 3, or 6 days. Transient GUS expression was histochemically quantified 2 days after bombardment as the percentages of GUS positive explants and the number of blue spots per GUS positive explant. Transient GUS expression was highest when 'PJM Hybrid' explants were incubated on regeneration medium 9 to 12 days before bombardment, whereas transient GUS expression by 'Iridon' explants was relatively high (>42%) for leaves incubated 3 to 18 days before bombardment. Exposure of 'PJM Hybrid' explants to media containing different concentrations of sucrose decreased transient GUS expression by 71% compared to those grown on sucrose-free medium. Alternatively, 27% more 'Iridon' explants transiently expressed GUS compared to those on sucrose-free medium, and the mean number of spots per GUS positive explant increased from 5 to 13. Transient GUS expression increased almost three fold when 'PJM Hybrid' explants were placed in the dark for six days compared those in the light. In contrast, the dark treatment had inconsistent effects on transient GUS expression by 'Iridon' explants. Manipulating explant culture conditions before biolistic bombardment influenced transient gene expression and illustrated that genotype-specific differences affect transformation potential.

SHOOT ORGANOGENESIS OF WATERMELON

Watermelon is a popular fruit originating from Africa, but cultivated in the south, southwest, and central states in the US. It is susceptible to many diseases, and thus is a good plant for genetic engineering. Somaclonal variation can produce new breeding lines, which is good for creating new cultivars. In the last assignment, we saw the negatives of somaclonal variation in organogenesis. Now we see an advantage. Seedling cotyledons are usually used for explants.

Exercises

Experiment 1 Difference in regeneration potential among distinct regions of watermelon cotyledons. In watermelon, the cotyledon gives the best outcome in organogenesis. Still, not all areas of the cotyledon give uniform results. Experiment 1 will locate the best areas of the cotyledon for adventitious shoot regeneration.

Procedure

- Imbibe seeds
- Remove seed coat and remove embryo
- Disinfect embryo
- Transfer embryos to vessels with germination medium and culture for 5-7 days
- Remove germinated seedlings for dissection. Remove cotyledons.
- Cut cotyledons into pieces and place on regeneration medium.
- Seal with parafilm. Place dishes randomly in growth chamber.
- Record # explants with shoots or vegetative buds. Transfer those explants to rooting medium.
- Once roots have formed, transfer to soil.
- Acclimatization: wash away agar from roots and transplant to soilless medium.
- Analyse data and transfer plants to greenhouse/garden when there are favourable conditions.

Anticipated Results

After 3-5 days explants should be 4-5x original size. After 3-4 weeks of culturing, buds, shoots, and leaves should be seen. Experiment 2 Effect of plant genotype on the ability of watermelon cotyledon explants to produce shoots Even within species, there is variation in response to a particular protocol. Thus it is important to choose those genotypes that are very regenerative for successful genetic transformation. Experiment 2 studies shoot production of 4 different cultivars.

Procedure

- Imbibe, disinfect, germinate seeds
- Remove seedlings and follow steps 5 and 6 from above procedure.
- Seal with parafilm. Place dishes randomly in growth chamber. At day 9-12, examine explants. Some should have bud formation.
- At week 4, record # explants with shoots or buds. Transfer to shoot elongation medium.
- Follow steps 9-11 above. When roots bigger than 1 cm have formed, record # that rooted.

Anticipated Results

After 3-5 days explants should be 4-5x original size. After 9-12 days, some should from buds. About 75-95% shoots will root on IBA medium. About 60-90% plants will acclimate.

Experiment 3 Influence of Darkness

In several plant species, dark treatment improves shoot regeneration. Although the role of dark treatment isn't understood, it is speculated that darkness conserves endogenous PGRs and reduces starch and oxidative polyphenols. Dark treatment shows reduced cell wall thickness and cell wall deposits, which may encourage organ regeneration.

Procedure

- Imbibe, disinfect, and culture on germination medium
- Remove seedlings and follow steps 5 and 6 from procedure 16.1. Place plates randomly in growth chamber.
- Examine explants at day 9-12. Some should have bud formation.
- At week 4, record # explants with shoots or bud formation. Transfer explants with shoots to shoot elongation medium.
- Acclimatize by following steps 9-11 in procedure.

DIRECT AND INDIRECT SHOOT ORGANOGENESIS

Torenia fournieri is also called the wishbone flower, bluewings, or torenia. It is easy to obtain and to grow.

GENERAL CONSIDERATIONS

Growth of Plants

Torenia can be grown from shot cuttings or seed. They should grow for a minimum of 8 weeks after germination. They should start flowering 12 weeks after germination. They are very vulnerable to white fly infestation.

Explant Preparation and Basal Medium

Remove leaves from young and hearty plants, which will give better results than older plants. Surface disinfect and rinse. Place only 3-4 leaves per beaker with bleach solution to increase success rate. Remove leaf margins and cut into pieces. Place with abaxial side on medium. Incubate and observe weekly. Look for shoot, leaf, and callus initiation. The text talks about "three types of organogenesis," which confuses me since I thought there were only two (direct and indirect). Experiment 1 Direct organogenesis from leaves of Torenia fournieri. Direct organogenesis is less frequent than indirect organogenesis. However, direct organogenesis can occur if placed on media with cytokinins and auxins. The purpose of experiment 1 is to show direct organogenesis of torenia.

Procedure

- Culture leaf sections on direct organogenesis medium. Place culture vessels randomly.

- In 4-6 weeks, count # shoots/culture vessel.
- If using for further experimentation, separate shoots aseptically when counting.
- In 3-5 weeks, remove plantlets and wash agar from roots. Transfer to soilless medium. Plants will be acclimatized in 2-3 weeks.

Anticipated Results

Contamination may be present 3-7 days after the culture. Discard those cultures. There may also be bleach damage, but the experiment will work if the damage isn't widespread. In 2-3 weeks, shoots should form. Each explants should have many shoots (except for damaged areas).

Experiment 2 Callus formation and indirect organogenesis. Indirect organogenesis has a callus formation step that is the precursor of roots/shoots. This experiment demonstrates indirect organogenesis from a callus. Given the serious problem of somaclonal variation in using indirect organogenesis, when would someone want to use indirect over directorganogenesis?

Procedure

- Culture leaf sections on callus induction medium. Place culture vessels randomly.
- In 3-4 weeks, subculture explants on new callus induction medium.
- After 5-6 on induction medium, remove callus and culture on indirect organogenesis medium.
- After 4-6 weeks on organogenesis medium, count # shoots/culture vessel.

A viable protocol has been developed for direct and indirect shoot regeneration of Vernonia cinerea. To establish a stable and high-frequency plant regeneration system, leaf and stem explants were tested with different combinations of á-naphthalene acetic acid (NAA), indole-3-acetic acid (IAA), and benzylaminopurine (BA). Lateral buds on nodal explants grew into shoots within 2 wk of culture in Murashige and Skoog (MS) basal medium supplemented with 20.9 ìM BA. Excision and culture of nodal segments from in vitro-raised shoots on fresh medium with the same concentration of BA facilitated development of more than 15 shoots per node.

Similarly leaf, nodal, and internodal explants were cultured on MS basal medium supplemented with different concentrations of BA, NAA, and IAA either alone or in combinations for callus induction and organogenesis. Shoot buds and/or roots were regenerated on callus. Shoot buds formed multiple shoots within 4 wk after incubation in induction medium. Adventitious buds and shoots proliferated when callus was cut into pieces and subcultured on MS basal medium containing 20.9 ìM BA and 5.3 μM NAA. This combination proved to be the best medium for enhanced adventitious shoot bud multiplication, generating a maximum of 50 shoots in 4 wk. This medium was also used successfully for shoot proliferation in liquid medium. Root formation was observed from callus induced in medium containing 8.05-13.4 µM NAA. Regenerated shoots exhibited flowering and root formation in MS basal medium without any growth regulators. Plantlets established in the field showed 85% survival and exhibited identical morphological characteristics as the donor plant.

SHOOT ORGANOGENESIS FROM PETUNIA LEAVES

The garden petunia is a popular plant is called a "cultigen" because this hybrid started in domestication. It is a hybrid of 3-4 petunia species. Garden petunias are easy to tissue culture.

GENERAL CONSIDERATIONS

Growth of Plants

Different petunia cultivars and genotypes will have different needs for PGRs, but almost all will produce shoots in vitro. Choose cultivars of the Grandiflora types. Avoid red flowering cultivars and mixed cultivars. Eight weeks before starting the experiment, plants seeds in a soilless medium. Do not use overhead watering, as this can cause fungal and bacterial contamination. White flies can be a problem as well. It's interesting that white flies. They were a problem as well. Seedlings should be growing rosettes when explanting. Flowering plants can be used, but may be contaminated. In each petri dish, place two leaf explants adaxial side up (upper surface up).

Experiment: The effects of cytokinin and auxin on petunia leaf explants. Experiment demonstrates the effect of PGRs on organogenesis. We will look at the effects of cytokinin and auxin separately, as well as the cytokinin to auxin ratio.

Procedure

- Surface disinfect leaves
- Remove bleach solution by swirling leaves in sterile/distilled water. Repeat once.
- Discard edges and tip. Dissect into 1-1.5 by 0.5 cm explants. Place 2 explants per petri dish with right adaxial side up. Incubate with fluorescent lamps. Arrange petri dishes randomly.
- Count # roots and number. Count # shoots that are at least 1 cm long. Measure callus formation with a relative scale.

Anticipated Results

After 1 week, roots will first start to appear on leaves growing on IBA medium. After 3 weeks, roots will mature. Shoots may be seen on leaves growing on medium with BA. After 7-10 days, callus will be seen. Best callus growth will be seen on medium with BA and IBA. Experiment 2 Regeneration of petunia from callus and monitoring somaclonal variation The purpose of this experiment is to contrast direct and indirect organogenesis.

Procedure

- Prepare explants as outlined above.
- After 3-4 weeks, remove callus in cubes of ~4mm. Also make new leaf explants.
- After 4 more weeks, remove shoots from callus and explants.
- Once shoots have rooted, transplant to conditions for seedling Acclimate to greenhouse conditions. When they are large enough, move to soilless medium.
- Arrange plants randomly
- Each week, record length of shoots, flower corolla length, width of flowers, number stamens, plant height, number of branches, length of branches, and any abnormalities.

Anticipated Results

Shoots will from explants and callus. There will be greater variability in plants from callus in comparison to plants from explants or stock plants.

PROCESS OF NONZYGOTIC EMBRYOGENESIS

Somatic embryogenesis is the process of a single cell or group of cells initiating the developmental pathway that leads to reproducible regeneration of nonzygotic embryos capable of germinating to form complete plants. Under natural conditions this pathway is not normally followed but from tissue cultures somatic embryogenesis occurs more often and as an alternative to organogenesis for regeneration of whole plants. According to somatic embryogenesis is, initiated either by preembryogenic determined cells or by induced embryogenic determined cells.

In preembryogenic determined cells, the embryogenic pathway is pre determined and the cells appear to only wait for the synthesis of an inducer (or removal of an inhibitor) to resume independent mitotic divisions in order to express their potential. Such cells are found in embryonic tissues (including scutellum of cereals), certain tissues of young in vitro grown plantlets, the nucellus, and the embryo sac (within ovules of mature plants).

Induced embryogenic determined cells, on the other hand, require redetermination to the embryogenic state by exposure to specific growth regulators such as 2, 4-D. These cells are dedifferentiated generally in microspore (anther) cultures and callus cultures. Once the embryogenic state has been reached both cell types proliferate in a similar manner as embryogenic determined cells. Plant lets are then directly produced by following the full embryogenic pathway as a coordinated group of embryogenic determined cells. Some times individual cell or cells from the group may escape and give rise to either embryoids or nodular embryogenic callus (e.g. scutellar callus) consisting of proembryoids. These are embryolike structures which are bipolar units and germinate into full plantlets under suitable culture conditions.

In dicotyledonous plants totipotent embryogenic cells have been most commonly obtained from explants of embryonic or young seedling tissues. Excised small tissues from young inflorescences (before maturation of floral primordia) are equally effective for the induction of somatic embryogenesis in cultures. Other explants used are the scutellum, young roots, petioles, immature leaf, and immature hypocotyl.

A high yield of embryogenic calli can also be obtained from isolated fully differentiated mesophyll cells or protoplasts in a defined culture medium. Somatic embryos germinate in situ or when they are excised and cultured individually on a fresh semisolid medium. The essential requirements for the induction and promotion of somatic embryos are primarily established by the use of suspension cultures. The in vitro development of somatic embryos was first observed in carrot suspension cultures was also able to induce somatic embryogenesis in a callus cultured on a semisolid medium. The presence of auxin in the medium is generally essential for embryo initiation.

Tissues or calli maintained continuously in an auxin free medium generally do not form embryos; therefore, somatic embryogenesis is achieved in two steps. First, the callus is initiated and multiplied on a medium rich in auxin (2, 4-D, 0.5 mg 1-1) which induces differentiation of localized groups of meristematic cells called embryogenic clumps. Second, the embryogenic clumps develop into mature embryos when transferred to a medium with a very low level of auxin (0.01-0.1 mg,-1) or no auxin at all.

Hence a medium with auxin is called a proliferation medium and one without auxin an embryo development medium. Mature embryos develop in embryo development medium. Mostly 2, 4- D has been used to induce somatic embryogenesis.

A substantial amount of nitrogen, usually in reduced form such as ammonium salts, is required for both embryo initiation and maturation. The source of nitrogen can be in the form of complex addenda such as coconut milk, casein hydrolysate, a mixture of amino acids or single amino acid, and ammonium ions. MS medium contains high levels of nitrogen in the form of ammonium nitrate. Other items, such as potassium, dissolved oxygen, activated charcoal are also found to playa role in somatic embryogenesis. In monocotyledonous plants, most of which are agriculturally and medicinally important, the vegetative parts do not readily proliferate in cultures. Hence explants are best taken from embryogenic or meristematic tissues (young inflorescences and leaves).

Germination of the somatic embryo can occur only when it is mature enough to have functional shoot and root apices capable of meristematic growth. High auxin levels can inhibit development and growth of the shoot meristem and often embryos mature when the embryogenic cell suspension is transferred to a medium lacking auxin. The addition of low level of cytokinin (zeatin 0.1 μm) in combination with ABA may prove beneficial for embryogenesis of low density cell cultures.

Various physical factors such as temperature also affect embryo maturation, depending on the requirement of the species. Maturation of somatic embryos proceeds more normally in complete darkness although light seems essential for somatic embryogenesis in some cultures. Somatic embryos germinate on agar medium without growth regulators. Single embryos may profit by the inclusion of low levels of zeatin (0.1 μm) in the medium.

Sometimes loss of morphogenic potential in embryogenic cultures is seen. This may be due to many reasons. Nuclear changes such as polyploidy, aneuploidy and chromosomal mutations in cultured cells may be responsible for the loss of organogenic or embryogenic potential in prolonged cultures. Altered hormonal balance within the cells or tissues may also be associated with decline in the embryogenic potential. Sometimes the non embryogenic cells of the explant increase under conditions favourable to their growth, resulting in a gradual loss of the embryogenic component during repeated subcultures.

After a prolonged period of uncertainty about the precise role of maternal genes in initiating embryogenesis in flowering plants, considerable evidence for the involvement of maternal genes in embryo and endosperm development in Arabidopsis thaliana has accumulated in recent years. Much attention has centred on a group of mutants known as fis, which display an ability to initiate partial embryogenesis and endosperm development in the absence of double fertilization. This article presents a brief overview of our current understanding of the role of non-zygotic parental genes in the development of these products of double fertilization in A. thaliana. Evidence shows that the expression of paternal alleles of some genes is frequently delayed during embryogenesis and endosperm development, and that the silencing occurs at the transcriptional level by genomic imprinting.

SOMATIC EMBRYOGENESIS FROM SEEDS OF MELON

Scoop out the seeds from a ripe melon and put them into a wire mesh sieve, then with running water over the seeds rub them gently against the mesh, using it to loosen and remove the stringy fibres. Next place the cleaned seeds in a bowl of water, stir it a few times. Some seeds will float to the top....these are immature or sterile melon seeds, they are hollow and/or light-weight and will float to the top of the water. Skim away these bad seeds and discard them. Stir a few more times and repeat the process until no more sterile seeds float to the the top. Drain the water from the remaining seeds.

Afterwards, line a heavy plate or baking pan with waxed paper, spread the seeds out in a single layer onto the waxed paper and place it in sunny spot to air-dry. Stir the seeds occasionally during the next few hours to make sure all sides are exposed to fresh air, this facilitates even drying. After a day in the sun bring the seeds into the house where they continue to dry for another week or two, stir them daily so they dry evenly. If you've got rainy weather the increased humidity can prolong the drying process another week or so. Melons have thick seeds so be sure they are thoroughly dry before packing them for storage.

Index

F

G

H

I

K

L

M

N

O

P

R

S